Praise for A

"[A] lovely, big-hearted book . . . brimming with compassion and the tales of the many, many humans who devote their days to making animals well."

—Emily Anthes, *The New York Times*

"This is a marvelous, smart, eloquent book—as much about human emotion as it is about animals and their inner lives. Braitman's research is fascinating, and she writes with the ease and engagement of a natural storyteller."

—Susan Orlean, bestselling author of *Rin Tin Tin*, *Saturday Night*, and *The Orchid Thief*

"*Animal Madness* is a gem."

—Marc Bekoff, *Psychology Today*

"Illuminating . . . Braitman's delightful balance of humor and poignancy brings each case to life. . . . [*Animal Madness*'s] continuous dose of hope should prove medicinal for humans and animals alike."

—*Publishers Weekly*

"[Written with] equal parts rigor and compassion, *Animal Madness* is a moving, pause-giving, and ultimately optimistic read."

—Maria Popova, *Brainpickings*

"This book should be required reading for veterinary and animal science students and for all who have any professional dealings with animals, wild and domesticated."

—Dr. Michael Fox, *St. Louis Post-Dispatch*

"In the hands of an observant and engaging writer like Braitman, this story is an outstanding example of a rigorous investigation presented in a most accessible way. Readers will also be rewarded by the deep compassion and gratitude she shows for all her subjects, both the animals and the humans who care for them."

—*The Bark*

"In the tradition of Marc Bekoff and Virginia Morell, Laurel Braitman deftly and elegantly makes the case that animals have complex emotional lives. This passionate, provocative, and insightful book deeply expands our knowledge and empathy for all species—especially, perhaps, our own."

—B. Natterson-Horowitz, M.D., and K. Bowers, coauthors of *Zoobiquity: Astonishing Connections Between Human and Animal Health*

"Where the BuzzFeed Animals page, for example, urges us to see animals as an undifferentiated mass of squee-worthy fluff, Braitman wants us to take animals seriously—to see them as individuals with life histories and psychologies as dramatic and intense as our own."

—*The New Yorker*

"Humane, insightful, and beautifully written, *Animal Madness* gives anthropomorphism a good name. Laurel Braitman's modern and nuanced definition of the word helps animals, helps people, and bolsters the connection between the two. Her thought-provoking book illuminates just how much we share with the creatures around us."

—Vicki Constantine Croke, author of *The Lady and the Panda* and *Elephant Company*

"*Animal Madness* is the sanest book I've read in a long time. Laurel Braitman irrefutably shows that animals think and feel, and experience the same emotions that we do. To deny this is crazy—which is why this fine book should be required reading."

—Sy Montgomery, author of *The Good Good Pig*

"Fascinating."

—*New York Post*

"Compulsively readable and thoroughly engaging. Laurel has the rare gift of being able to combine ideas, research and personal experience into a compelling narrative. Yet behind the engaging tone and the lightness of touch there is a deep seriousness, as indeed there should be. For the ideas that animate *Animal Madness* are of the greatest urgency and importance."

—Amitav Ghosh, author of *River of Smoke, The Glass Palace,* and *The Hungry Tide*

"There is much here that will remind readers of Jeffrey Moussaieff Masson—a gift for storytelling, strong observational talents, an easy familiarity with the background material and a warm level of empathy. . . . Engaging . . . Sparks curiosity."

—*Kirkus Reviews*

"Loving animals is easy. Thinking clearly about them can be almost impossible. Only a writer as earnestly curious as Laurel Braitman—so irrepressibly game to understand the animal mind—could draw this elegantly on both the findings of academic scientists and the observations of a used-elephant salesman in Thailand, on the sorrows of a famous captive grizzly bear in nineteenth-century San Francisco, and the anxieties of her own dog. *Animal Madness* is a big-hearted and wildly intelligent book. Braitman rigorously demystifies so much about the other animals of our world while simultaneously generating even greater feelings of wonder."

—Jon Mooallem, author of *Wild Ones*

"The wonderful thing [Braitman] discovered is that it is possible for animals to heal, a message crystallized by her encounters with 'friendly' gray whales who sought out human contact, even though they still bore harpoon scars from the whaling days."

—*Booklist*

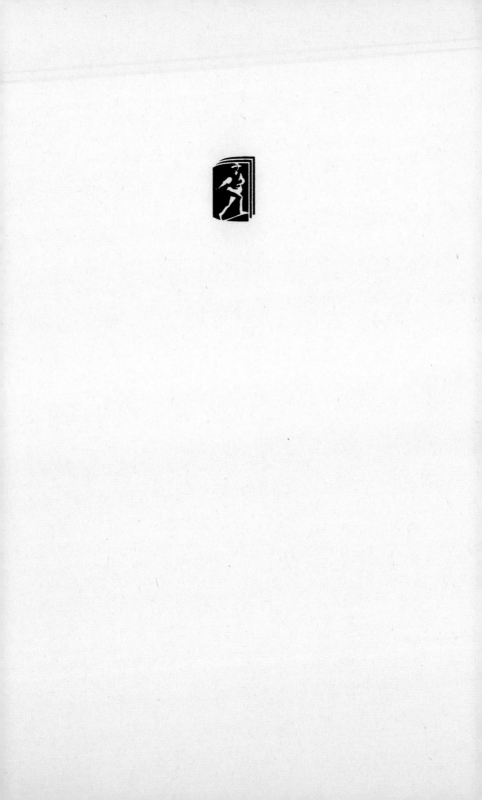

Animal Madness

Inside Their Minds

Laurel Braitman

Simon & Schuster Paperbacks
New York London Toronto Sydney New Delhi

Simon & Schuster Paperbacks
An Imprint of Simon & Schuster, Inc.
1230 Avenue of the Americas
New York, NY 10020

Note: Certain names and identifying characteristics have been changed.

Previously published as *Animal Madness: How Anxious Dogs, Compulsive Parrots, and Elephants in Recovery Help Us Understand Ourselves.*

First Simon & Schuster paperback edition October 2015

For information about special discounts for bulk purchases, please contact Simon & Schuster Special Sales at 1-866-506-1949 or business@simonandschuster.com.

The Simon & Schuster Speakers Bureau can bring authors to your live event. For more information or to book an event, contact the Simon & Schuster Speakers Bureau at 1-866-248-3049 or visit our website at www.simonspeakers.com.

Interior design by Ruth Lee-Mui

Manufactured in the United States of America

10 9 8 7 6 5 4 3

The Library of Congress has cataloged the hardcover edition as follows:

Braitman, Laurel, author.
Animal madness : how anxious dogs, compulsive parrots, and elephants in recovery help us understand ourselves / Laurel Braitman. — First Simon & Schuster hardcover edition.
pages cm
1. Animal behavior. 2. Animal psychology. I. Title.
QL751.B6834 2014
591.5—dc2
2014000791

ISBN 978-1-4516-2700-8
ISBN 978-1-4516-2701-5 (pbk)
ISBN 978-1-4516-2702-2 (ebook)

To all the animals I've loved before,
especially Lynn, Howard, and Dr. Mel

You see a dog growls when it's angry, and wags its tail when it's pleased. Now *I* growl when I'm pleased and wag my tail when I'm angry. Therefore I'm mad.

The Cheshire Cat to Alice, in Lewis Carroll, *The Annotated Alice*

Some day I'll join him right there,
but now he's gone with his shaggy coat,
his bad manners and his cold nose.

Pablo Neruda, "A Dog Has Died"

Contents

Foreword

A few months ago I was walking down a coastal trail north of San Francisco when I realized I was in love. The object of my affection was trotting ahead of me on a mysteriously urgent errand to smell a blackberry bush.

His name is Cedar and I am his human and he is my dog and he looks, depending on whether he's curling his tail up over his back, like a tiny Akita or a dark fox.

It was my friend Vanessa who made sure I adopted him. I'd been in Portland less than two hours when she loaded me into her van and drove me to the Oregon Humane Society. I'd been talking about adopting a dog for a while but I was hesitant. When you've known loss—messy, broken, snot-sobbing loss—it takes bravery and more than a smidgen of self-delusion to let yourself become completely and totally enamored with someone else again. Even if, maybe *especially* if, that someone is a dog. For reasons that this book will make clear, opening myself up to another canine took me awhile. I was hesitant, cagey, and a little nervous. I was not, however, cynical.

I didn't notice Cedar on our first visit to the shelter. Instead I'd asked to see a glossy black Labrador that turned out to radiate anger at other dogs simply for existing. It was on our second visit that we saw Cedar. He was mopey and sharing his cage with a shepherd mix. I took this as a sign that he wasn't inherently angry. He also had soft pointed ears and looked like he was wearing white athletic socks on his front paws. When Vanessa pressed her hand against the chain-link of his enclosure, he padded over and leaned his weight into her, casting his eyes mournfully upward and sneezing. In the cement meet-and-greet area Vanessa gave him bits of cheese from a tube while I followed him around asking pointedly: "Are you my dog? Can you tell me if you have separation anxiety? Please don't have separation anxiety. Also, are you housebroken? Please be housebroken. How do you feel about boats? Cats? Strangers?"

According to his chart, Cedar had been given up twice, once at six weeks old, and then again two years later. He'd been microchipped and when the shelter called the number linked to his chip they were told, "We don't want him anymore." No one would tell me why. It's possible that no one knew but I think that it's more likely no one was talking because they wanted this furry, sad little creature to find a home where people expected the best of him.

For some reason I cannot entirely explain, I did.

So far it's working out really well. This is mostly because I hired a dog trainer to teach us how to behave. Her name is Lisa Caper and when she pulled up to my house for the first time, wearing a T-shirt with a version of Shepard Fairey's iconic Obama/Hope image reimagined as a terrier with his head cocked and the word *adopt* written in giant block letters—I felt like everything was going to be okay. Things are also working out simply because Cedar is being himself. He has an inner calm and sturdy athletic confidence that my last dog did not. Most of Cedar's problems, or I should say, my problems with Cedar, stem from the fact that living with him is sometimes a bit like living with a very tall raccoon. He loves to get his paws wet and then put them all over everything and he will eat anything and everything left out on the counter. He also hates speeding road bikes and their spandex-clad riders, though I don't really blame him, and he thrills at the scent of cat, turkey sausage, and the dead seabirds that sometimes wind up on the beach near our house. He rolls on their remains until the wet feathers and tiny bones stick to his coat like a dead-bird costume. Already, after just a few months, I can't imagine my life without him.

The truth is that I wouldn't have adopted Cedar if not for this book and the readers who wrote to me with their own stories of dogs, cats, and other creatures they loved who tested their patience, the limits of their affection, and their preconceptions of animal minds. The stories of people helping phobic horses confidently face pedestrians with umbrellas, dogs cheering up elephants mourning lost companions, or goats rousing donkeys from deepest depression, blanketed me with hope not just for the ability of creatures to heal from emotional suffering but also the lengths that people and other animals will go to mend each other's broken spirits.

One Texan rancher called into a public radio station in Houston to tell me that all of his dogs have personality quirks, some of them verging on mental illness but that "a dog is just God coming at us ass-backwards." Whether you believe in God with a capital G doesn't really matter. In every dog—and possibly every donkey, kangaroo, or dolphin, there is a chance, often far more than one, for grace, forgiveness, and recovery.

Introduction

Mac the miniature donkey can be kind of a jerk. He bats his eyelashes, angles his long furred ears toward you, flatteringly, like TV antennas, and pushes his belly up against your thighs. Then, just as you've grown comfortable with his small, stocky presence, his burro smell of sagebrush and sweet alfalfa, something dark and confusing stirs within him. He stiffens, whips his head back, and bites down hard on the bony part of your shin and doesn't let go. Or he rears to stamp his hooves on your toes, or kicks his back legs like sharp springs in the direction of your kneecaps or into your actual kneecaps. If this wasn't painful, it would be funny. Mac is, after all, the size of a goat. But because you can't predict when it will happen, he is also a little scary. Mac shifts so suddenly from being affectionate and needy to violent and aggressive, transformations that don't seem to be triggered by anything in particular, that some people have taken to calling him "schizo donkey."

I am not one of these people. But I believe that he's disturbed. This, however, is not Mac's fault. Not entirely anyway. His mother, a stoic Sardinian miniature donkey, lived on the ranch where I grew up. She died within days of giving birth to Mac, and he was given to me to raise. I was twelve years old and saw this tiny donkey as a living stuffed toy. I spent hours bottle-feeding him and playing with him, until I got distracted by *Anne of Green Gables* books and my seventh-grade crush, a tan boy who skateboarded behind the local McDonald's. Mac was weaned too quickly, exiled to a corral without a donkey mother to show him the ropes—a small, unself-confident creature among indifferent adults. Another donkey may have been fine, but Mac wasn't another donkey. Eventually he began to turn his attacks on himself, biting his own fur off in chunks when he became frustrated or erupting in violent outbursts against people and other animals, outbursts that kept him from receiving the affection he also seemed to crave. Now, more than twenty years later, I know that Mac's experience and the disturbing behavior that resulted from it, is far from unique.

Humans aren't the only animals to suffer from emotional thunderstorms that make our lives more difficult, and sometimes impossible. Like Charles Darwin, who came to this realization more than a century ago, I believe that nonhuman animals can suffer from mental illnesses that are quite similar to human disorders. I was convinced by the experiences of many creatures I came to know, from Mac to a series of Asian elephants, but none more persuasively than a Bernese Mountain Dog named Oliver that my husband and I adopted. Oliver's extreme fear, anxiety, and compulsions cracked open my world and prompted me to investigate whether other animals could be mentally ill. This book is the tale of what I found: the story of my own struggle to help Oliver and the journey it inspired, a search to understand what identifying insanity in other animals might tell us about ourselves.

There isn't a branch of veterinary science, psychology, ethology (the science of animal behavior), neuroscience, or wildlife ecology dedicated to investigating whether animals can be mentally ill. What

I have done in this book is draw together evidence from the veterinary sciences and pharmaceutical and psychological studies; first-person accounts of zookeepers, animal trainers, psychiatrists, neuroscientists, and pet owners; observations made by nineteenth-century naturalists and contemporary biologists and wildlife scientists; and many ordinary people who simply had something to say about animals doing odd things around them. All of these threads, when pulled together, suggest that humans and other animals are more similar than many of us might think when it comes to mental states and behaviors gone awry—experiencing churning fear, for example, in situations that don't call for it, feeling unable to shake a paralyzing sadness, or being haunted by a ceaseless compulsion to wash our hands or paws. Abnormal behaviors like these tip into the territory of mental illness when they keep creatures—human or not—from engaging in what is normal for *them*. This is true for a dog single-mindedly focused on licking his tail until it's bare and oozy, a sea lion fixated on swimming in endless circles, a gorilla too sad and withdrawn to play with her troop members, or a human so petrified of escalators he avoids department stores.*

Every animal with a mind has the capacity to lose hold of it from time to time. Sometimes the trigger is abuse or mistreatment, but not always. I've come across depressed and anxious gorillas, compulsive horses, rats, donkeys, and seals, obsessive parrots, self-harming dolphins, and dogs with dementia, many of whom share their exhibits, homes, or habitats with other creatures who don't suffer from the same problems. I've also gotten to know curious whales, confident

---

* In this book I refer to abnormal behavior as the people who spend time with these animals do: as madness, mental illness, evidence of mental disorders, insanity, and more. These are generic words unfurled like leaky umbrellas over a whole host of behaviors considered abnormal. They're obviously unable to describe the ever-shifting patterns of the animal mind, not to mention the social expectations of what is *normal* in humans and other animals. Madness is a mirror that needs normalcy to exist. This distinction can be a murky one.

bonobos, thrilled elephants, contented tigers, and grateful orangutans. There is plenty of abnormal behavior in the animal world, captive, domestic, and wild, and plenty of evidence of recovery; you simply need to know where and how to find it. Oliver was my guide, even if he was too busy compulsively licking his paws to notice.

Acknowledging parallels between human and other animal mental health is a bit like recognizing capacities for language, tool use, and culture in other creatures. That is, it's a blow to the idea that humans are the only animals to feel or express emotion in complex and surprising ways. It is also anthropomorphic, the projection of human emotions, characteristics, and desires onto nonhuman beings or things. We can choose, though, to anthropomorphize *well* and, by doing so, make more accurate interpretations of animals' behavior and emotional lives. Instead of self-centered projection, anthropomorphism can be a recognition of bits and pieces of our human selves in other animals and vice versa.

Identifying mental illness in other creatures and helping them recover also sheds light on our humanity. Our relationships with suffering animals often make us better versions of ourselves, helping us empathize with our dogs, cats, and guinea pigs, turning us into bonobo or gorilla psychiatrists, or inspiring the most dedicated among us to found cat shelters or elephant sanctuaries.

For me, the realization that mental illness and the capacity to recover from it is something we share with many other animals is comforting news. When, as humans, we feel our most anxious, compulsive, scared, depressed, or enraged, we're also revealing ourselves to be surprisingly like the other creatures with whom we share the planet. As Darwin's father told him, "There is a perfect gradation between sound people and insane. . . . Everybody is insane at some time." As with people, so with everyone else too.

Chapter One

The Tail Tip of the Iceberg

> A bluetick hound bays out there in the fog, running scared and lost because he can't see. No tracks on the ground but the ones he's making, and he sniffs in every direction with his cold red-rubber nose and picks up no scent but his own fear, fear burning down into him like steam.
>
> Ken Kesey, *One Flew over the Cuckoo's Nest*

If a Dog Falls When No One Is Home

On a warm May afternoon in 2003, a little boy I'd never met was doing his homework in the sunroom off his family's kitchen in Mount Pleasant, a leafy neighborhood in Washington, D.C. The back of our apartment building faced the boy's house, and as he worked, he looked out to the row of urban yards along the alley, separated by chain link or small planks of sagging wooden fencing. He happened to look up that Saturday just as Oliver, our dark-eyed Bernese Mountain Dog, jumped through the kitchen window of our fourth-floor apartment.

No one had seen Oliver at the window, even though it must have taken him a long time to push the air-conditioning unit out of the way and rip a hole through the wire mesh of the screen that was big enough for his 120-pound body to fit through. The pet sitter that we'd left him with had gone to the farmer's market, leaving Oliver by himself for two hours. He must have begun to slash and chew through the screen as soon as he realized he was alone. Once he made the hole large enough, Oliver hauled himself through the opening, more than fifty feet above the ground.

"Mom!" the boy screamed. "A dog fell out of the sky!"

Later the boy's mom would tell us that she thought her son was making up a story, but there was fear in his voice that made her think otherwise. They found Oliver in the backyard of our building. He'd landed inside the cement stairwell of the basement apartment.

I'll never forget the phone call that followed. I was clutching a gin and tonic and had, until that moment, been worrying about underarm stains on my new chiffon dress. Jude was drinking a beer and sweating through the knees of his pants. We were milling about, uncomfortable in the heat, at a wedding reception for one of Jude's cousins in South Carolina. The wait staff had just announced the opening of the buffet when his cell phone rang.

The woman told us that she found Oliver lying in a heap. When he noticed her and her son pushing the backyard gate open, he'd tried to get up, wagging his tail weakly. Oliver's lips and gums were bloody and raw from gnawing at the metal screen, and he couldn't walk. The mother and son carried him to their car and rushed to the local animal hospital. In order to begin treatment the hospital required a $600 deposit; the woman gave them a check and then drove home to knock on the doors of our building to find out who this odd, broken dog belonged to.

"The vets didn't know the extent of his injuries when I left him," she told Jude and me when she reached us at the wedding, "but they did say that they'd never seen a dog survive a fall like this."

Overwhelmed, we thanked the woman for her generosity and hung up. I begged Jude to leave with me immediately. But it was almost evening in South Carolina and we couldn't make the last flight out in time. So we called the animal hospital to ask for any news (there wasn't any yet) and sat through the rest of the wedding, distracted and scared.

When I was twenty-one and on my way into the bathroom of a bar in upstate New York, I met Jude. We fell for each other in a way that felt like head injury—wholly and completely, with the sort of blurred vision that seemed to make anything possible. Before long we had a list of top-ten future pets. After a trip to China and Tibet, it grew to include a pair of yaks, and from the beginning I wanted to live with a capybara, but mostly we dreamt of dogs. At the very top of the wish list was a Bernese Mountain Dog. Bred to guard livestock and pull carts of cheese and milk through the Swiss Alps, Berners are handsome, broad, and regal, with an air of accessible friendship. Dog food companies know this. So do automakers. Bernese are the supermodels of the canine world, popping up in advertisements for organic kibble, paper towels, perfume, SUVs, and phone plans.

When Jude and I moved into an apartment in Washington that allowed dogs and was located just off Rock Creek Park's pools of water and walking trails, I started looking for puppies.

I found them. But I was crushed to learn that purebred Bernese Mountain Dogs sold for nearly $2,000 each. I was working for an environmental conservation organization at the time, and Jude, a government geologist, wasn't earning much more than I was. We couldn't afford a puppy that expensive, and even if we could, I couldn't justify spending that much on a dog. So a few months went by during which we felt like perverts at the dog park—dogless people who came to look at dogs, luring other people's pets over to be petted with clandestine pockets of treats. "Heeeeere doggie doggie."

And then one day I received an email from a breeder I'd contacted a few months earlier. One of his adult dogs was available now, "for free!" He told me that this Berner, named Oliver, was four years old and wasn't getting the attention he needed from his current family. He said that since Oliver was an adult dog he required slightly less exercise than a puppy and would be more easygoing.

I scheduled our first meeting to take place within twenty-four hours. When we pulled up to the veterinary office to meet Oliver and his current family, we saw a young girl walking a gigantic dog on the clinic's front lawn. He carried his white-tipped tail like a flag, raised high and arching over his back. His white paws were lionlike, huge and spreading, and his coat glossy and feathered as a 1970s shag. He looked happy to be walking with the small girl, and his gait was jaunty as she led him back and forth across the lawn.

When I think about it now, it's striking how much I didn't notice. Adopting a family pet from a veterinary office and not the family's home was perhaps the first clue. There were many others but I was blind to all of them.

Oliver was being boarded at the vet because he wasn't legally allowed to remain in the family's neighborhood. He'd had an altercation with a neighbor and her dog, and they were threatening to sue. While it sounds quite serious to me now, it didn't at the time. The mother of the family, Oliver's primary human, explained that he'd "just gotten so excited about the neighbor's new dog that he dashed through their electric fence to say hello." The dogs began to fight and the woman tried to break it up with her hands. Oliver bit the woman while she was trying to separate them. I didn't need to hear more. Everyone knows you shouldn't break up a dogfight with your bare hands; that's what garden hoses are for. Plus, this neighbor must have been unreasonable. Jude and I would be able to control our dog. He just needed some training.

In retrospect I know the biting story was the tip of the iceberg, or really the tip of the tail on a very large dog, but at the moment I didn't, I couldn't, absorb it.

We'd fallen for Oliver at first sight. It felt more like a physical sensation than a conscious decision. It certainly wasn't rational. We brought him home that same afternoon.

After a few days of cool appraisal, Oliver settled into a routine with Jude and me and became very affectionate. We spent hours playing hide and seek in our apartment and the park, playfully tweaking his whiskers, wondering aloud what his voice might sound like if he could talk, and filling endless trash bags with the fur we brushed from his coat. It wasn't until a few months into our relationship with Oliver that his truly bizarre behavior started to manifest. But once it did, it spread like spilled molasses: sticky, inexorably expansive, and difficult to contain.

The first real sign of trouble I discovered by accident. Jude had already left for work. I said goodbye to Oliver and locked the house, only to realize as soon as I reached my car that I'd left the keys in our apartment. As I headed back up the block to our building I heard a plaintive yowling—not feline or human and not from the National Zoo, a few blocks away. It was a bark that sounded like the squeak of an animal too large to squeak (this was before I knew any elephants), and it was coming from our apartment.

When I stepped onto the front porch the barking stopped and was replaced by a loud skittering sound. As I climbed the steps to the top floor, the crablike skittering got louder. It was, I realized, the sound of Oliver's toenails on the wooden floor as he sprinted back and forth along the length of the apartment. When I opened the door he was panting and wild-eyed. He bounded up to me as if I'd just returned from a months-long expedition, not a five-minute trip to the car. I picked up my keys, walked Oliver back to his dog bed, petted him a bit, and then got up to leave. When I reached the sidewalk I sat on the porch and waited. After about ten minutes of quiet, I stood up in relief. Then suddenly, after only a few steps, there it was—the yowlingsqueakbark. Again and again and again. I looked up and saw Oliver's giant head pressed against our bedroom window, his paws on the sill. He was looking down at me with his tongue lolling. He'd waited to bark until he saw me leave the porch. I was already late for work. As I walked down the sidewalk I kept turning around. Oliver had moved to the living-room window so that he could watch me walk farther down the street. The barking increased when I turned the corner, and the whole drive to my office I could hear it inside my head.

That evening, when Jude got back from work, he discovered that Oliver had gnawed through the center of two bath towels and turned the pillows on our bed into a pile of goose down and shredded cases. There was also a mysterious pile of wood shavings in the hallway and toenail tracks in the floors, like ghost tracings on a chalkboard, in front of all the windows in the apartment. Strangely, his front paws were also quite wet.

Later that night, as Jude and I lay in bed, our heads resting on folded sweaters, he slid close to me and said, "Do you think there's anything that his old family didn't tell us?"

I could feel Oliver's presence next to us in the dark. He always began the evening curled into a large oval in the doorway to our bedroom and then, after we'd fallen asleep, moved to his dog bed, a round cushion with the footprint of a Smart car, next to the sofa. He was breathing softly.

"I can't imagine they would have lied."

And yet, even as I said the words I could feel the doubt coming loose within me like disturbed sediment on the bottom of a pond.

What Darwin Knew

Trying to understand what was happening between Oliver's furry ears while he savaged our towels or yowled alone at the window was confusing. In many ways, attempting to understand the relationship between what animals are thinking and what they are doing always has been.

In 1649 the French philosopher René Descartes argued that animals were automatons, lacking in feeling and self-awareness and operated unconsciously, like living machines. For Descartes and many other philosophers, capacities for self-consciousness and feeling were the sole province of humanity, the rational and moral tethers that tied humans to God and proved we were made in his image. This idea of animals as machines proved to be sturdy and enduring, revisited time and again for hundreds of years to prop up arguments for humanity's superior intelligence, reasoning, morality, and more. Well into the twentieth century, identifying humanlike emotions or consciousness in other animals tended to be seen as childish or irrational.

The most resounding blow to this idea of human exceptionalism, at least in Western scientific circles, was delivered by Charles Darwin, first in *On the Origin of Species*, then in *Descent of Man*, and quite richly detailed in *On the Expression of the Emotions in Man and Animals*, published in 1872. *Expression* was one of Darwin's last published arguments in support of his larger theory that humans were just another kind of animal. He believed that the similar emotional experiences of people and other creatures were additional proof that we shared animal ancestors.

In *Expression* Darwin described surliness, contempt, and disgust in chimps, astonishment among Paraguayan monkeys, love among dogs, between dogs and cats, and between dogs and humans. Perhaps most surprisingly he argued that many of these creatures were capable of enacting revenge, behaving courageously, and expressing their impatience or suspicion. A female terrier of Darwin's, after having her puppies taken away and killed, impressed him so much "with the manner in which she then tried to satisfy her instinctive maternal love by expending it on [Darwin]; and her desire to lick [his] hands rose to an insatiable passion." He was also convinced dogs experienced disappointment and dejection.

"Not far from my house," he wrote, "a path branches off to the right, leading to the hot-house, which I used often to visit for a few moments, to look at my experimental plants. This was always a great disappointment to the dog, as he did not know whether I should continue my walk; and the instantaneous and complete change of expression which came over him, as soon as my body swerved in the least towards the path (and I sometimes tried this as an experiment) was laughable. His look of dejection was known to every member of the family and was called his *hot-house face*."

According to Darwin this doggish disappointment was unmistakable—his head would droop, his "whole body sinking a little and remaining motionless; the ears and tail falling suddenly down, but the tail was by no means wagged. . . . His aspect was that of piteous, hopeless dejection." And yet, "hot-house face" was really only the beginning for Darwin.

He went on to document grief-stricken elephants, contented house cats, pumas, cheetahs, and ocelots (who expressed their satisfaction with purring), as well as tigers, whom he believed did not purr at all but instead emitted "a peculiar short snuffle, accompanied by the closure of the eyelids" when happy. He wrote about deer at the London Zoo—who approached him because, he believed, they were curious. And he talked about fear and anger in musk-ox, goats, horses, and porcupines. He was also interested in laughter. "Young Orangs, when tickled," reported Darwin, ". . . grin and make a chuckling sound" and "their eyes grow brighter."

It wasn't until he published a revised edition of *Descent of Man* in 1874 that Darwin opined on insanity in other animals directly. He wrote:

> Man and the higher animals especially the Primates, have some few instincts in common. All have the same senses, intuitions, and sensations—similar passions, affections and emotions, even the more complex ones, such as jealousy, suspicion, emulation, gratitude, and magnanimity; they practise deceit and are revengeful; they are sometimes susceptible to ridicule, and even have a sense of humour; they feel wonder and curiosity; they posses the same faculties of imitation, attention, deliberation, choice, memory, imagination, the association of ideas, and reason, though in very

> different degrees. The individuals of the same species graduate in intellect from absolute imbecility to high excellence. They are also liable to insanity, though far less often than in the case of man.

Darwin doesn't seem to have done any original research on the topic; instead he cites William Lauder Lindsay, a Scottish physician and natural historian who believed nonhuman animals could lose their minds. In a paper Lindsay published in 1871 in the *Journal of Mental Science,* he wrote, "I hope to prove that, both in its normal and abnormal operations, mind is essentially the same in man and other animals."

Lindsay knew a fair bit about both, particularly the human insane. He'd been appointed medical officer to Murray's Royal Institution for the Insane at Perth in 1854 and held the job for twenty-five years. Meanwhile he kept up with his botanical interests, publishing a popular book on British lichen in 1870, and like Darwin, he was a member of the Royal Society, which awarded him a medal for "eminence in natural history." Lindsay combined his interest in natural history and his experience treating the mentally ill in a two-volume masterwork published in 1880 titled *Mind in the Lower Animals.* It covered morality and religion, language, the mental condition of children and "savages," and more. But it is the second volume, *Mind in Disease,* that is truly remarkable.

Like Darwin, Lindsay believed that the minds of insane people, criminals, non-Europeans, and animals were similar. Insane people could be recognized by "their use of teeth for vicious biting" and their "filthy habits." Lindsay wrote that many of these insane people "'eat and drink like beasts,' tearing raw flesh and lapping water; they bolt their food and gorge themselves as certain carnivora do." He also believed many preferred to spend time with other animals instead of people, acquiring something like animal language that allowed them to communicate with their nonhuman companions. Lindsay noted that an Italian "idiot" known as the Bird Man would leap on one leg, stretch his arms out like wings, and hide his head in his armpit. He also chirped when frightened or at the sight of strangers.

Lindsay also wrote about feral children like the Wolf Children of India, said to be raised by wolves. He classified them as a subtype of lunatic that walked on all fours, climbed trees, prowled around at night, lapped water like oxen, smelled food before eating it, gnawed on bones, refused clothing, and had no language, sense of shame, or ability to smile. Like generations of physicians before him, Lindsay understood his patients by analogy to other animals.

Insane humans were also compared to—and treated like—animals at the famous Bethlem Royal Hospital in London, the place that inspired the word *bedlam* for the chaos so often found within. Until the hospital outlawed visits by the general public in 1770, Bethlem was a popular spectacle. Watching the mentally ill, like the patient supposed to crow all day long like a rooster, was considered good entertainment, along with other pursuits, like prostitution, that flourished in and around the hospital. Despite serving as a human menagerie of the insane, Bethlem almost certainly housed sane people too, who had been committed because they were inconvenient or too eccentric for their families. As in an animal menagerie, the more uncontrollable patients were chained by the neck or foot to the wall and stripped naked. It's not surprising that the stench and brutal conditions of the hospital, as well as the weird behavior of so many of its patients, tended to remind people of dog kennels or circuses. Conditions improved over time, but one 1811 visitor reported that chains and handcuffs were still being used, and some of the incurables "are kept as wild beasts constantly in fetters."

Lindsay is intriguing because, despite working as the medical officer at another British insane asylum, he didn't limit his studies to crazy humans acting like animals. He also refused to see animals themselves as dumb beasts. Instead Lindsay believed that animals *themselves* could go insane. He was even convinced that some human lunatics were more mentally degenerate than sane dogs or horses. In *Mind in Disease*, a sort of Victorian mental illness field guide, Lindsay posited many forms of animal insanity, from dementia and nymphomania to delusions and melancholia.

Lindsay was also convinced that animals exhibited what he called "wounded feelings" of many kinds, and he tells story after story on the subject. There was a mother stork who "let herself" be burned alive rather than desert her young and a Newfoundland dog who was so sad after being scolded, then ceremoniously beaten with a handkerchief, and finally having a door shut in his face when about to leave the room with the nurse and the family children (his usual companions) that he "tried twice to drown himself in a ditch but survived . . . only to stop eating." He died soon thereafter.

All of this was a worthy course of study for Lindsay not simply because he was convinced that insanity in other animals was a lot like insanity in humans but because it was also dangerous. "Mental defect or disorder," as he called it, in horses, oxen, or dogs could be terrifying. The cause of violence or aggression in these animals was often puzzling and mysterious, and it inspired fear because there were so many horses, dogs, and cattle living in close proximity to people in his day, even in big cities. Angry oxen bent on murder or horses mad with the desire to kick or stomp were actual public health risks during Lindsay's lifetime and for a long while afterward.

Silver Lining

The day after Oliver jumped out of our apartment window, Jude and I caught the first flight back to D.C. and drove straight to the animal hospital. A tech ushered us into the back area of the clinic and said, "We honestly have never seen a dog survive a fall like this. We've been bringing all of the vet students by to see him." She led us to a bank of cages along a far wall and said that Oliver was a bit groggy but awake.

He was curled into a sleepy lump inside a cage that was barely big enough for him to turn around in. A rectangular area of his front left leg was shaved clean, and his freckled muzzle was marked with jagged cuts and scratches. "Beast!" I called, his nickname.

Oliver raised his head and looked straight at Jude and me. His tail thumped awkwardly against the floor of the cage and he tried to get up. I felt relieved and also useless, unable to figure out how to stroke him through the wire mesh.

The attending veterinarian approached and asked if we had a moment to talk. "The silver lining," he told us, "is that Oliver is too sore to try and jump out of your apartment again any time soon."

Even though he had fallen fifty-five feet onto cement, to the shock of every vet and vet tech at the hospital, he hadn't broken a single bone. He was bruised and sore and wouldn't be able to walk for weeks, but the clinic staff told us that he'd make a full recovery, at least physically. "Make a sling from a bed sheet and carry him downstairs to use the bathroom every few hours," the vet told us. "Also, you are going to have to see a veterinary behaviorist. I will give you some Valium you can dose him with now, but that is not a long-term solution."

"What is the long-term solution?" I asked.

"Move to a first-floor apartment," he said and left the room.

Had we known what to look for, Jude and I might have noticed the full extent of Oliver's anxiety before he jumped out the window. Looking back, I was distressed by his distress, and humbled by it, but I'm not sure I ever completely understood what he was capable of.

As our first year with Oliver wore on, Jude and I had begun to notice ever stranger behavior and continued to wonder if Oliver had experienced something traumatizing when he lived with his previous family. His anxiety accreted steadily whenever we left the house. He then exploded in a slobbery, excited fiesta of return, even if we'd only gone downstairs to take out the trash. In the evenings he'd snap at flies that didn't exist. Training his gaze on what seemed like invisible insects, he tracked them like a pointer. Oliver was in a kind of trance as he did this and couldn't be distracted with cheese, bits of meat, or affection. He was also becoming something of a liability at the dog park; he had begun to approach the place as a sort of canine buffet, the smallest Dachshunds and pugs like unattended snacks. He hadn't bitten another dog yet, but he would catch sight of a creature that piqued his interest and take off at a sprint, no matter how far away the other animal was, his large bulk stopping just short of bowling the dog over, terrifying their human companions. This did not seem to be done playfully.

Oliver also ate a variety of inedibles with gusto, things like plastic and sometimes hand towels; since he was years out of puppydom, Jude and I found this troubling. One night, after watching him retch for hours and produce nothing, we made a late-night trip to the vet hospital, where the staff took an X-ray and found a large obstruction in his lower intestine.

"Surgery is likely the only solution," the vet told us, "but first we can try something else. It's a long shot, but a doggy enema might work."

An hour later a tech appeared in the waiting room and presented us with what I thought was a small brown, plastic accordion. "This is a first for us," she said, "but we think it is an intact sleeve of Saltines."

Oliver had not only eaten the sleeve of crackers whole; he'd also eaten the ziploc bag that they were stored in. His intestinal tract had compressed the plastic into what looked like a bile-cured musical instrument.

Then there were the wet paws. The soggy feet that Jude and I noticed early on were quickly traced to a habit in which Oliver licked his front paws for hours at a time. We tried changing his diet, washing him with different shampoo, and walking him along different trails, just to make sure he wasn't suffering from an allergy, to no avail. The licking continued, to such an extent that tongued spots on his once lushly furred front paws turned bare and oozy. Sometimes he gave up on his paws and focused instead on his tail, chewing open a sore that he licked until it looked like pastrami and smelled worse. The vet told us that this was a compulsive behavior and to make him wear a plastic cone collar. Oliver, like most dogs, hated that thing. At first he tried to outrun it. He could see the cone out of the corner of his eyes, looming uncomfortably just out of reach. He would rush around the house, running a few steps and then looking anxiously side to side. But no matter how fast he dashed to and fro, the cone stayed in his side vision. We felt embarrassed for him and took it off.

By this point Oliver's anxiety was beginning to wear on me. If we didn't return home by five or six in the evening, we knew he would have destroyed pillows and towels or chewed on wooden moldings. He scratched so hard at our floorboards that it looked as if we lived with giant termites. Hiring a dog walker to come in the afternoons helped but didn't fix the problem, and one afternoon when the dog walker took Oliver back to his own house and left him alone for an hour, Oliver clawed and chewed his couch upholstery into damp shreds. Jude and I ended up coordinating our schedules so that one of us went into work late and one of us came home early. If we were with him, outside of the fly-snapping and prey drive at the dog park, Oliver was the picture of calm. Alone he was a tornado.

I found this out because I filmed him. Jude and I were curious why some days were worse than others on this new Richter scale of destruction, so I borrowed a video camera and set it up to film the apartment when we left. There was, it turned out, something else besides being left alone that could send Oliver beyond the brink of composure: thunderstorms. If those two events were combined, it was as if someone had tossed an anxiety grenade into the apartment. He frothed at the mouth, paced, quivered, and settled down in the crack between the bed and the wall, only to get up again seconds later and try to wedge his large body underneath the coffee table. Unfortunately it seemed as though every other day in the summer the humidity built into a thunderstorm that crested a few hours before we returned home. Sitting in my office across town I'd see the flashes of light through the window, feel the thunder in my chest, and worry about Oliver, a quaking fur ball of nerves, back at the house.

In his beautiful book *Dog Years,* Mark Doty writes, "Being in love is our most common version of the unsayable; everyone seems to recognize that you can't experience it from the outside, not quite. . . . Maybe the experience of loving an animal is actually more resistant to language, since animals cannot speak back to us, cannot characterize themselves or correct our assumptions about them." Caring for animals like Oliver happens outside of verbal language, but it's a descriptive language all the same. Dogs in particular make us more expressive in all kinds of ways. They make us act more like dogs, rolling on the floor or hopping side to side to get them excited, a sort of transspecies basketball drill. They make us stop at good places to pee. They make us go to the park and notice the weather, mouldering bits of trash, entrances to the burrows of small animals. In short, they make us pay attention to what we might otherwise miss.

Dogs are also good barometers for relationships and often act like the third corner of a triangle connecting two people who otherwise would look only at each other. Oliver was no exception.

As his anxiety grew, and with it his need for structure, exercise, companionship, and routine, life became more stressful for Jude and me. We also had different ideas about what structure and routine actually meant. Jude had raised a guide dog for the blind, and while he knew a lot about training confident, calm dogs, I thought he lacked compassion for Oliver's idiosyncrasies. Once, he'd taken Oliver on a work trip out of town and left him alone for the day at a friend's house—something that would not have been a problem for an easy-going dog. Oliver, however, jumped out of the living-room window (luckily on the first floor) and brought the friend's two dogs with him. It took hours for all three to be rounded up again. Jude, feeling that he couldn't leave Oliver at his friend's house again lest he make another jailbreak, took him to a nearby kennel and left him there for the rest of the week. When they came home, I felt that Oliver's anxiety over being left alone had only increased. He jumped out of our apartment a few weeks later.

In general, of the two of us, Jude was much more likely to say, "He's a dog. He can handle it." Looking back, I don't know who was right. I think we were both alone at sea in our particular ways. But I was beginning to think of Jude as more callous than he should be. And Jude thought I was becoming the kind of person who spent too much time and money worrying about something that we couldn't fix and blamed him unfairly. I suspected that Jude lacked compassion not just for Oliver but for me too. Our leash was fraying.

My preconceptions about nonhuman minds were fraying too. I was suddenly seeing Olivers and potential Olivers everywhere. It was as if my own dog's crisis had given me canine-anxiety-tinted goggles. I still noted dogs doing dog things, but I was beginning to regard them as individuals with their own emotional weather systems that guided their behavior as they whizzed, panted, lolled, and humped. These weather systems could also compel them to do odd things. As I talked about Oliver's puzzling behavior with other dog owners at the park, at dinner parties, with people I'd just met and others I'd known for years, I started to collect their stories too.

It turns out that almost everyone has come across a disturbed animal at some point, and most people want to tell you about it. I've been pulled aside at almost every social gathering I've attended in the past six years to be regaled with tales of cats peeing only on left shoes or plucking their bellies bald while hidden under the bed, other dogs who've jumped from apartment buildings or reacted with mortal fear to stop signs or anything that makes a flapping sound, hamsters who wouldn't get off their wheels, and parrots who developed violent fixations on people who wear baseball caps or have long hair.

Just how similar *are* these experiences to human ones? Extrapolating from a monkey's seeming depression to a human's, may, because of our many primate similarities, be relatively easy. But what about the emotional experiences of other animals? Of dogs like Oliver? Was what he felt when left alone anything like the terror I remember feeling when I woke from a nightmare in the middle of the night at a friend's sleepover party, unable for the first few minutes to remember where I was or find my mother?

Returns and Arrivals

In many ways the past forty to fifty years of research on animal emotions and behavior represents a long, slow, scientific U-turn back to Darwin and his arguments for the shared nature of emotional experience. Researchers like Nikolaas Tinbergen and Konrad Lorenz laid the foundation for this U-turn. Tinbergen was a renowned behaviorist, working from the 1930s through the 1960s, who studied birds and insects. Lorenz experimented over the same years on innate versus learned behavior, in fighting fish and in birds who followed him about like a mother goose. Their studies represented an alternative to the research of B. F. Skinner and the radical behaviorists, who tended to see animal behavior more like Descartes did, a disembodied series of responses. Lorenz even described one of his geese as depressed when she refused food and stopped waddling around after one of her wings was clipped.

The work of these researchers and their peers created the field of ethology, or animal behavior, as we know it today and cleared a path for others like Jane Goodall. When Goodall shared tales of expressive chimps welcoming her into their social lives in the Gombe in the 1960s, she helped shift public opinion of what nonhuman animals were capable of. Books like Rachel Carson's *Silent Spring*, published in 1962, also helped galvanize the new environmental movement, contributing to what would turn out to be a decades-long fertile environment for recognizing animal minds, feelings, and kinship.

The sea change gathered force in 1976 when the zoologist Donald Griffin published *Animal Awareness*, positing that animals have conscious minds. It was helped by Roger Payne and Scott McVay's recordings of humpback whale songs, which made them seem not like instinctual automatons but musicians, Dian Fossey's work with gorillas in Rwanda, and reports by elephant researchers such as Cynthia Moss, Joyce Poole, and Katy Payne on the conscious, emotional, and communicative creatures in Africa in the 1970s, 1980s, and 1990s. All of this suggests that Descartes is now in the doghouse and the dogs have left.

The neuroscientist Jaak Panksepp holds the Baily Endowed Chair of Animal Well-Being Science at Washington State University's College of Veterinary Medicine. He's also a distinguished research professor emeritus of psychology at Bowling Green State University and the head of Affective Neuroscience Research at the Falk Center for Molecular Therapeutics at Northwestern. He has another, slightly less sonorous title as well: rat tickler. One of my favorite YouTube videos is of Dr. Panksepp stirring an open-topped cage of chubby rats with his hand as they roll over to be tickled. "We obtained these transducers that are called bat detectors that can bring very high frequency sounds down to our auditory range," he says as the camera pans over the apparently joyful rodents chittering away. "And when we did this and listened in, we could tickle animals and generate a LOT of vocal activity that appeared to be laughter." The rats emit this same sound when they're mating, when they're about to receive food, when a lactating mother is reunited with her baby, and most of all, when two friendly rats are playing with each other. The rats make a totally different sound, also inaudible to humans, when they're scared, fighting, or have just been defeated in a tussle with another rat. Baby rats make a version of this same sound when they're abandoned or kept from their mothers. Panksepp believes the happy sound roughly corresponds to human laughter and the lower sound signals distress or psychic pain. He compares it to human moaning.

Like Lauder Lindsay, Panksepp began his career in a mental institution. One of his last college summers he took a job as a night orderly in the psychiatric unit of a Pittsburgh hospital. The position gave him time to get to know individual patients, from those with relatively minor problems to the most violent and psychotic patients kept in padded cells. He spent his free time reading about their life histories and watching how the patients responded to the newly available pyschiatric drugs of the 1960s. "Toward the end of my undergraduate days," he wrote, "I increasingly wanted to understand how the human mind, especially emotions, could become so imbalanced as to wreak seemingly endless havoc upon one's ability to live a happy life in the outside world." He became a clinical psychologist, and eventually, a neuroscientist focused on plumbing emotional states.

After decades of research, Panksepp is convinced that most animal brains, from Oliver's to a ticklish mouse's, likely have the capacity for dreaming, for taking pleasure in eating, for feeling anger, fear, love, lust, grief, and acceptance from their mothers, for being playful, and for some conception of selfhood, an argument that might have seemed painfully unscientific just forty years ago. Panksepp believes that emotional capacity evolved in mammals long before the emergence of the human neocortex and its massive powers of cognition. He is careful to say that this doesn't mean that all animal or even mammalian emotions are the same. And when it comes to complex cognitive skills, he believes that the human brain puts all others to shame. But he is convinced that other animals have many special abilities that we don't have and this may extend to emotional states. Rats, for example, have richer olfactory lives, eagles have impressive eyesight, and dolphins can sense the world via sight, sound, sonar, and touch. These abilities may translate into more and different feelings associated with their various sensory or cognitive experiences. Panksepp believes that rabbits, for example, may have bigger or different capacities for fear while cats may have larger capacities for aggression and anger.

Over the past fifteen years the cognitive ethologist Marc Bekoff has published accounts of many types of animal emotions, from compassionate chimps to contrite hyenas. The primatologist Frans de Waal has written of altruism, empathy, and morality in bonobos and other apes. An explosion of recent research on dogs plumbs their ability to mirror the emotions of their owners, and studies of hormonal fluctuations in baboons after the death of their troops' babies have shown monthlong spikes of glucocorticoid stress hormones in the mothers, chemical surges that point toward a long grieving process. A number of recent studies have gone far beyond our closest relatives to argue for the possible emotional capacities of honeybees, octopi, chickens, and even fruit flies. The results of these studies are changing debates about animal minds from "Do they have emotions?" to "What sorts of emotions do they have and why?"

Perhaps this shouldn't be too surprising. As the neurologist Antonio Damasio has argued, emotions are a necessary part of animal social behavior. Consciously or not, they guide our behavior, helping us to flee from danger, seek pleasure, avoid pain, or bond with the right fellow creatures. Both dolphins and parrots, for example, can exhibit symptoms similar to human sadness and depression after the loss of a companion. They might ignore food or refuse to play with others. Other social animals, like dogs, often do the same. These emotions are consequences of a very helpful evolutionary process: attaching to others who protect you, feed you, play with you, groom you, hunt or forage with you, or otherwise make your life more enjoyable or productive. Affective states, as the emotional expressions of animals are known, are useful whether you're a prairie dog collaborating with other prairie dogs on a tunnel extension or a harried human negotiating who is going to pick up dinner on the way home from work.

Lori Marino is a senior lecturer in the Neuroscience and Behavioral Biology Program at Emory University and has researched primate, dolphin, and whale intelligence and brain evolution for decades. She has also worked on key studies of dolphin cognition, proving, along with Diana Reiss, that dolphins can recognize themselves in mirrors. "I think that emotions—although they are subject to selection—are one of the oldest parts of psychology, laid down in the first animals," Marino told me. "This is because without emotions an individual cannot act or make the kinds of decisions that are key to survival. Of course, some emotions are basic and others are tied into cognitive processes, so some are more complex than others. But every animal has emotions."

The ethologist Jonathan Balcombe believes that emotions likely evolved with consciousness, as the two serve each other. Today, researchers are no longer debating whether other animals are conscious but to what *degree*. Recent studies have attempted to show that consciousness isn't limited to humans and the other great apes, mammals, or even, perhaps, vertebrates. A subset of these animals has also been shown to be self-conscious in the context of cognitive and behavioral experiments; that is, they were able to conceive of themselves as beings independent from other animals and from the rest of their environment. Mirror recognition tests are the stock in trade of animal cognition research; they consist of drawing or dyeing a mark on an animal's body and then placing a mirror in front of them. If while looking in the mirror the animal touches the marked spot in a statistically significant manner, he or she is demonstrating self-awareness. That is, the animals are using the mirror as a tool to explore the mark that wasn't there before, something the researchers consider proof that the animals conceive of themselves as the beings in the mirror.

As of this writing, the only animals to have been proven self-aware in such a way are chimpanzees, orangutans, elephants, orcas, belugas, bottlenose dolphins, magpies, and humans, but only after the age of two. Pigs have been tested but the results were inconclusive. One pig looked behind the mirror to find the food reflected in it. And while African Grey parrots used the mirrors as tools to find food in cupboards, it was not obvious that they recognized themselves. These experiments, while helpful, demonstrate only which animals are *interested* in looking at themselves in mirrors. The actual list of self-aware animals may be much longer. The African Greys, for example, might have known that they were looking at themselves but may have found the mirrors more worthwhile to use as tools for finding snacks. Not caring about what you look like isn't the same as not *knowing* what you look like.

In 2012 a group of prominent neuroanatomists, cognitive neuroscientists, neurophysiologists, and ethologists released the Cambridge Declaration on Consciousness. The declaration sought to establish, once and for all, that mammals, birds, and even some cephalopods, like octopi, are conscious creatures with the capacity to experience emotions. The authors argued that convergent evolution in animals gave many creatures the capacity for emotional experiences, even if they don't have a cortex, or at least one as complex as the human neocortex.

And yet, despite the pronouncements on consciousness and the flowering of new research, debates surrounding animal emotion and feeling are as lively as ever. Researchers studying animal cognition, emotion, and intelligence often disagree about what capacities nonhuman animals have, as well as the best ways to evaluate them. The burgeoning field of affective neuroscience, or the neuroscience of emotion, has not simplified the topic. If anything, it has made it far more complex. Neuroscientists, behaviorists, and psychologists at many of the world's top research institutions have varying theories on how humans process emotions, how many emotions we share with other animals, and even what emotions actually *are*.

Despite centuries of investigation by everyone from natural historians, psychologists, and psychiatrists, to ethicists, neuroscientists, and philosophers, there is still no universal definition of emotion or consciousness. As I mentioned earlier, a number of researchers have agreed that animals share the capacity for the emotions of fear and enjoyment. It's highly likely, however, as the neuroscientist Jaak Panksepp suggests, that animals experience many more than these. What, for example, is the bee emotion associated with seeing a particularly pleasing ultraviolet pattern inside a flower? What does the dolphin emotion for sensing a sonar ping from a long-lost companion feel like? The octopus emotion associated with performing a sudden, flushing change of skin color? Other animals have different physiological experiences than we do and those may come with their own emotional experiences. Because of this, it's difficult to make a finite list. There isn't consensus even on the universal human emotions. The psychologist Paul Ekman put forth the most famous list of what he called "basic" human emotions: anger, fear, sadness, enjoyment, disgust, and surprise. But what about excitement, shame, awe, relief, jealousy, love, or joy? Attempting to reduce all of these complex states to a grocery list of experiences may be beside the point, especially since we know how useful they can be.

Humans have to be especially careful when ascribing emotional states to other animals' behavior. Consider the waterlogged possum I found a few winters ago, crouching in a metal trashcan near my house in Boston. It was a cold morning and I heard a scratching sound as I walked by. The possum, a female, was huddling beneath a piece of cardboard. I assumed that she'd fallen in the night before and then couldn't climb the smooth-sided walls to freedom. But how did this possum feel when I peered down into the trashcan? I was a giant, silhouetted by the bright morning sun, wearing a fuzzy hat and talking at her in human-speak. It's tempting to conclude that the possum hid underneath the piece of cardboard because she was scared of me, and we know she was scared of me because she hid underneath the piece of cardboard. This sort of circular reasoning is an alluring trap to fall into. Interpreting the possum's emotional state from her behavior would be much more accurate if we knew this particular possum's natural history and perhaps even her own past experiences. (Did she do this often? Did she have a thing for cardboard or human trash? Was she raised by a wildlife rehabilitator and therefore not very scared of people?) The benefits that come with knowing an individual animal, his or her normal and abnormal behavior, is why so many of us first learn about other animals' emotional lives from our pets. We spend a lot of time with dogs and cats and we come to know them not at the species level but as individuals. Oliver's fear, anxiety, and compulsions were noticeable to me only because I knew what he was like when he wasn't feeling fearful, anxious, or compulsive. When he hid from me, for example, it wasn't because he was feeling scared; it was because we were playing hide and seek.

As I watched Oliver's disturbing behavior grow more intense, his nightly relentless paw licking, for example, or his frenzied concern over being left by himself, I puzzled over what was going on in his mind. Like so many other animals, he was a furry enigma. And yet discovering the particularities of what he was actually thinking didn't matter that much when it came to helping him. The reality of Oliver's raw, self-inflicted sores and my inability to distract him from making them worse was enough to tell me that he was too focused on something that was doing him harm. On one particularly bad evening, he gnawed on the base of his tail until he'd made a hole the size of a tennis ball. But he would choose other body parts too, taking a break from his tail to lick some other limb into hairlessness or injury. What I didn't know, what I feared no one may know, was exactly why he was doing this, but I wanted to find out.

Anxiety, Alzheimer's, and Other Animal Problems

The first person I turned to for help in understanding Oliver's mind was a physician named Phil Weinstein. A professor of neurosurgery and the president emeritus of the Society of Neurological Surgeons at the University of California, San Francisco, Phil has taught dozens of UCSF residents and pioneered a slew of neurosurgeries to correct spinal cord injuries. He also spends every morning and evening walking with Alf, his and his wife Jill's sixteen-year-old Australian shepherd. Alf is independent, thoughtful, and prone to burying his head in the crotches of visitors. He has never submitted to the indignities of a leash; he has never needed it. For years, he paused and looked both ways before crossing the streets of his neighborhood and never trotted too far ahead of Phil and Jill, constantly circling back to make sure that his humans were where they should be. When he sits, he folds his front paws on top of each other and cocks his head to listen to the people around him. As Phil and I talked one morning across their kitchen table, Alf hurried into the room and then stopped, looking from side to side, appearing confused. It was as if he'd forgotten why he'd come into the kitchen in the first place. Then he began to turn in wide circles. Phil told me that Alf recently developed canine Alzheimer's. His athleticism had given way to herky-jerky movements, and from time to time he failed to recognize people he knew.

Behaviorally, the disease is similar in aging dogs and aging humans. We become confused, the familiar turns foreign and scary, we may be grouchier or more easily frustrated than we were previously, and before we know it, we don't recognize the postman or remember where we left our bones or keys. Physiologically, there are similarities too, primarily that Alzheimer's is a result of nerve cell death and tissue loss. But the way the damage unfurls in dogs and humans is a bit different. In people, the cortex and hippocampus shrink and plaque, or abnormal clusters of proteins, builds up between the nerve cells, reducing the mind to a shadow of its former self. Because dogs have shorter lives, the damage that gives rise to their confusion and other signs of dementia is less advanced: that is, there isn't enough time for plaque to build up to the extent it does in people. Instead, canine Alzheimer's seems to be due to a hardening and narrowing of the arteries that supply blood to their brains. Starved for oxygen and nutrients, the brain withers and shrinks. Because of these similarities, a few recent studies have used dogs with dementia to try to understand the effects of a diet rich in antioxidants on cognitive function in both species. Perhaps veterinarians will soon be urging dog owners to add blueberries and leafy greens to their pets' kibble. There's also the option of training elderly dogs to do new things, just as aging humans are being encouraged to fill in crossword puzzles and learn new languages to stave off dementia.

As Phil urged Alf to stop turning in busy circles, I asked him about other potential similarities—namely, how similar my anxiety might be to Oliver's.

"The underlying brain structures that are involved in these responses are really not that different at all," Phil said. He went on to explain that the basic neurological hardware for emotional states exists across animal species, and with these similarities comes the possibility of malfunction.

Learning about fear and responding to it involve neural pathways that send information about a certain fear-inducing triggers to brain regions that determine an emotional response and behavior: freezing, fleeing, attempting to defend oneself, or, in Oliver's case, hopping out of a window or gnawing through a wooden door.

These neurological processes work similarly in almost every species, including birds and even reptiles. That is, fear responses aren't coordinated by the parts of the brain that allow us to achieve particularly human cognitive acts, such as writing novels or solving crossword puzzles—the frontal, temporal, and parietal lobes of the neocortex. This wrinkled layer of gray matter that's highly developed in humans and other great apes, as well as whales, dolphins, and elephants, helps coordinate complex cognitive processes. Our responses to fear and anxiety are different and probably originate in the subcortical regions of the brain, shared by most vertebrates and perhaps other creatures as well. Animals capable of complex thought may have more nuanced and coordinated responses to danger, perceived or real, once we sense it. Humans and other animals with a lot of brainpower can construct elaborate escape plans, for example, or develop sophisticated ideas about whatever is agitating or scaring us. But the *emotional* experience of the anxiety or fear might be similar regardless of intelligence.

These similarities are one set of reasons that nonhuman animals have been used for more than a century as neurophysiology research subjects in the quest to develop therapies for people. In the mid 1930s, the Yale neurophysiologist John Fulton performed the first frontal lobotomies on two anxious and angry chimps named Becky and Lucy. After the operation Fulton reported that Becky in particular looked like she'd joined a "happiness cult." His results helped inspire other researchers to try the surgery on people. Electroconvulsive "shock" therapy was first developed in other creatures as well, not as a treatment for animal schizophrenia but rather to determine safe voltage levels for humans. Italian researchers induced seizures in dogs and, in 1937, visited a pig slaughterhouse in Rome where the animals were stunned into unconsciousness before their throats were cut. If the pigs weren't immediately killed, they experienced the kind of convulsions that the researchers hoped would function as psychiatric cures in human patients. By 1938, a schizophrenic man known as Enrico X was given eighty volts of electricity that caused him to seize, go pale, and, oddly enough, start singing. After two more sets of shocks he called out in clear Italian, "Attention! Another time is murderous!" Within a few years, ECT had taken hold of psychiatry, first in Switzerland, then sweeping through Germany, France, the United Kingdom, Latin America, and, finally, the United States. By 1947, nine out of ten American mental hospitals were using some form of electroshock therapy on patients.

I asked Phil if a dog with a shock collar could be considered to be undergoing ECT. He laughed but said some of the psychosurgeries he's observed may work in other animals. "Many of the most common human mental disorders have to do with inappropriate fear and anxiety responses. It isn't likely that humans are the only animals to occasionally feel scared or anxious in situations that don't call for these emotions. It's also quite possible that other animals develop obsessive-compulsive disorders and other forms of mental illness." When neurosurgeons operate on people with extreme cases of OCD, for example, they destroy a small region of gray matter. The patients are conscious during surgery, and when the surgeon stimulates an area that floods the person with desire to, say, wash his hands or check a lock, the surgeon singes the corresponding bit of tissue. Often the OCD symptoms fade after the operation. No one has tried these surgeries on compulsively paw-licking or tail-chasing dogs, but perhaps they should.

They would be difficult in animals like Oliver, though, since the surgeon wouldn't be able to ask him about his desire to lick himself, then cauterize the corresponding brain region. Much of animal mental health is like this; that is, we can't definitively know what they're feeling. Studying the neurophysiology of animal emotions is possible in a limited way by mapping the firing of neural networks as the animals act fearfully or seem to be experiencing pleasure. Recent magnetic resonance imaging (MRI) of dogs as they're reunited with their owners or discover food is coming suggests that the neuro-networks that process these positive emotional experiences function similarly in them and us.

Most animals cannot narrate their emotional experiences for humans, and even if they could (signing apes, say, or talking parrots), this isn't necessarily the best measure of what they're actually experiencing. There's something of a parallel with people who can't, or won't, articulate their emotional responses or feelings when asked about them. The complex process of making sense of our racing heartbeat, sweaty palms, and surges of good or bad feelings is what undergirds much of psychotherapy and analysis. We simply don't always *know* what we're feeling while we're feeling it. And yet there's so much value in making educated guesses about animal emotions, especially when the outcome could be restoring their mental health. We know, for example, as Phil said, that fear and anxiety give rise to the majority of mental illnesses in humans, from debilitating phobias to post-traumatic stress disorder (PTSD). According to a recent estimate by the National Public Health Service, half of all mental problems in the United States, besides those related to drug or alcohol addiction, are made up of anxiety disorders. These include phobias, panic attacks, PTSD, obsessive-compulsive disorders, and generalized anxiety.

One researcher attempting to understand the physiological processes underlying human mental illness is Joseph LeDoux, a sort of neo-Skinnerian neurophysiologist at New York University. LeDoux is the recipient of dozens of awards (including the American Psychological Association's Distinguished Scientific Contributions Award for "reinvigorating the field of emotion"). He also plays in a rock band called the Amygdaloids, named for the almond-shaped set of neurons, or amygdala, associated with emotional memory in the brain. LeDoux is the author of *The Emotional Brain* and *Synaptic Self*, and he researches the processing and storage of emotional memories, particularly traumatic ones, in the brain. But he doesn't do his research on humans.

On LeDoux's office door at NYU is a newspaper clipping of a guinea pig dressed up as a Christmas elf, with tiny felt antlers. There is also a yellowing newspaper article titled "Rats: From Pests to Pets" and a *Peanuts* cartoon in which Snoopy is talking about heartbreak. His bookshelves are lined with titles like *Extreme Shyness and Social*

Phobia, *Readings in Animal Behavior*, and *The End of Stress as We Know It*. There are also a lot of neuroscience textbooks, encyclopedias of cognitive science, and a copy of *On the Expression of the Emotions* by Darwin.

LeDoux's insights into the human brain in his books and his many journal articles are based on more than thirty years of research on rodents. Recent experiments in his lab have focused on understanding the noradrenaline system in the amygdala. His research suggests that changing levels of regulatory transmitters in the brain (such as norepinephrine) may affect whether the kinds of memories that, in humans, tend to give rise to anxiety disorders like PTSD, actually end up as traumatizing.

Since he works in mice and rats as opposed to people, I asked LeDoux if it made sense to think of himself not solely as one of the world's preeminent fear experts but as a *rat* fear expert. He told me that whether he was looking at a rodent or a person was actually beside the point. "It's not the rat part of the rat," he said, that makes it a good study animal. "It is their amygdalas. Because theirs are so similar to ours."

LeDoux believes that feelings, as humans think of them, are a product of language. Other animals may have feelings, he argues, but we will never know them and that is not the goal of his research.

LeDoux is correct that the ways we describe our feelings are particular to our species, products of language, culture, individual brain chemistry, and our own learned experiences of what we find fun, satisfying, or terrifying. This is true even at the individual level. Old wooden roller coasters terrify me, but my brother, a firefighter paramedic who hops out of helicopters to rescue injured hikers and drags people from car wrecks, finds them boring. As neuroscientists like Panksepp argue, the fact that we use language to describe these sensations and experiences doesn't mean that feelings are limited to human beings. And they may be just as individualized. Perhaps the dog version of my brother prefers standing in the open bed of a pickup truck to riding inside the cab, even on the freeway. We can't know what other animals feel, but that doesn't prove that they aren't feeling *something*. The trick is to attempt to understand what they may be experiencing without projecting all of our own feelings onto theirs.

On my last visit to LeDoux's office he told me that using rats and mice has allowed him and other researchers to make detailed maps of how animals learn about and respond to danger, real or imagined. LeDoux and I don't agree, however, on whether his research on fear in rats demonstrates that rats can have fear or anxiety disorders. He told me that you'd have to observe animals in their natural habitat in order to find out how fear and stress changed their behavior. Yet his work depends upon making rats fearful enough to alter some aspect of their behavior so that it can be studied scientifically. If these behaviors occur with a frequency or intensity that disrupts their normal life (a relative concept for a lab rodent), then it fits the definition of mental illness in humans. Rats who have been shocked enough times to lose interest in food, for example, or in playfully interacting with their cage mates may be exhibiting a rodent version of induced depression or a depression-like state. At its most extreme, this state is considered "learned helplessness," a phrase coined by the psychologists Martin Seligman and Steven Maier in 1967. The researchers shocked a group of dogs into a state of such indifference that they could no longer muster the energy to escape the pain or react to it, even when all they had to do was jump over a low partition to safety. They simply gave up, resigned to their fate. Seligman saw parallels in humans caught in horrible circumstances that were outside of their control. "Such uncontrollable events can significantly debilitate organisms," he wrote, "they produce passivity in the face of trauma, inability to learn that responding is effective, and emotional stress in animals, and possible depression in man."

Le Doux may shy away from equating human and rat depression because he is leery of anthropomorphizing (although he did admit to me that he could tell when his pet cat was happy). Like a heavy leash that drags along behind nearly all twentieth-century efforts to understand the emotional lives of other animals, anthropomorphism was resented and feared. Radical behaviorists like B. F. Skinner, comparative psychologists, ecologists, and many ethologists warned against sentimentalizing other animals and rejected Darwin's ideas on animal emotions, working to suppress what they considered subpar science. For a long time anthropomorphism was a dirty word in the behavioral sciences, despite the fact that experimental animals were busy acting as models for human psychobiological phenomena inside laboratories worldwide.

And yet, no one has quite been able to do away with the practice. Millions of people a year watch films that feature talking animals wearing chef's hats or swim trunks while they cook or drive cars. We read animal fables to our children that function like human morality lessons and many pet owners' most embarrassing pleasure is speaking for their cats and dogs. A few months ago I watched a man greet a friend at his front door while holding on to a salivating, excited spaniel by the collar. "Spooky is just *so* happy to meet you!" the man said to his visitor. "Aren't you, Spooky?" He lowered his voice and continued, "Yessss, I wuv new people!"

There is a reason anthropomorphism has stubbornly refused to go extinct. In and of itself it's not problematic. In fact all human thinking about animals is, in some sense, anthropomorphic since we're the ones doing the thinking. The challenge is to anthropomorphize well. The psychologist and cognitive researcher Diana Reiss, who has worked on dolphin communication and cognition for more than thirty years, argues that we should avoid *anthropocentrism:* the belief that humans are unique in our abilities and that our intelligence is the only one that counts. Diana's own dog, whom she shares with her husband, the neuroscientist Stuart Firestein, is a black Newfoundland named Orson who is yak-like, sweet, shy, and prone to magical thinking. Every time Orson returns to their apartment on an upper floor of a Columbia University faculty building, he enters in the exact same way. "The elevator opens and instead of going straight to the front door," Diana said, "he always walks a few steps down the hallway in the opposite direction." When he reaches a low window, he glances out. When the door clicks open, he turns around and walks back into to the apartment. "At some time Orson must have experienced the apartment door opening right after he looked out of the window. He's gotten it into his head that in order for him to enter the apartment, he has to first perform this ritualized act."

B. F. Skinner wrote about superstitious animal behavior in 1947, when, after putting a group of pigeons in a cage with a machine that dropped food pellets at regular intervals, the birds began to behave oddly. A few of them turned in circles but only a specific number of times, or swung their heads to and fro like dizzy pendulums in sequence, apparently convinced that by repeating whatever they were doing when they were last fed, the pellets would again drop into view. Animal magical thinking is, of course, not limited to Newfoundlands and pigeons. Professional athletes may be the most fitting equivalent to Skinner's birds: the Olympic swimmer Michael Phelps swings his arms exactly three times before hopping in the water to race, Michael Jordan wore his college basketball shorts under his pro basketball shorts on game days, and tennis star Serena Williams refuses to change her socks once a tournament has begun. These lucky charms may work for the athletes because they keep their confidence up and make them more comfortable. Superstitious acts by human and nonhuman animals are a function of otherwise unrelated events becoming associated in meaningful ways. This is similar, in some sense, to the faulty logic behind uncritical anthropomorphism. That is, relying too heavily on one's own limited perspective can encourage someone to ascribe meaning where it might not exist.

We can avoid this by refusing to see other animals as extensions of ourselves. Being humble helps too. In 1906 the naturalist William J. Long wrote in *Briar Patch Philosophy by Peter Rabbit*, "It is possible . . . that your simple man, who lives close to nature and speaks in enduring human terms, is nearer to the truth of animal life than is your psychologist who lives in a library and today speaks a language that is tomorrow forgotten." He may have been onto something. The best interpreters of animal behavior are often those people whose colleagues are nonhumans. Zookeepers, exterminators, trainers, sanctuary workers, dog walkers, breeders, and the staff and volunteers at animal shelters spend their working lives with animals, often the very same ones, day in and day out. In order to achieve even the most basic tasks their jobs require, they must convince other animals to do things they may not want to do, such as urging a gorilla to walk into a travel crate on her own, stopping an altercation among giraffes hell-bent on annoying one another, or encouraging a snarly dog to let his his toenails be clipped. These keepers, trainers, and groomers become intimately familiar with the animals' tastes and individual preferences, their behavioral quirks, the other animals they prefer to spend time with, and what sorts of treats will entice them and which ones won't make them budge.

Jose Luis Becerra is a wildlife removal expert in demand. His business card identifies him as a "humane critter trapper" alongside a photo of a raccoon atop a telephone pole. Jose has de-skunked and de-possummed Nicolas Cage's Malibu mansion and knows all of the best ways to extricate families of raccoons from attics. (Most involve cans of tuna or cat food.) He told me that he considers the animals that he sets traps for and later releases in dry riverbeds, canyons, and other secret locations around southern California to be his colleagues.

"I'm only good at my job if I can learn to think like they do, if I can literally imagine myself in their place, with their desires," he told me while pulling a skunk out from underneath my childhood bedroom. He had lured it into his trap with tuna-flavored Fancy Feast and covered himself up with a large plastic trash bag so he wouldn't get sprayed.

The cognitive ethologist Marc Bekoff makes a similar argument.

When he is trying to determine what a dog is thinking or feeling, he says that he has to be anthropomorphic but he tries to do it from the dog's point of view. "Just because I say a dog is happy or jealous," Bekoff wrote, "this doesn't mean he's happy or jealous as humans are. . . . Being anthropomorphic is a linguistic tool to make the thoughts and feelings of other animals accessible to humans."

Robert Sapolsky, the wild-haired and charismatic Stanford neuroscientist and author of *Why Zebras Don't Get Ulcers* and *A Primate's Memoir,* studies baboons living in the wild in Kenya. His research has demonstrated that changes in their social hierarchy affect not only their behavior but also their physiology. Lower-ranking baboons are often bullied and lead far more stressful lives than the troop's higher-ranking baboons. The brains of these antagonized primates are steeped in a nearly constant stream of stress hormones that, with long-term exposure, cause neurological damage. Sapolsky sees the baboons he studies as individuals, writing extensively about their personal quirks and the various ways their shifting ranks affect their emotional and physical health. His attention to their psychodramas, along with his efforts to gauge personalities of individual baboons, may have helped him conclude that their physiologic responses to stress approximated our own. His research has revolutionized how we think of the affects of chronic and acute stress on the human brain.

"I'm not anthropomorphizing," Sapolsky has written. "Part of the challenge in understanding the behavior of a species is that they look like us for a reason. That's not projecting human values. That's primatizing the generalities that we share with them."

With all due respect to Sapolsky, I think he *is* anthropomorphizing, and that's fine because, as he says, his conclusions are based on shared generalities, not unfounded projections. We've inherited a bias against identifying with other animals that isn't useful, and it's high time we discarded it.

Men, Monkeys, Mothers

One of the most famous instances of disturbed creatures drafted into helping humans better understand themselves unfurled throughout the 1950s and 1960s inside Dr. Harry Harlow's comparative psychology lab at the University of Wisconsin, Madison. There, a series of chilling experiments forever changed how we understand the role of touch and affection in the healthy development of primate infants—human and otherwise.

Harlow wrote more than three hundred scientific books and articles, founded two different research laboratories, and created and oversaw one of the first productive breeding colonies of monkeys in the United States. Between 1955 and 1960 he and his team bred enough baby rhesus monkeys to make massive psychological testing experiments possible. He was the recipient of both the National Medal of Science (1967) and the Gold Medal from the American Psychological Foundation (1973). He was also a dark lord of monkey torture.

In a now infamous series of experiments, rhesus monkeys were separated from their mothers at birth and caged alone inside the lab. Able to see other monkeys and the human lab staff, they were allowed no physical contact with them. These isolated infant monkeys quickly began staring into space, clutching themselves, repetitively rocking and biting their own bodies and their cages. Harlow performed a variety of experiments on the infants. One series involved offering baby monkeys a choice between a fake monkey mother made from wire, with a terrifyingly crocodilian head, that dispensed milk, and a second fake monkey mother who had no milk but was covered in terrycloth and had a round head with two eyes, a mouth, and vaguely simian ears. The baby monkeys clung to the cloth-covered mother, even though doing so meant they stayed hungry. This type of experiment was repeated endlessly, with all sorts of mothers built to repel infants (some blasted air, others had hidden spikes), to plumb the connection between maternal rejection and psychopathology.

Another of Harlow's experiments demonstrated that depriving an infant of touch and social contact caused irrevocable psychological damage. He called the experimental apparatus he used for these tests the "pit of despair." In this inescapable, smooth-sided, stainless steel chamber, monkeys would usually cease all movement, curl into a tight ball and stay that way. Harlow called this behavior "induced depression." He would then remove the monkeys from the pit and attempt to make them less depressed.

To do this, he took the now extremely abnormally behaving monkeys—they rocked, bit and mouthed themselves, did not groom or play, and were prone to aggression—and put them into separate cages, each with a single "therapist" monkey. According to Harlow, the therapist monkeys (who were not raised in pits of despair) clung to the fearful ones and offered comfort and warmth. After a few weeks many of the once bizarrely behaving monkeys began to play with their therapists. According to the researchers, after a year many of the previously abnormal monkeys were indistinguishable from the others.

At the same time Harlow was experimenting on monkey infants, institutionalized human infants were sometimes suffering similar fates, not in pits of despair per se but in orphanages and hospital wards where they were rarely touched. Children who were left in these institutions with masked and gloved caregivers who didn't caress, rock, kiss, or hug them but who were given plenty of food, kept clean, and offered medical care nevertheless failed to put on weight. They also didn't learn to walk, talk, or sit. Like Harlow's monkeys, they developed odd behaviors such as staring into space and moving their hands bizarrely. As Deborah Blum writes in *Love at Goon Park*, the only object the children saw for any length of time was the ceiling.

Throughout the 1940s and 1950s, the psychoanalyst and psychiatrist René Spitz observed many of these institutionalized children and documented how they withered and failed to develop. The problem, Spitz was convinced, was not that the sterile environments of these places were boring, static, or lacked cognitive stimulation, though that was true and terrible. It was that there was no one to love the children. Or as Blum writes, even to like them, smile at them, or give them a careless hug. Spitz believed that the lack of human touch and affection left the children vulnerable to infection and disease. More than a third of the babies he studied died, and many of those who survived remained institutionalized forty years later, unable to care for themselves.

John Bowlby, a British psychologist, psychiatrist, analyst, and frequent correspondent of Harlow's, investigated the importance of affection for children isolated in hospitals during the same time period and discovered similar results. He was extremely interested in animal behavior, and not only exchanged letters with Harlow but also with famous ethologists like Konrad Lorenz, Robert Hinde, and Niko Tinbergen. Bowlby was convinced that babies who weren't held or played with during their hospital stays, like the isolated infant monkeys, eventually developed life-threatening apathy and depression. He predicted that these children would grow up with stunted cognitive and language skills as well as attention problems and difficulty forming relationships with others.

Bowlby's and Spitz's research, combined with Harlow's experimental results, eventually helped change what people thought it meant to provide for an infant. In a way it was Harlow's benighted, suffering monkeys who taught us that some things are more important than food and shelter and that touch and affection are crucial to the healthy development of primates—human or otherwise. Over time, at least in the United States, orphanages were replaced by foster care families and group homes. Bowlby became known for contributing to the development of attachment theory, which he described as a "lasting psychological connectedness between human beings."

Harlow's monkeys also ended up helping, in a roundabout way, other institutionalized primates. In many zoos, monkey and ape mothers are now allowed to raise their own infants because of what Harlow, Bowlby, and Spitz established more than fifty years ago.

Captive apes learn to be good mothers by watching others or remembering their own experiences growing up. In at least one case, zoo gorillas, who were part of a troop in which there'd been no recent births, were shown videos of other gorillas giving birth so that when the time came, the females wouldn't be terrified of what was happening to their bodies. This same zoo also brought in the wife of a groundskeeper who had recently given birth to demonstrate nursing. She sat quietly feeding her human infant, while the gorillas watched with interest through the cage wall. Other zoos have hired midwives and lactation consultants to help teach their apes how to nurse and to be affectionate with their babies.

Zoos do this because primates who are raised by dysfunctional or fearful mothers, like Harlow's infant monkeys, can develop cognitive, linguistic, and emotional problems that make it difficult for them to interact with their own babies and other troop members when they grow up. At orphanages for young gorillas, orangutans, and bonobos whose parents or other troop members have been killed in the bush meat trade or by poachers, the most successful rehabilitations are due to the youngsters' relationships with surrogate ape mothers who cuddle, groom, and play with them. At one of these places, a bonobo sanctuary called Lola Ya Bonobo located just outside of Kinshasa in the Democratic Republic of Congo, the surrogate mothers are human women, whose nearly twenty-four-hour presence and constant physical contact with the babies helps the bonobos grow into confident, well-adjusted adults who can eventually spend all of their time with their own species inside the sanctuary's protected forest.

Pavlov, Personality, and PTSD

As word spread among my friends and family about Oliver's weird behavior, they began forwarding me articles about dogs who roused themselves from depression by making friends with orangutans or links to stories about Scotland's Overtoun Bridge, otherwise known as "the dog suicide bridge," where a number of dogs are said to have mysteriously plunged to their deaths (but who may have merely been following the scent of rabbits or foxes). Most of these articles I filed away in a rumpled folder labeled "Animal Crackers," but a few caught my eye.

In the wake of the Iraq War and during the ongoing armed conflict in Afghanistan, stories of anxious canines, such as "Military Dogs of War Also Suffer Post Traumatic Stress Disorder," "War Dogs Are Taking Xanax for Puppy PTSD," "More Military Dogs Show Signs of Combat Stress," and "Four Legged Warriors Show Signs of PTSD," popped up in the popular media. Reporters marveled at the novelty of these walking psychologically wounded, but all I could think about was how this news was perhaps not quite as novel as the journalists suggested. Ivan Pavlov, whose work almost a century ago focused on disordered dogs, wouldn't have been surprised in the least.

The Russian physiologist was interested in far more than the conditioned responses to cues that famously sent his dogs salivating. Pavlov spent decades exploring the physiological basis for human neuroses and the relationship between disordered human and canine minds. He even lived out the last years of his life as a researcher inside a clinic for nervous diseases attempting to help disturbed humans. Pavlov's life's work set the stage for much of our contemporary understanding of the effects of trauma on human behavior, memory, and mental health and is one of the main reasons that today's dogs of war can be said to have PTSD.

Pavlov became interested in canine neurosis after reading Freud's writings on the patient Anna O. She put on a happy face while caring for her terminally ill father, whom she loved deeply. Anna did this for his benefit, hiding her own feelings of despair and loss. Freud believed this internal conflict gave rise to her neurosis.

Pavlov wanted to simulate this conflict in his dogs to better understand the mechanisms of neurosis. The first of his canine experiments, conducted in 1914, went like this: A woman in the lab conditioned a dog by shocking the skin on his hip while he ate to make him associate the shock with food. Eventually he began salivating upon being shocked on the hip. When more shocks were applied to the dog, but this time on different body parts, the dog's behavior suddenly changed. He became listless instead of alert. He drooled and lowered his head, tail, and eyelids. He was apathetic and unresponsive and began salivating to strange cues, like loud noises in the lab. At other times the dog wasn't lethargic at all but became so agitated that he broke the straps that held him in place. Pavlov was convinced that he had created the perfect experimental model for human neurosis like Anna O's, resulting from a deep-seated conflict between signals that excited (bell = yum, food coming) and inhibited (shock = ouch, pain). These signals were so confusing that they drove the dogs to distraction.

The lab performed endless variations of these sorts of experiments. One, which involved not dogs but cats, consisted of a hungry cat with an electrode affixed to her tail. The cat, who had been fed mice for weeks, was placed inside a chamber with a mouse. The instant the cat pounced she received a shock and immediately spat out the rodent, running as far away as she could. For weeks afterward, the cat was scared and immobile, her heart racing, whenever a mouse was in sight. A photo of one of these experiments shows a crouching black and white cat, holding perfectly still while white mice loll about on her back and head like sunning cruise ship passengers.

In 1924, Pavlov's views on the similarities between disordered humans and other animals were confirmed when a loud and violent storm sent floodwaters gushing into his Leningrad lab. He and his fellow researchers were able to save the dogs, but once the water receded and work resumed, a few of them became much more agitated during the behavioral tests than they had before. Pavlov came to the conclusion that these dogs had "weaker nervous systems" than the other, hardier dogs who seemed unaffected by the storm. To test his theory, he simulated the flood, sending a stream of water onto the lab-room floor while he watched the dogs' behavior. At least one reacted anxiously, glancing at his feet, barking, and dashing in circles as soon as he spied the seeping water. Others were able to perform the behavioral tests as they had before, even in the presence of the simulated flood. Pavlov concluded that an individual's personality, dog or human, informed his or her responses to potentially traumatic experiences. Today, we may take issue with his loaded terms, "weak" or "strong," to denote individual personality, but Pavlov intended these terms as descriptors for a kind of emotional resilience that is now largely taken for granted.

Pavlov had his critics, especially when it came to his work comparing disordered dogs to disordered people. Many psychoanalysts and physiologists were rightfully skeptical that dogs who were shocked into confusion, catatonia, and other disturbed mental states were comparable to humans suffering from nervous disorders. They doubted that dogs made neurotic in the lab were sufficiently similar to patients like Anna O. These critics suggested that what Pavlov was mirroring in his experiments was not *human* neurosis originating from within but a kind of tension that came from being in a stressful environment or in close proximity to something unpleasant.

Other psychoanalysts argued that Pavlov's work was inferior to analysis. For Pavlov, though, talking about one's problems was unnecessary. Knowledge of mental life could come from observation. For him, dogs were just simple versions of humans; the main thing separating canine neurosis from human neurosis was that the latter stemmed from more complex situations. Furthermore he felt confident that his ability to return his dogs to their normal, nonneurotic state, using caffeine to rouse them from catatonia or reconditioning them back from the brink of insanity with a more logical series of stimuli and rewards, could act as a map for curing neuroses in people.

Pavlov's research laid the foundation for a spectrum of experiments on nervous diseases in other animals. His contemporaries and successors attempted to induce neurosis in sheep, goats, pigs, pigeons, rats, and cats, many of whom, like Pavlov's dogs, eventually broke down in the face of repeated random shocks or confusing cues. The animals who recovered were held up as models for soldiers suffering from war neurosis, a disorder that, in the wake of the Vietnam War, would morph into PTSD.

During World War II, military physicians and psychiatrists noticed that soldiers could display symptoms that were quite similar to those of neurotic experimental animals. The men's hearts raced, they sweated, they felt their anxiety levels swell, and they were easily startled. In 1943 one American psychiatrist suggested that acute war neurosis should be treated with the same deconditioning procedures used on experimental animals like Pavlov's dogs. The canines were exposed to a trigger they'd learned to associate with a shock, but no shock would come. Eventually they learned that the painful stimuli was no longer something to fear. These ideas were adopted by the military, and a selection of nervous troops was sent to "Battle Noise School" in the South Pacific, where they were exposed to mock gunfire, controlled land-mine explosions, and simulated dive-bombing attacks in the hope that they'd learn to respond without crippling fear. I don't know the extent to which the school helped the men who were sent there, but ever since, Pavlovian ideas of conditioning and deconditioning have informed our understanding of nervous disorders and the many therapies used to treat them, especially PTSD.

Today, the disorder is characterized by anxiety that arises in the wake of a traumatic event, causing sufferers to experience some combination of flashbacks or upsetting memories that interfere with their daily life. People may also have terrifying nightmares, unsettling reactions to situations that serve as reminders of the traumatic event, and feelings of detachment or emotional numbness. PTSD sufferers also experience a variety of symptoms that mirror many of those documented in Pavlov's dogs: difficulty concentrating, startling easily, acting hypervigilant, feeling irritable, expressing outsize anger, feeling dizzy, fainting, racing heart rates, sweating, headaches, and more. While diagnosing anxiety disorders like PTSD in humans is now a largely verbal process, it wasn't always so. Physical symptoms were once the signposts that physicians frequently followed to diagnosis. Nineteenth-century and early-twentieth-century physicians who treated survivors of traumatic events like bloody train accidents or tramplings often used their patients' physiological symptoms to gauge their emotional suffering.

Doctors treating soldiers in the wake of World War I, for example, occasionally compared those who fell mute after being sent to the front with animals who froze in the presence of predators. Men who stopped speaking after traumatizing battles were, according to the English anthropologist, neurologist, and psychiatrist William H. Rivers, exhibiting the same kind of immobility and silence adopted by prey animals to avoid being eaten or attacked.

Recently, in a sort of reversal of these earlier observations, other animals have been compared to humans with PTSD. In the past few years African elephant calves who witness violent culling campaigns that kill their elders, dogs trained for search and rescue work who survived explosions, the deaths of their handlers, or were forced to work long hours under stressful conditions, and laboratory-dwelling apes who were kept in cramped cages for years on end have all been said to be suffering from trauma disorders.

Chimps who have spent time at testing facilities can have nightmares and what appear to be flashbacks of painful or scary experiences. They can become either more aggressive or more withdrawn, startle too easily, and have problems forming new, healthy relationships with other chimps or their human keepers, even after they've been retired to sanctuaries. The ethologist Jonathan Balcombe shared an account of this sort of suffering at the Fauna Sanctuary in Quebec, Canada, a refuge for chimps previously used in research. One afternoon, keepers loaded a shipment of materials onto a metal trolley. The unknowing staff members pushed the trolley past the enclosure of two different chimps, Tom and Pablo. As soon as the chimps caught sight of it they let out a frightened shriek. Hearing them, the other sanctuary chimps rushed to the bars of their enclosures, rocking back and forth and shrieking along with Tom and Pablo. The staff later realized that the same trolley, or one that looked just like it, had been used to transport unconscious chimps to the surgery room at the research facility where Tom and Pablo had lived, and were experimented upon, two years earlier.

Whether these animals were indeed experiencing the same sorts of feelings that humans diagnosed with PTSD do is impossible to prove, but then again there is no single human experience of PTSD. Human sufferers all have varying degrees and types of symptoms. How people decide to label these feelings and behaviors—symptoms like acting fearfully; feeling anxious, depressed, or aggressive; or not wanting to be social—is less important than the fact that they can make someone miserable. Signs of suffering are visible to the careful, compassionate observer. You don't need to be able to talk to someone to notice these symptoms in another person—which is why disorders such as "shell shock" and "war neuroses" could be identified by observation as opposed to in talk therapy at the turn of the twentieth century and earlier. Even today PTSD is sometimes diagnosed in humans using observation as opposed to interviews. Traumatized infants and pre-school-age children, for example, may be diagnosed when psychiatrists see warning signs in their play behavior or in the ways they interact with family members, social workers, or the therapists themselves.

Oliver definitely had abnormally high levels of anxiety, from startling too easily to his vigilant and then panicked reaction to the sight of work bags or suitcases. That being said, I don't think he had PTSD. His anxiety was actually mild in comparison to many canine survivors of natural disasters like Hurricane Katrina, dogs who huddle under tables or transform from friendly and approachable to vicious and timid. The behaviorists and trainers who have treated these dogs believe their symptoms are comparable to PTSD. They attribute the dogs' difficulties to their forced abandonment during and after the storm, having been trapped by floodwaters, having gone without food for days or weeks at a time, having had to negotiate new and frightening environments, or having lost the companionship of their humans.

A few search and rescue dogs exposed to the loud, dangerous, and unfamiliar environment of the World Trade Center site after the September 11 attacks also became agitated, depressed, irritable, and uninterested in playing. Others grew hypervigilant and aggressive. Some no longer do search and rescue work.

Lee Charles Kelley is a dog trainer interested in canine trauma disorders. He is also a writer of detective novels, including *Twas the Bite Before Christmas* and *To Collar a Killer*, about an NYPD cop turned dog trainer. Kelley's website has sections called "Neo-Freudian dog training" and "Support Group/Canine PTSD." He offers a checklist for people interested in diagnosing their own pets, with questions like "Has the dog been in a serious accident, fire, or explosion?" If the answer is yes, then Kelley asks others, such as "Does the dog react as if he's actually reexperiencing a traumatic event or events?" or "Does the dog seem to be having vivid dreams or possible nightmares?" Kelley believes that the number of dogs who've survived traumatic experiences reaches into the millions—from those injured in fights with other dogs to those mistreated by humans and left at shelters to survivors of car accidents or deployments. His own dog, a Dalmatian named Fred, suffers from extreme panic attacks triggered by new noises. Kelley found that asking him to bark when he was descending into panic was helpful (it distracted him), as was giving him something to carry in his mouth when they went for walks. A tennis ball is Fred's transitional object.

The dogs most frequently diagnosed with PTSD, however, are those canine soldiers whose appearance in the headlines first caught my eye. Of the roughly 650 American military dogs deployed in Iraq and Afghanistan, more than 5 percent were assumed by the military to be suffering from canine PTSD. Dr. Walter Burghardt, a retired air force colonel and the current chief of behavioral medicine at the military working dog hospital at Lackland Air Force Base in San Antonio, Texas, believes the disorder applies to many dogs exposed to gunfire, explosions, and combat-related violence. Like human soldiers, not all of these seemingly traumatized dogs have the same symptoms, but many exhibit some variation of hypervigilance and drastic personality and behavioral changes, becoming more likely to bite, for example, or more timid and easily startled than they were predeployment. These dogs sometimes avoid buildings they once entered with no hesitation, refuse to sniff cars at checkpoints, or balk when approached by men in foreign uniforms.

Because dog noses are still the most effective tools for finding chemical improvised explosive devices, especially in Afghanistan, where IEDs made from fertilizers are common, the number of deployed dogs has spiked in recent years. There has been a corresponding increase in the number of dogs sent home from war for psychological problems. In an attempt to forestall these problems, Burghardt made a series of instructional videos to help soldiers spot PTSD in their dogs. He communicates with the deployed dog handlers on Skype, advising them on their dogs' behavior and sometimes prescribing Xanax or antidepressants. If the dogs refuse to work it can be dangerous for them and their humans, so the government has also started sending the dogs at risk of emotional breakdown back to the States for therapy, a deconditioning process based, at least in part, on the same ideas behind Pavlov's research and the Battle Noise School for men. If after three months of behavioral reconditioning and training the dog soldiers still hide under cots at the sound of gunfire or refuse to hop into a car to sniff its interior, they're retired to civilian life. As with human veterans attempting to assimilate to life back home, this process is often difficult. The dogs' adoptive families struggle to deal with the behavioral and emotional problems they have inherited.

Goodbye, Beast

Two years after Oliver jumped from our apartment, Jude and I traveled to my family's house in southern California for Christmas. We left him at a boarding kennel outside of Boston. By that point leaving Oliver by himself anywhere but the canine equivalent of a padded cell was untenable. He could hurt himself and destroy furniture, floors, windows, and doors if we left him with friends. Since his jump we knew we couldn't leave him in our house with a pet sitter. Honestly, if we could have left Oliver in a car for a week with sufficient food and water and someone to come by to walk him a few times a day, he probably would have been fine. He loved the car and was never anxious there. He knew that being in the backseat meant that he was not going to be left behind. But you can't leave a dog alone in your car for a week, even if that's where he's happiest. So we took Oliver to a kennel, where they put him in a large dog run with his bed and a few toys stuffed with treats (not that toys interested him, even ones filled with cheese). The kennel staff walked him twice a day, and we felt confident that there was no way he would try to escape and hurt himself. We were wrong.

Despite our efforts to help him, Oliver's anxiety at being left alone only increased in the years he lived with us. His storm phobia reduced him to a shaking, inconsolable mess, and it took him hours, sometimes days to recover. He continued to eat things that weren't food if we left him alone past 5 p.m., and with every passing night he seemed to hunt for invisible flies for longer periods of time. Whenever Oliver was agitated—which was often—he gnawed things in the apartment. He also became more aggressive at the dog park and snapped at a few young children. We were tired. By that time Jude and I had tried virtually every means of therapy and treatment available to American pet owners. We'd taken him to a veterinary behaviorist, given him first Valium, then Prozac, then both. We practiced behavioral modification and training in an attempt to manage his anxiety. We played him recorded sounds of storms to desensitize him to thunder and jingled our keys even when we weren't planning on leaving the house. We took him on long walks, then long hikes. We tried to socialize him with other dogs. We gave him toys and treats. We gave him affection. We thought about getting him another animal companion and then decided against it. We tried, and failed, to give him certainty.

When we left Oliver at the kennel that December, Jude and I planned to be away less than a week. My family in California are farmers, and one afternoon, less than three days into our trip, Jude, my mom, and I walked to the very top of the hill behind the house. We stood at the property line where rusted barbed wire dipped between the posts like bunting and lemon orchards spread out beneath us. My cell phone rang. And then Jude's. I don't remember who picked up first, but I remember what we were told. "You are going to have to act fast." "We're not sure if he is going to make it." "It happened so quickly." "We are so sorry." "No, we don't know why."

Oliver had worked himself into a panic after his afternoon walk and began to anxiously chew on a piece of wood on the door in his dog run. By the time someone noticed what he was doing, it was too late.

"It couldn't have been going on for more than a half hour," the manager of the kennel told me. "But he was panting and making wheezing sounds. And then I noticed how he was standing."

Oliver was suffering from bloat. This horrid and probably excruciatingly painful predicament comes about when a dog's stomach fills with air, fluid, or food and twists, putting pressure on other internal organs and possibly cutting off their blood supply. You have about forty-five minutes to perform surgery and untwist the stomach before there's irreparable damage. Bloat is notorious for affecting deep, barrel-chested dogs like Bernese Mountain Dogs, Saint Bernards, and Basset Hounds. There's no single thing that brings it on and I couldn't find any research linking anxiety to bloat. But I believe that's what happened in Oliver's case. He was in a frenzy. He was gulping air and chunks of wood. He was agitated and scared. He was alone.

When we reached the attending veterinarian on the phone, she told us that Oliver was in the operating room. They had opened him up as soon as he arrived and unwound his twisted intestines, tacked them aside, and surveyed the damage. She said that it was bad and she couldn't guarantee that further surgery would help him. She also said that if we went ahead and performed the surgery, we were looking at expenses, including the procedures that they'd already done, of between $10,000 and $15,000.

"You might want to think this over," she said to Jude, "but don't think for long because we can't keep him here on the operating table."

I looked at the neat geometry of tree rows below us and began to cry. I thought of Oliver's soft body on a steel table, his flank splayed open, his heavy unknowing head.

Jude put his arms around me and said something, but I don't know what it was. I heard only the blood in my ears and felt a sudden, thudding grief.

We called the veterinarian back and told her to put Oliver down. She assured us that he wouldn't feel any pain, that he was already unconscious. I made her promise that she would cradle his head and stroke him while he died, that she would call him "Beast" and tell him that we loved him. And then I asked, lamely, "Do you think we're bad people?"

I was remembering the story of a friend's friend in veterinary school who treated a Labrador for extensive injuries after he was hit by a car. The family who brought the dog in loved him but couldn't afford the recommended surgeries and decided to put him down. The vet student knew that the dog would survive if the costly treatment was done. She let the family say goodbye to him, and then, after they'd left the clinic, she treated the dog and adopted him herself. I found this story chilling. It was good, after all, that the dog had survived. But shouldn't she have offered to do the work for free and sent the dog home with his family? As we waited for the vet to call us back and tell us when it was over, I had visions of Oliver bounding out of the hospital with other, richer people. Or with the soft-spoken lady veterinarian. I thought of him turning around, scanning the sidewalk for us, and then getting into someone else's car.

"No," the vet said to us, "I understand."

There's a greeting card that I found in the gift shop of the Sigmund Freud Museum in London and keep tacked above my desk next to a drawing of a squirrel in a T-shirt shooting heroin. The card is black with bright yellow type and reads, "Blessed are the cracked for they let in the light." Supposedly Groucho Marx said it, though I can't find proof. If he did say it, he probably wasn't referring to neurotic dogs. But he could have been.

Oliver died more than six years ago now, and when I think about him, I ache. I bet Jude does too, but we don't talk about things like that anymore. We don't talk much at all actually. We divorced the year after Oliver died and a few years after that he stopped taking my calls. I can't say that we broke up because of what happened with Oliver. That would be a lie, or at least it wouldn't be the whole truth. I do believe, however, that if Oliver had lived, we may not have broken up when we did. Dogs have a way of gluing people together, even ones who are already coming unglued.

Now it feels like I walk around with a few different drafty spaces in my chest. One is in the shape of a dog, and there's at least one more in the shape of a man. And yet, in the years since Oliver died, I've fallen in love again anyway—with a half dozen elephants, a few elephant seals, a troop of gorillas, one young whale, a couple of long-dead squirrels, and a handful of men and women who came into my life as if they'd been tugged there by invisible leashes. I'm not sure I would have found any of these creatures otherwise. Losses and disappointment can do that if you're lucky. Before you know it your pain has welcomed the world. That's what happened to me, anyway. One anxious dog brought me the entire animal kingdom. I owe him everything.

Chapter Two

Diagnosing the Elephant

Abnormal is the new normal.

Jon Ronson

Mel Richardson mentioned orangutan masturbation within fifteen minutes of our first meeting. We were standing in the dusty gravel parking lot of the Performing Animal Welfare Society sanctuary. Mel told me that if I ever saw an orangutan sitting cross-legged and rocking back and forth on her heels, she was probably pleasuring herself. He would know.

PAWS, as the sanctuary is known to the people who work there or visit for the occasional fund-raising dinner alongside the elephant corrals, is a refuge and retirement center in a particularly lush part of the Sierra Nevada foothills of California for tigers, bears, elephants, and other animal actors once used in film or television and for those rescued from circuses and zoos. Mel, a tall man with a neat gray goatee and a cell phone clipped to his belt, was the sanctuary's consulting veterinarian. He is one of the most experienced exotic animal vets in the world, having spent more than thirty years tending to hundreds of different species, from free-living gorillas in Congo and Rwanda to Pablo Escobar's hippos, zebras, and ostriches at the Colombian drug lord's private zoo in Medellín, along with the dogs, cats, and birds that came through Mel's private practice in Chico, California.

He treated all of these animals not only for physical problems like infections and broken bones but also for emotional ones. He's seen almost every conceivable abnormal behavior: from phobia-addled dogs and traumatized horses to depressed lions and compulsively self-pleasuring apes and walruses. He frequently serves as an expert witness in abuse cases. I'd contacted Mel because I wanted to know how he goes about diagnosing an animal with a mental disorder.

"Well," he said as we wandered past the sanctuary's elephant Jacuzzi, "it's not exactly like mental illness in humans, but I believe other animals experience similar things all the time." To make a diagnosis, Mel first looks at the animal's environment; he says that a creature living in bad conditions or one that is being abused will often have both physical and mental problems. He also talks to people. "With pets, I depend on a detailed interview with my human client. Zoo animals are actually easier to diagnose because you don't have to depend on a pet owner to describe the problem or worry about what they might not be telling you. In the zoo, there's no intermediary."

For humans with psychiatric troubles, the diagnostic process is usually verbal. As I mentioned earlier, there are exceptions for children and adults who can't or won't speak, but most often a person explains her symptoms to a therapist, social worker, or psychiatrist. These self-reported symptoms, combined with a mental health professional's own observations of the patient, lead to a diagnosis. Today's diagnoses correspond to more than thirteen thousand codes in the *Diagnostic and Statistical Manual of Mental Disorders*, the atlas of recognized human mental problems first published in 1952. The *DSM* codes are used as a guide for practitioners and required by insurance companies, though very rarely does a person fit neatly into a single category. The *DSM* is also a historical document that is constantly reinterpreted to fit the times, with new disorders being included and others removed. Post-traumatic stress disorder was added in 1980, for example, while homosexuality was taken out, but sadly not completely until 1982. Premenstrual syndrome, once known as "menstrual insanity," is another disorder that has gone through many classificatory changes over the years. It wasn't included in the *DSM* in 1980, but in 1999 the USDA accepted premenstrual dysphoric disorder, a slightly different term for PMS, as a legitimate reason to prescribe Prozac. And so, like many other psychiatric conditions, the disorder was defined by the pyschopharmaceuticals used to treat it.

What this means for an animal who can't speak to us and thus falls outside of the most common mode of human diagnosis is that the diagnostic process relies almost entirely on observation, and sometimes, as with premenstrual dysphoric disorder, on how the animals respond to drug treatment.

Unfortunately there is no *DSM* for animals. The closest thing I could find was Mel. After touring the grassy fields at PAWS, where a few female elephants dozed beneath oak trees, one snoring audibly, he led me to what looked like a large dog run surrounded by a tall, chain-link fence. It contained a pacing and restless blur of stripes, a small female tiger named Sunita. She looked at Mel with what seemed like annoyance, boredom, and a deep, abiding suspicion.

Sunita was born in a residential house in the southern California town of Glen Avon, in San Bernardino County. The home belonged to a man named John Weinhart, who lived there with his wife, his young son, and his collection of tigers. When animal control agents raided Weinhart's home in 2003, they found fifty-eight dead tiger cubs stuffed into freezers, dozens of rotting and desiccated tiger carcasses spread about the property, a few alligators swimming in a bathtub, and ten live tigers, one of whom was on the back patio swiping at the door to the kitchen. Weinhart's son's Easter candy was inside the refrigerator next to tiger tranquilizers. He also kept dozens of tigers at a former sewage plant in the city of Colton, about ten miles away. He called his ragged menagerie a "rescue operation."

"I live with them," Weinhart told a newspaper reporter three years before his property was raided. "The pores of my skin smell like a tiger. So when I go around one . . . they accept me as a tiger."

After their rescue from San Bernardino, the tigers were sent to PAWS. The cats now live in ample cages and outdoor enclosures with pools, completely inaccessible to the public. The sanctuary makes sure the tigers receive enough exercise by moving them from a large sunny yard to smaller enclosures with their own dens every two hours. They are enticed to move with treats of chicken necks and drumsticks, beef hearts, ground turkey, and paper bags. The meat they eat; the paper bags they relish tearing to bits. This new life is the opposite of the cramped, dark quarters they were used to in Riverside. But unlike many of the other tigers who quickly assimilated to the sanctuary, Sunita took a much longer time to relax. She enjoyed her gory meals and treats, but she wouldn't touch her food in the presence of people or other tigers. She also howled and whined and refused to lie down when certain people were present. Because she's smaller than many of the other tigers, the PAWS keepers believe she may have been picked on by the larger cats back at Weinhart's house.

Mel brought me to see Sunita because of a disorder that she shares with roughly 10 percent of American schoolchildren. Sunita blinks her eyes and twitches her muzzle, repeatedly, like a human with an extreme facial twitch. Mel is convinced that she has a stress-related tic disorder. In humans, tic disorders are divided by type: chronic, transient, Tourette's syndrome, and "not otherwise specified." They can be vocal or motor or both and affect children and adults, often worsening when someone is feeling stressed. Sunita's facial twitch seems to grow more intense and frequent when she's stressed—particularly around veterinarians like Mel, who have given her vaccinations, or a few keepers she just doesn't like. When Sunita first arrived at the sanctuary she would throw her body at the chain-link walls of her enclosure anytime a human walked past, her face twitching all the while. Mel diagnosed a tic disorder and he hoped that she would grow out of it, like many human children whose tics often lessen and disappear as they age.

Two years later Sunita *is* calmer and more confident. She only occasionally paces along the fenced far wall of her enclosure and she's put on weight. Her coat is thick and full and she no longer waits until she's alone to eat. The tics haven't gone away entirely but Mel believes that Sunita's may be with her forever, a response to stressful situations that she can't seem to leave behind. As we stood alongside her enclosure, watching the keepers prepare the next meal of chicken and beef parts, Mel asked me why I was so interested in Sunita. I told him about my feelings of guilt over what happened to Oliver and how powerless I'd felt in the face of his compulsions and phobias.

"It sounds like Oliver was disturbed," Mel said, "and you did everything you could. Sometimes that's not enough. Sometimes it is."

Since he retired from the sanctuary and his small animal practice, Mel works primarily as a consulting veterinarian in animal rescue operations and as an advisor in facilities like PAWS that care for these animals long-term. Because of these animals' past experiences, the problems he sees are drastic: depressed elephants kept for years in cramped and isolated corrals, mutilated horses inside Canadian slaughterhouses, and traumatized chimps previously used as test subjects for hepatitis and other infectious disease research. He believes that most of the psychological problems in these animals are a function of their lives in captivity. Mel's work with companion animals, however, has convinced him that creatures whose natural environments are our homes, barns, and yards and who are accustomed to living with humans can still develop obsessive-compulsive disorders, weirdly specific fears, extreme anxiety, pica (eating inedible objects), self-mutilating habits, and depression, even if they haven't been mistreated.

"Your dog," he said, "may prove the point. You and Jude offered him kindness, love, stability, exercise, and still his problems got worse."

Doctor Do-a-lot

After Oliver's jump I was scared to leave him alone in the apartment. We couldn't ignore what he was capable of. Before going to work in the mornings, Jude and I dragged our wooden kitchen chairs in front of the windows and pulled the shades down. I knew this was illogical. Oliver clearly understood where the windows were and could have easily pushed the chairs aside. But even a solely visual barrier seemed important at the time. I also started calling veterinary behavior clinics in earnest, begging the receptionists for an appointment. I imagined these men and women lording over waiting rooms full of recently well-adjusted dogs who no longer gave a damn that they were sometimes left alone, the kind of carefree retrievers that trot across beaches in television commercials for human arthritis drugs. The receptionists were the professional gatekeepers to a mental peace I hadn't had in months, or at least I hoped they were. Most behaviorists had long waiting lists for new clients, but I eventually found a woman with a practice in rural Maryland who agreed to see us. Before the appointment I filled out a detailed questionnaire, answering questions about Oliver's snack preferences and whether he'd bitten anyone before.

On the day of the meeting I paid $350 and waited patiently for a diagnosis, a plan, and salvation while I sat flipping through copies of *Dog Fancy* in the waiting room. The first thing I noticed about the vet was that she spoke softly and warmly and wore pants without any dog hair on them. She offered Oliver a treat as soon as we entered her office. He immediately curled up at my feet and closed his eyes, the picture of canine calm, while I described his panic and anxiety. I felt as if I'd taken a car to the mechanic to complain about a strange noise only to have it hum along perfectly. The only noise Oliver was making was his snores, pleasant and regular. The vet, to her credit, still seemed interested and asked me questions about his behavior: what he did when Jude and I came home, the specifics of his diet, how long his walks were and where we took him, the layout of our apartment and how he used it, the long list of the things he'd destroyed, his responses to certain people and other animals. I rattled off the depth and breadth of his behavioral quirks, and eventually she stopped asking me questions. Instead she looked at Oliver and sighed. "You are going to have to do a lot of work."

"So it's possible to help him?" I asked.

"Yes," she said, "it should be."

It occurred to me then that animal behaviorists may do a brisker business selling hope rather than advice. Could the behavioral training portion of veterinary school actually be a course in human psychology? Was this woman *my* therapist too? As if in answer to my silent questions, she pulled out a prescription pad from her desk and wrote two, one for Prozac and one for Valium, and then she printed out a sheaf of papers and handed them to me.

"I believe your dog has a severe case of separation anxiety," she said, "and a thunderstorm phobia. And also perhaps, acral lick dermatitis, which is another way of saying he compulsively licks himself, like people with OCD wash their hands." The papers contained a variety of exercises that Jude and I were to do with Oliver in order to help him dissociate certain cues from his fears of our leaving him alone and also of thunderstorms. The vet also wrote down the name of a website where I could buy a CD of thunderstorm sounds to use to desensitize him to the booming thunder that sent him panicking. I woke Oliver up and we left the office. I felt better, more encouraged, than I had been when we went in and I could swear Oliver did too.

Receiving his diagnoses of separation anxiety, thunderstorm phobia, and a strain of canine OCD also gave me something to Google. I discovered that, like attention deficit disorder, separation anxiety has not always been an applicable diagnosis. Dogs can be diagnosed with the disorder today because it's a recognizable human affliction, currently defined in the *DSM* as "developmentally inappropriate and excessive anxiety concerning separation from home or from those to whom the individual is attached" for at least a month. It became a viable diagnosis in 1978; before that, children overly anxious about attending school, being left alone at home, or their parents dying were either not diagnosed or were simply considered sensitive. A similar process happened with dogs.

Many people, at least those who could afford it, became more distanced from livestock and working animals in the late nineteenth century and closer to animals who didn't have to work, like pet dogs and birds. By the early twentieth century dogs were beginning to be considered, at least a bit, like children. The historian Katherine Grier argues that Victorian prints, small decorative statues, cards, and other widely circulated products started to portray animals as friends. Illustrations of babies and puppies playing together as equals, or nursing cats and their kittens alongside nursing human mothers, helped encourage people to use the same kind of endearments to describe their pets that they used for their own children. This shift laid the foundation for the idea of shared emotional problems, evidence that many people were comfortable equating certain animals (pet dogs, for example, as opposed to foxes or coyotes) with humans, not merely as friendly companions but as beings with similar emotional lives, and eventually, similar brain chemistry.

The night after my veterinary behavior appointment, I spent the first of many hours inside virtual dog parks like bernertalk.com, reading other people's tales of canine woe. The diagnoses also gave me something to tell my mom when she asked why all of our kitchen chairs were piled up against the living-room windows. It was a sanctioned excuse I could give my coworkers when I left the office at 5:30 p.m. sharp. "My dog has a disorder," I would say, while heading toward the door. "A few of them. I have to go home on a schedule or he is going to fall to pieces." Perhaps literally.

Despite my enthusiastic defense of Oliver's diagnoses, I was still a bit conflicted. I felt that they were a bit too one-size-fits-all, a horse blanket tossed over him in a manner that didn't account for his individual responses and behavior, all of which, I was convinced, stemmed from one thing: fear and anxiety over being abandoned. His repetitive behaviors, such as his incessant licking, were a self-destructive way of calming himself and an outlet for his anxiety. Since his base level of agitation was already so high, his fear of thunderstorms was perhaps more extreme than it would have been otherwise. Indeed, Oliver's fears and anxieties were so severe that they colored his entire life and, by extension, mine and Jude's too.

I quickly realized I wasn't alone in my late-night Internet searches or in my hope that with Oliver's diagnoses would come peace of mind in the form of clinical intervention and acknowledgment from someone who knew more than I did. The reason it took me a while to secure an appointment with a behaviorist was that their practices are so busy. The American College of Veterinary Behaviorists currently certifies fifty-seven vets specializing in behavioral and emotional issues, all of whom churn out diagnoses for drastic self-destructive behaviors like Oliver's but also for annoying or annoyingly consistent activities—not just a single episode of pooping on the couch, for example, but regular pooping on the couch, perhaps combined with eating the poop. This small number of behaviorists is somewhat misleading, however, because the ability to make diagnoses like coprophagy (the poop eating), separation anxiety, thunderstorm phobia, or pica isn't limited to behaviorists; any veterinarian can diagnose mental illness or behavioral disorders and prescribe psychopharmaceuticals. The *actual* number of vets diagnosing emotional problems is probably closer to 90,200, the number of certified and actively practicing veterinarians in the United States.

The director of one of the busiest clinics specializing in behavioral problems, the Animal Behavior Clinic at Tufts University's Veterinary School, is the veterinarian Nicholas Dodman. He is the author of books such as *The Dog Who Loved Too Much* and *The Cat*

Who Cried for Help and has written dozens of scholarly papers about nonhuman disorders, from compulsive licking in canines to equine self-mutilation syndrome, which is similar, he says, to Tourette's syndrome in humans. Dodman primarily treats pet dogs, cats, and occasionally, horses and parrots. Most of these animals have not been abused or abandoned; after all, their human companions are paying quite handsomely for help.

The first time I visited the Tufts Animal Behavior Clinic it was to meet with Dodman's colleague, Nicole Cottam. The waiting room smelled animally and was full of people and their creatures on leashes, in carriers, or, in the case of one cat, a plastic laundry basket covered with a dishtowel. The room was divided down the middle by a four-foot-high partition, one side for dogs and the other for cats, each with its own television set. The dog side was watching QVC, the cat side, a talk show. With no animal myself, I didn't know where to sit. I spotted a man holding a Tupperware container with a chinchilla inside, and sat down next to him.

After Cottam retrieved me, she took me on a tour through the hospital. There were the cages of blood-donor cats, strays who now live at Tufts to supply blood to felines who need it; the orthopedic wing, where dozens of animals in brightly colored casts lean up against the walls of their cages waiting, long-eyed and droopy, to go home; and the main office of the behavioral clinic.

The first thing I noticed was the sheer volume of VHS tapes. Walls of shelves were lined with black plastic tapes in cardboard sleeves. The labels down their spines were handwritten, all in different script, some in pen, some in pencil, some neat, some scrawled. There was "Roxie" and "Chip" and "Snooker" and "Bill" and "Ralphie." It looked like a video rental shop from the 1980s, but instead of John Candy movies these were documentary features starring Poodles, Labs, Rottweilers, and cats.

Cottam saw me staring at the shelves. "For those people who

don't live close enough to the clinic, we do a remote consulting service over the phone," she said. "We ask people to document the problems they're seeing at home. Now everyone sends pictures and video to us via email, but before, you had to set up a camera when you left the house and then send us a tape." I was looking at an immense visual archive of animal emotional problems.

Cottam and Dodman both see hundreds of behavioral cases a year. They are also busy with their own research projects, focusing on many of the more common disorders they treat at the clinic. Cottam is currently investigating thunderstorm phobias in dogs, but she's seen all manner of other emotional issues too. After 9/11 she treated a few of the search and rescue dogs who had worked at the World Trade Center site for fear and anxiety problems that turned them into hesitant and unstable versions of their former selves. She believes that the dogs' extreme anxiety and fear was triggered by the sights, sounds, and long hours they spent working in the rubble. Later she treated canine survivors of Hurricane Katrina who had been adopted after the storm and then brought into the clinic when they started reacting fearfully to sounds or sights that reminded them of the flooding or the experience of being left by their previous families. She has also seen many strange animal fixations, from cats bent on eating only small shiny things to her own dog, who runs, shaking, from billowing fabric. "He freaks out when he sees sheets blowing on the line," she said, "or flags, or my neighbor's tarp in the wind."

Like the military's canine behaviorist, Cottam is convinced that dogs can suffer from a canine version of PTSD and mentioned the 9/11 search and rescue dogs, as well as canine survivors of Hurricane Katrina as examples. "It's possible that it's just extreme shyness that keeps them cowering under tables, beds, couches, too afraid to come out," she told me, "but I believe it's something closer to PTSD. When they do come out, they hug the wall. They appear traumatized."

At Tufts they treat these dogs by decreasing their stress levels and putting them on medication. "We also try and get to the bottom of their specific fears and try and treat those," she said. One dog's fear response may be triggered by loud sounds, for example, while another may react to men in uniform.

Like Mel Richardson, Cottam believes that other animals exhibit their own versions of almost every human psychiatric problem, but she's careful about how she says so.

"Take obsessive-compulsive disorder, for example," she said. "Now, I *think* that animals are having obsessive thoughts when they lick the leg of a chair without pause for hours on end. But I can't *prove* that they're thinking obsessively, so when Nick and I publish on the topic, we use the term *compulsive disorder.* We leave out the *obsessive* because using that term would imply that we know what the animal is thinking. A human obsessive hand-washer can describe his obsession to us, but an obsessively licking Doberman cannot."

I thought about Oliver's compulsive paw-licking and his weirdly vacant fly snapping and asked if she thought that our behaviorist had been right when she said that Oliver had a form of canine OCD.

"It's possible. Compulsions are a major problem. You should see the crop circles in some people's yards. They happen because a dog just won't stop chasing its tail."

Cottam and Dodman help their patients with their obsessions and fears, some of which are extremely specific. "I've seen dogs that are scared of all kinds of strange things, like shadows, bright sunlight, even contrails in the sky, or beeping noises from alarms or microwaves. The beeps can be really scary because the dogs don't understand where they're coming from. Another time we treated an older dog for his fear of flies. He had been swarmed when he was a puppy."

Treating phobias, especially a thunderstorm one like Oliver had, is especially tough since the dogs are reacting not only to noise but also to changes in atmospheric pressure or flashes of lightning. Desensitizing dogs by simulating all of these at once is nearly impossible.

The most common problem they see at the Tufts clinic, however, is extreme aggression. "I think it may be similar to impulse control disorders in humans, and it's usually related to jealousy," Cottam told me. "Certain dogs will see two of their humans hugging and just lose it. Or they will be jealous of another dog."

As Cottam walked me back to the waiting room I paused in front of a gigantic scale ("We had a giraffe in last week") and asked why she thought Oliver might have had the mental troubles he did. I tried to cover the whine in my voice with an earnest researcher tone, but she still looked at me pityingly.

"I wouldn't be that upset if pure-bred dogs disappeared," she said. "Look up the Carolina swamp dog. It's a forty-pound tan dog. It's what dogs would look like if we gave up on breeds. Eventually they'd all look a lot more like dingoes and coyotes."

I asked why this was better.

"If you're breeding a dog like a Bernese Mountain Dog, you are breeding for very particular physical traits, such as coat color, body shape, and more, and you end up with whatever behavioral traits are associated with the physical ones."

I thought about the perfect white stripe down Oliver's nose, the white tip of his tail, the identical brown eyebrows, the black coat like a lush tablecloth, and the lineages of award-winning show Berners with names that call to mind Viking kings (Igor Vom Eck-Manns-Hof) or yachts (Glory V Legacy). Almost every Berner has these exact markings, and their lineages are celebrated as if they were four-legged Daughters of the American Revolution. The dogs look so much alike, in fact, that ever since Oliver's death, whenever I run into one on the street I feel as if I've seen a panting, tricolored ghost.

A parallel in human breeding would be deeply disturbing. What would happen, for example, if a small group of people were made to have children with another small group solely based on the length of their forearms, the color of their leg hair, the shape of their ears, the shade of their palms or backside, or the size of their feet? It would be reminiscent of the misguided, racist, and often terrifying eugenics programs of the early twentieth century in the United States and later in Nazi Germany. Now imagine this group's children were forced to make the same mating decisions, and their children's children, maybe even their children's children's children. Pretty soon you would have a human version of a Bernese Mountain Dog.

Many breeders will tell you that they don't breed for solely physical traits, that they raise family dogs with good personalities, who are easygoing and as sane as possible. But to simply meet a breed standard (the precise requirements designated by the American Kennel Club), a dog must have specific markings and proportions. As Cottam suggested, some of these markings may be connected to other traits that don't make for a well-adjusted creature, traits like anxiety, fearfulness, or aggression. The AKC's breed standard for Bernese Mountain Dogs has descriptive subsections for "forequarters" and a thick paragraph on markings but only two sentences on the dogs' temperament, which, according to the AKC, should be "self-confident, alert and good natured but never sharp or shy."

Certain breeds are notoriously prone to specific disorders. Mel Richardson has encountered so many Bull Terriers with tail-chasing compulsions that he believes the behavior is in some way genetic. Dodman has treated Terriers and Border Collies and many other breeds for tail chasing too, as well as dogs unhealthily fixated on following shadows, chewing rocks, and licking all sorts of surfaces that shouldn't taste good. Chasing patches of light is reportedly most common in Old English Sheepdogs, Wirehaired Fox Terriers, and Rottweilers. Snapping at flies that don't exist is prevalent in German Shepherds, Cavalier King Charles Spaniels, and Norwich Terriers—though I can attest to its presence in Berners. As for cats, Siamese, Burmese, Tonkinese, Singapura, exotic breeds like the Ocicat, a domestic cat spotted a bit like an ocelot, and Munchkins, bred for disproportionately short legs, a sort of feline version of a Dachshund, have each been known for their compulsive tendencies.

In Oliver's case I believe his particular distress cocktail derived from a mix of his constricted gene pool, his past experiences, and his neurophysiology. I was never able to discover the exact trigger that set him on his path to madness, but I have a few guesses. Figuring out what was wrong with Oliver was a process of determining exactly what was bothering him and attempting to understand where and when his anxiety might have begun in the first place.

After visiting the veterinary behaviorist, I wrote to the breeder who introduced Jude and me to Oliver's previous family. I asked him if he knew anything about our dog's quirks and he told us, for the first time, a bit of Oliver's life story.

From the time Oliver left the breeder's house, where he rolled around as a playful pup with his brothers and sisters, throughout his first four years with his new family, he was adored and received loads of affection. He went on lots of walks and liked to lie in the living room alongside the family's other dog, an Old English Sheepdog. Life was easy and peaceful, the dogs spent their days gazing out the glass sliding door into the yard. When the family's youngest daughter, a high school student, became pregnant and decided to keep the baby, everything changed. Suddenly Oliver was no longer the fuzzy sun of his family's solar system. He was unseated by the teenage pregnancy and then a new baby, and he didn't like it one bit. He reacted by trying to wend his way back into the center of family life, doing so as a dog does. He pooped where he wasn't supposed to, he bit the neighbor after going after her dog, he broke through an electric fence. He also started gnawing on things he knew he wasn't supposed to gnaw on. Throughout it all he probably only wanted the affection he was used to. But no matter what he did, he wasn't getting it.

I'm convinced that his family meant well. They loved Oliver, but they were overwhelmed. Their daughter and the new baby came first. The more attention Oliver demanded, the more frustrated they became and the more often they locked him away. First they put him in the garage, but he chewed the moldings off the windows in his attempts to break out. Then they tried putting him in a crate. But they didn't first teach him first that the crate was a nice, safe place to be. Being locked up in a small space without any of his people around probably just made Oliver more upset. He destroyed the plastic and wire of the crate in his efforts to break free and rejoin the family. This, I believe, is when they started looking for someone to take him off their hands.

A glossy, wry brunette with a taste for sparkly earrings, Dr. E'Lise Christensen is New York City's only certified veterinary behaviorist. As such, I expected her to be older. Instead she looks like she's in her mid- to late thirties and is the kind of self-aware vet with whom you want to get drunk and then beg for stories about animal psychopaths. Or maybe that's just me.

The first time I saw Christensen in person she was sharing a stage with Dr. Richard Friedman, a professor of clinical psychiatry and the director of the Psychopharmacology Clinic at Cornell's Weill Medical College. She was giving a talk about anxiety in dogs and Friedman was discussing panic and anxiety disorders in his human patients. They were speaking at Rockefeller University in New York, as part of a conference on the overlaps between human and nonhuman medicine. "It's shocking how similar anxiety disorders are in humans and in dogs," Friedman said. "[My patients] walk around in a state of constant semi-dread, as if the world is a very dangerous place." Christensen nodded in agreement. "Anxiety disorders," he continued, "are far and away the most common disorders in the United States today. There's a ten to twenty percent risk that someone will suffer from extreme anxiety or panic attacks over the course of their lifetime."

Christensen countered, somewhat wistfully, that unlike psychiatrists, veterinary behaviorists seem to see fewer cases of extreme anxiety than they once did. This is because general veterinarians have learned, in the last decade or so, to treat panicked and anxious pets with medication and behavioral therapy, therefore sending fewer of their clients on to specialists. She believes that around 40 percent of the behavioral issues propelling pet owners into these vet clinics are due to separation anxiety. "But I only see the toughest cases, the dogs who don't respond to the first line of treatment." These animals are so distressed, their suffering so extreme, that Christensen is sure that if the dogs were humans, they would be hospitalized in psychiatric wards. "By the time they see me, these animals can be on five different meds, at really high dosages, and they're still panicking," she said.

Many of Christensen's separation anxiety patients come in with underlying issues related to what she considers impulse control, a phenomenon that's common at the Tufts Animal Behavior Clinic as well. When I asked Christensen how a dog with an impulse control disorder is different from a normal dog, she said that it comes down to reactivity. That is, a dog who used to growl before biting but was punished for the growling and now bites without warning isn't a dog with an impulse control disorder. He's a dog that has learned not to growl. A dog who has no warning behaviors of any kind, who never snaps or growls but just bites, is a dog with an impulse control problem. "It's like a person who shoots first and asks questions later," Christensen told me. "If they're more anxious than the average bear and don't stop to think and consider the consquences before biting, this is an issue." She says that for a dog or a person choosing aggression is a risk. You can be injured or your social world can be disrupted when you lash out. "Dogs make choices. They don't always make good choices, however, and sometimes their choices are too fast." These animals will struggle with impulse control and anxiety for the rest of their lives. "You can't guarantee a cure and if anyone does, they're lying."

Another veterinary behaviorist interested in the causes and treatment of separation anxiety is Dr. Karen Overall, the former director of the Behavior Clinic at University of Pennsylvania's School of Veterinary Medicine. Overall has spent years researching mental disorders in companion animals and believes that no animal disorder is a perfect mirror of a human condition, but like so many veterinarians

I spoke to, she's convinced that dogs develop issues similar to many human psychiatric disorders. These include generalized anxiety disorder, attachment disorders, social phobias, OCD, PTSD, panic disorder, the aggressive impulse-control disorders, and, like Alf the Australian Shepherd, Alzheimer's disease. She makes diagnoses just as E'Lise Christensen, Mel Richardson, Nick Dodman, and Nicole Cottam do, by interviewing pet owners in detail, compiling a history of the animal in question, and observing their behavior.

Richard Friedman believes that the experiences of his human patients with generalized panic and anxiety disorders are similar to Christensen's anxious dogs because the overwhelming impulse for people with panic disorders is to flee. Friedman treats his patients with pharmaceuticals and cognitive behavior therapy to help them overcome their impulses to bolt. "The treatment outcome for my patients is really good," he said, "but it's a chronic disease. The recurrence rate over a twelve-year period is really high."

Unlike these generalized panic and anxiety disorders that Friedman sees in his patients at Cornell, the human form of separation anxiety is somewhat different from the dog version. In people the disorder tends to manifest before the age of eighteen and is usually triggered by being away from a parent or loved one. In the dogs that Christensen, Overall, and the other vets treat, the affliction seems to appear at any age and tends to come from being left alone, not by being separated from a specific person. The presence of a human, any human, often helps. Some canines, like Oliver, react when they are left all alone in the house, while others become distressed if they're locked in one particular room or a crate. Most express their anxiety by trying to escape. This is what canine separation anxiety has in common with human panic disorders and generalized anxiety disorders, diagnosed in people who feel excessively anxious and worried for more than six months, can't control their worrying, are restless or tense, can't sleep, are grouchier than normal, have trouble concentrating, or are occasionally overcome with a desire to run away.

Separation anxiety in dogs probably feels like a life-or-death predicament, their bodies coursing with flat-out panic; this could explain their extreme actions. When Oliver was left alone, he may have felt that it was forever, that no one he cared for was ever coming back for him. It didn't matter that Jude and I always came home eventually. Oliver's fear was a tsunami gathering on the horizon, threatening to crush him, and it activated every impulse within him to fight his way to safety. I suspect that he tried to escape so mightily not because he wanted to find us per se but because he was trying to run from the terrible discomfort he felt at being left alone. When Oliver was busily reducing a door to sawdust or digging up a hardwood floor to try and burrow his way outside, I don't think he was engaged in a specific, Lassie-like effort to locate me and Jude. Instead he was mad with his own anxiety, and dogs in extremis do what dogs are capable of. Chewing, digging, pacing, and frothing are a few of the things in their repertoire.

During her residency in upstate New York, E'Lise Christensen treated a German Shepherd whose anxiety compelled him to destroy a kitchen window anytime he was left alone. Once he'd gone through the window, his panic subsided and he curled up and waited in the yard for his humans to come home. After replacing the window multiple times, the family decided to simply leave it open. "Since he was a German Shepherd," said Christensen, "they didn't have to worry about burglars." She's also treated dogs who jump out of apartment buildings in panic like Oliver did; one survived by landing on the air-conditioning unit of an apartment a few floors below.

Oliver didn't just jump from great heights, however, he also whined and barked for help; he clawed and gnawed on furniture, floors, doors, sheets, towels, pillows, and anything else within reach; he panted and salivated, licked himself raw, and tried to bolt. Other dogs poop and pee where they're not supposed to. A few others express their anxiety by withdrawing deeply into themselves and growing less active; these are the drooly, quiet martyrs. Sometimes they stop eating. Christensen believes that anorexia—when a dog won't eat or drink when she is alone—can be a sign of separation anxiety. "Emotional eating in dogs," she said, "usually means not eating."

The Cat Whisperer

One of the largest animal shelters in the United States, the San Francisco Society for the Prevention of Cruelty to Animals (SFSPCA) houses between 230 and 300 cats and almost as many dogs. The cats live in tile-floored, glass-walled "condominiums" with fuzzy cat trees, chairs for potential adopters to sit in and visit with the cats, and televisions playing videos of squirrels dashing across green lawns or birds preening in birdbaths. From his post in the Cat Behavior Office, Daniel Quagliozzi helps diagnose problems in cats when they first arrive at the shelter. He also acts like an extended warranty for the shelter cats, answering questions from people about their newly adopted felines once they take them home.

Daniel has two sleeves of swirling tattoos, his forearm features an angel cat and his right, a devilish one. His knuckles spell out CAT'S MEOW in red and black ink. Daniel has treated everything from fear aggression that makes cats into snarling whiskered weapons to pica. On top of his desk in the Cat Behavior Office is a small Mexican folk-art box with the name Diablito written on it in glitter. Inside are the ashes of a shelter cat who used to hang out in their office. He was prone to explosive diarrhea.

"A large part of what makes a cat happy," Daniel told me, "is routine. They like having their expectations met and knowing how the day will unfold. They're well behaved when nothing is different. When things change they often go haywire." I couldn't help but think of so many humans I know with the same problem.

The biggest challenge for cats new to the shelter, Daniel believes, is that most of them come from comfortable homes and are then thrust into a strange environment with different smells, people, food, and routines.

"We like to say here that we manage a three-lane highway," Daniel said. "There are the fast-lane cats that come into the shelter and are unfazed by all of the new and strange things around them. They eat anything and are sociable from the beginning. Then there are the slow-lane cats. These cats come in and will literally hide under a blanket for two months. It takes a lot of work to get them to come out, and because it's a shelter, we may have to move them to a new room, and the whole process starts over. The middle-lane cats are, you know, somewhere in the middle."

To identify whether a given cat is depressed, Daniel asks himself a series of questions while carefully observing their behavior: Is the cat eating? Has she used the litter box? Has she moved at all? After three days, if the cat still hasn't touched her food but is otherwise healthy, Daniel says that it may be a sign of depression.

"I'll approach them and see what they do. Do they nuzzle into my hand, or do they not even move their chin? A depressed cat simply won't respond. The fearful cats are very responsive. . . . They hiss, swat. . . . The depressed cats are really just these little lifeless lumps."

Outside of his work at the SFSPCA Daniel runs a behavior consultancy called Go, Cat, Go. He makes house visits to people with cat problems throughout the San Francisco Bay area. Every week Daniel receives dozens of messages on his cell phone from frustrated and desperate people who want deliverance or simply insight into their cat's mind. When I last visited him at the office, Daniel had just finished speaking to a distraught woman who was convinced her cat had a split personality disorder. The once sweet and cuddly cat was attacking their French exchange student for no apparent reason and had scratched the young man's leg so badly that he had to be taken to the emergency room.

"Usually," Daniel said, "people call and want to know, Why does my cat hate me? Why is he scratching me? Or crapping in my shoe? Or eating my dress shirts? Part of my work is making them realize that it isn't about them."

In the case of the anti-Francophone feline, Daniel felt that the cat may have been reacting to the presence of a stranger in the house. The exchange student was sleeping in the recently vacated bedroom of his favorite human, a teenage boy who had just left for college. Perhaps the cat felt the exchange student was some sort of invader who had done away with his loved one.

Daniel sees himself as an interpreter of these sorts of interspecies mysteries. He asks lots of questions, as nonjudgmentally as possible, and pays careful attention to the way a client's house is set up, the dynamics between the people who live there, and how a cat spends his time.

"They will tell me everything that I need to know," he says. "For example, how do the cats like to use their environment, and are they being provided for in a way that makes them feel at ease? Do they have their own little areas that feel safe and controlled? Do they have the food and the cat litter they like? These things may sound small, but they have a huge impact on their mental health."

To create an environment that encourages cat sanity, Daniel suggests his clients reserve places that are cat-only, such as cat trees. "They're ugly, but cats like having things that are just theirs. This makes them feel protected. It's best if these places are also tall, like the top of a bookcase or refrigerator, because being able to look down on people and other animals in the house makes them feel secure." This was not particularly surprising.

"Also, these additions to their territory should *not* be tucked away from the action. They want to be part of everything that's going on." Daniel also encourages his clients to engage in play therapy with their cats, which is really just play. One of the most recommended cat toys for this is something called "Da Bird," a miniature fishing pole dangling a garishly colored feather clump. You're meant to wave Da Bird in the air like a demented conductor or someone who's smoked too much of da herb as your cat chases it to and fro. If the original lure becomes boring, you can swap it out for an even more sparkly option that looks like it's been plucked from a Vegas showgirl.

Still, no matter how many Da Birds a cat receives or how many scenic vistas they have to look down upon humans and dogs, they can still develop odd behaviors. Daniel's own cat, a Seal Point Siamese Munchkin named Cubby, has his own issues. He also has the watercolored face of a Siamese and the stubby paws of a Munchkin. Because of his short legs, Cubby can't swat, but he hisses, usually at other cats. To Daniel's dismay, Cubby suffers from feline hyperesthesia, a disorder characterized by a sudden, intermittent desire to savagely attack his own tail. Cats with hyperesthesia stalk their twitching tails as if they are menacing objects or invaders and then they pounce so hard that they sometimes rip their own flesh.

Daniel didn't know why Cubby was attacking himself. Their house, where Cubby rules the bedroom and sometimes the hallway and kitchen, has multiple cat trees, a tunnel for running back and forth, and private sleeping quarters in a closet. It is, in short, an ideal cat habitat, and Cubby could not find a human more attuned to his needs. Daniel decided to medicate Cubby. After thirty days on Prozac, the cat stopped acting as if he was possessed. A few years later, Cubby has recovered. He continues to take a small maintenance dose of Prozac, which limits his self-mutilating episodes to a mere thirty seconds or so per week. The rest of the time he sleeps in a sunny window, waiting for Daniel to come home and play Da Bird, or watch him as he runs on his short little legs through his cat tunnel.

The Elephant on the Beach

Daniel discovered with Cubby and the other cats he's helped that careful observation is necessary to diagnose them correctly and begin the healing process. But there's something else that's often key to understanding why a creature is disturbed: the animal's individual history. Oliver's experiences with his first family, for example, affected the ways he related to us. This is especially the case for animals who have lived through big changes, such as the shelter cats who were removed from their family homes, or animals who were raised under odd circumstances, like Sunita the tiger, or Rara, an elephant who grew up at a Sheraton Hotel.

When Rara was a year-old calf, she was taken from her mother and sold to a Sheraton hotel in Krabi, a luxurious beachfront resort in southern Thailand. She spent most of every day chained inside a cement-floored, open-air pavilion near the thatch-roofed house of her mahout and his family on the hotel grounds. Once or twice a day Rara was brought to the hotel lobby or grassy lawn to pose for photos with guests, to be stroked and cooed over, and to be hand-fed bananas. In the heat of the afternoon her mahout walked her to the hotel beach for a swim and to play in the warm water with the tourists, many of whom splashed around with her in the shallows and took photos of her while she sprayed seawater out of her trunk or dug wet holes in the sand. In those first few years Rara was goofy and charismatic and bonded easily and often with the hotel guests. She liked to use her trunk to play the harmonica and would dash to the hotel's outdoor showers where she knew how to turn on the taps. She drank and played under the spout until her mahout made her stop.

Rara was a growing elephant though, and over time becoming more of a liability than a photo-op. By her sixth birthday she weighed thousands of pounds and, because of her size, was too dangerous for the tourists to swim with. Her enthusiastic trunk grabs on guests' forearms, charming when she was small, were now so strong that even though she was only being playful she could easily knock people down or accidentally trample them. She was also becoming more opinionated about how she wanted to spend her time and was more difficult to control when she didn't feel like doing something her mahout asked of her. As a result, Rara was chained more often and for longer periods. A few concerned hotel guests, some of whom came back to the Sheraton year after year to visit Rara, called themselves "Rara's Fan Club" and shared photos and videos of her on Facebook and YouTube, began to worry for her future.

One hotel guest, a generous banker from Hong Kong named Silke Preussker, hatched a plan to buy Rara from her owners and bring her to an elephant ecotourism park where she would be around other elephants and chained only at night. Within a year, Silke had joined forces with a number of other Rara advocates and convinced the Sheraton Hotel (which promised never to lease another hotel elephant) to bring the young elephant north to Elephant Nature Park, a verdant valley outside of Chiang Mai, where the elephants are not made to perform and often form lasting friendships with one another.

I met Rara shortly after her arrival and immediately understood why she had so many fans. She was mischievous, affectionate, silly, constantly getting in the way of construction crews, tasting the landscaping plants, and causing her mahout, a slight Burmese man named Gawn, to chase after her ceaselessly. "She reminds me of my four-year-old son," he said one morning, watching nervously as Rara tried to balance on top of a pile of logs using only two of her feet. She continually glanced over at us, as though to make sure that we were paying attention to her antics.

Having been separated from her mother and all other elephants at such a young age, Rara was scared of the park herd. She had no basic elephant culture; she was at a loss when it came to approaching new elephants and didn't know how to show affection or express herself in a nonthreatening way. Because of this, the other elephants were skeptical of her. Rara preferred to spend time with the park's human guests, particularly white women, who had been the font of bananas and affection at the Sheraton. She disliked Thai men, except Gawn, whom she loved fiercely. The rest of the park's male staff gave her a wide berth. Once, when Gawn was unable to come to work and Rara was given a new mahout for the day, she terrified the park employees by throwing an elephant-size tantrum that resulted in a smashed car and overturned baskets of produce.

This behavior isn't particularly surprising if Rara's life history is taken into account. Elephants learn from their mothers, aunties, and other herd members how to be elephants: how to show joy and anger, what to eat and how to eat it, the best ways to stroke a companion, and how to physically protect themselves. Like humans, they're not born knowing how to behave. In the herd Rara also would have been disciplined when she acted inappropriately. After she was taken from her mother, the only teachers she had were humans. She spent most of her time confined, and when she was free it was only to be patted by tourists and given treats. She interacted with new humans all the time, and each of these people responded to her differently—some with affection and others with fear. The most important relationships, those that would have taught her how to be an elephant, were taken from her. As a result, Rara grew into a sort of human-elephant hybrid, an outsider in both worlds.

And yet she was lovely. I learned to rumble like she did, a sort of rolled-R throaty hum, and she would respond in kind. If I was gone for just a few hours and then ran into her and Gawn in the park, she treated me like a long-lost friend, running her trunk over my head and face, blowing air onto my crotch, rumbling and squeaking, ready to begin whatever game we'd last played. I hoped that she would learn to be an elephant among elephants, but I admit I also enjoyed the fact that she liked me. It's wonderful to make a new human friend, but it's even better to be friends with an elephant. It was also a bit depressing. Didn't human-elephant friendships usually end poorly, with the elephants winding up in circuses or as crop raiders? Shouldn't Rara be less fond of the species that took her from her mother and kept her chained for years? Why on earth did she still like people?

It may have been because she didn't have much of a choice. Elephants like Rara exist in a complex emotional world where they must balance their own needs with the humans who attempt to control them—a tightrope walk of interspecies expectations that can be both harmful and healing. In order to learn more about these relationships I left the park for the city of Chiang Mai to meet someone I'd been told could teach me about the emotional health of captive, working elephants.

Pi Som Sak is Karen, an ethnic group from northern Thailand and southern Burma with a long history of working with Asian elephants. He is also an elephant trader who buys and sells pachyderms of various ages and abilities, a bit like a used car dealer who specializes in different makes and models of the same brand. His family has had elephants for as long as anyone can remember. Until relatively recently their elephants lived in forests that the Karen protected for this purpose. They were captives, but weren't severely constrained. When they weren't busy doing heavy labor in the village the elephants lived more or less as they chose. Adults dragged a heavy, unsecured leg chain that, as it trailed behind them through the undergrowth, left a path that showed their mahouts where they'd gone when it was time to round them up again. Many of these elephants were born to mothers who had been part of Karen communities for generations. They gave birth, had sex, and basically made their own choices during most of the year. They were trained to respond to only a few commands, such as *Stop*, *Go*, *Open your mouth*, and *Lift your foot*.

Before World War II, roughly two-thirds of Thailand was covered in thick forests that were home to wild elephants, tigers, rhinoceros, wild cattle, leopards, wild dogs, and monkeys. But by the 1950s these forests were being cut at an increasing pace. The Thai government began granting large-scale logging concessions to foreign companies. Many Karen men found jobs in the logging industry working as mahouts. Since many of the prime logging regions had no roads, elephants were used in place of trucks to transport logs, men, and supplies. They also stacked the harvested logs, dragged them to rivers to be floated elsewhere, and helped pull stumps from the ground.

Today there is very little forest left in Thailand, and what remains is protected. Logging is illegal; this means that the roughly 2,500 elephants who still live there are now out of work, finding themselves in the sad position of having been forced to log their own habitat out of existence. There is still a thriving elephant market, however, and it exists to serve tourists willing to pay for elephant rides and treks or to see the animals paint pictures of flowers or the word "Love" in English with their trunks, play soccer or basketball, twirl a hula hoop, and throw darts.

I visited Pi Som Sak at the Chiang Mai Zoo, where he lives in a small house on a forested hill overlooking the city and the flat dirt spots where his elephants are chained at night. He is a wealthy man who could live in a nice house in town, but he chooses to stay here with his family and his elephants, some of whom give rides at the zoo. "I like the view," he told me, motioning to three female elephants staked in front of his house, calmly eating the wet tubular hearts of banana trees.

Pi Som Sak is often on his cell phone, talking in hushed tones about prices, and then driving off to a small village to visit a calf or adult whose owners want to sell. It is his job to determine how resilient an elephant is. Emotionally distraught elephants are cheap; aggressive elephants are even cheaper. Cheapest of all is an elephant who has killed one or more people. Som Sak wants to buy well-adjusted elephants because they're the best investments, so he has spent his entire life figuring out how to judge their emotional health as quickly as possible.

Certain behaviors signal a disturbed elephant. "I look for particular movements," he said. "If they 'pound the rice'"—meaning that they nod their head up and down without stopping—"that is bad luck and not an elephant you should buy. Also if their ears are not flapping in unison, but one at a time, that means they could be very dangerous. And their tail should look beautiful, like a lion's tail. If the end is missing, it is bad luck. Maybe they fight with others and had it bitten off. Also when they look at you, they should blink. If they just stare at you blankly, that is bad."

When Som Sak buys a new elephant and brings him or her back to Chiang Mai, a long truck ride is often involved. This can be stressful for the elephants, especially if they have never ridden in the back of a truck before. Som Sak's process is in a way similar to those early flood experiments on emotional resiliency. An elephant road trip is his version of Pavlov's fake flood.

When a new elephant arrives at Som Sak's house, he takes her off into the forest, gives her plenty of food, and leaves her alone. "Later, I sneak back and hide behind a tree where I can see the elephant but she can't see me. If she is flapping her ears and breaking off tree branches to scratch herself or use as flyswatters, then I know the elephant is okay. If she is just standing there, she is not."

Pi Som Sak believes that the emotional stability of an adult elephant is largely due to what his or her life was like as a calf. "It has to do with the mother," he told me. "If they have a good mother, they will usually be good because she will treat them well."

I asked Som Sak about Rara, telling him a bit about her early years. He agreed that living in a hotel is no way for an elephant to grow up, but he also pointed out that she must have had a good mahout or she would have already become violent. Som Sak is also convinced that base temperament can be inherited. A kind and gentle mother is more likely to give birth to a kind and gentle calf, and an aggressive mother is more likely to have an aggressive calf. But when it comes to the emotional health of captive elephants who have to interact with humans, the most important thing is how they are treated by these humans, and their mahout in particular. This is not as easy as simply treating one's own elephant well though, because an elephant's relationships with other elephants are so important that they can become upset if they see *another* elephant being beaten or ignored or harassed by their mahout.

* * *

Unfortunately, despite Gawn's careful, loving attention and the possibilities of her new life at the park, Rara died a few months after she arrived. One morning, after a night of refusing food, she had a heart attack and collapsed. Her autopsy showed that she had herpes, a disease that, in elephants, causes heart problems instead of sores. Her heart was so enlarged that the park veterinarian, Dr. Grishda Langka, couldn't believe that she had lived as long as she did. Her fans and friends were devastated. Silke Preussker, who had worked so hard to bring her to Elephant Nature Park, flew in from Hong Kong to pay her respects. We stood next to Rara's grave, a giant mound of earth, and made a pile of young coconuts, Rara's favorite snack.

For a long time after I returned from Thailand I thought of Rara every day. I carried with me, and still do, a small wooden carving of her that Gawn made while she grazed nearby. The fact that Rara stood still long enough for him to carve her likeness amazes me. And when I turn the tiny, smiling wooden elephant over in my hands I always think about other conflicted, difficult, and charming animals. Oliver had been fragile when he came to Jude and me because, perhaps, that was simply the sort of canine he was, but he was also a collection of his experiences as a young dog. Rara was the same. Now we would never know if she would have eventually gotten over her fear of other elephants. I wanted to understand more about the role of animals' early life experience on their long-term mental health. And so I turned to the creatures that generations of scientists have used to unravel the mysteries of the mind: rats, mice, and children.

Bruce Perry is a child psychiatrist, neuroscientist, and the former chief of psychiatry at the Texas Children's Hospital. He specializes in helping traumatized kids, treating the survivors of the Branch Davidian siege in Waco, Texas, for example, and many others who have survived genocide, rape, neglect, and abandonment. He has also helped various organizations plan trauma responses in the wake of tragedies such as Hurricane Katrina, the Columbine school shootings, and September 11. In his book (with coauthor Maia Szalavitz), *The Boy Who Was Raised as a Dog,* Perry describes a few children he treated who grew up in highly abnormal environments, many of whom suffered early life traumas that affected them well into adulthood. There was the boy who was left alone all day as an infant while his mother took long walks around the city, whose cries brought no help and who grew up to be a rapist with an inability to feel emotions as others do. Another boy was raised in a kennel like a dog, among other dogs, by a well-meaning but completely incapable male guardian.

Perry has also done research on rats. Part of his doctoral work focused on understanding the role of neurotransmitters like norepinephrine (noradrenaline) and epinephrine (adrenaline) in the fight or flight response. Rats in his neuropharmacology lab were exposed to stressful stimuli, such as shocks or sharp sounds, as they tried to negotiate a maze. Some of the rats were still able to solve the maze easily, while others fell apart when exposed to even the tiniest stressor, forgetting everything that they already knew. The researchers determined that the rats most sensitive to stress had extremely overactive adrenaline and noradrenaline systems (that is, a more sensitive fight-or-flight response). The overabundance of these stress hormones caused an avalanche of changes in other parts of their brain, hampering the rats' ability to respond to stress. The developmental stage during which these rats were exposed to stressors also affected the extent of their neurological changes. Earlier studies demonstrated that if rat pups were handled for even a few minutes by a human in the lab, something they found especially stressful, the resulting changes in their stress hormone levels and their behavior would last into adulthood.

Human stress response systems, like those of rats, can become too easily triggered by potentially disturbing things: plane turbulence, heights, people who resemble a person who hurt them, insects, or the millions of other sights, sounds, or experiences people can find worrisome. There may also be lasting effects on the function of the brainstem, limbic system, and cortex—the parts of the brain responsible for everything from controlling heart rate and blood pressure to the capacity for abstract thinking and decision making—and, tellingly, emotional states such as sadness, love, and happiness. In 2009, the U.S. Department of Health and Human Services published a report, "Understanding the Effects of Maltreatment on Brain Development." They didn't have nonhuman animals in mind, but the processes the report describes are similar. During fetal development, neurons in the animal brain are created and migrate to the various parts of the brain where they will stay. The development of synapses between these neurons, the neuronal pathways that give rise to memories, decisions, emotions, and other mental experiences, happens a bit in the womb and then takes off after birth in response to a young animal's experience in the world. If some synapses and neuronal pathways are not used, or if they are bathed in high levels of stress hormones, they can atrophy; this can result in serious emotional problems as the animal grows. Research on both human and other animal brains has shown that they may be more susceptible to damage at certain times than others, as in Perry's rat pups, and this damage may lead to emotional problems later.

Perry's very first patient, a seven-year-old girl named Tina, reminded him of his earlier work on the stressed young rodents. Tina was sexually abused between the ages of four and six by the teenage son of her babysitter. At least once a week for two years, the boy tied Tina and her younger brother up while he raped and sodomized them with various objects. He threatened to kill both children if they told anyone. When Tina appeared in Perry's office a year after the abuse ended, she was having trouble sleeping and paying attention, difficulty with fine motor control, coordination, and aspects of her speech, and sometimes she misunderstood social cues from people around her. Like the baby rats exposed to stressors that affected the function and development of different brain regions, Tina's abuse affected her neurological development during a key period in her growth. Perry was convinced that the years of stress caused a string of changes in Tina: altered stress hormone receptors and increased sensitivity and dysfunction, which were responsible for her developmental difficulties. This, along with her memories of the abuse, made learning and paying attention harder. She also acted aggressively in school. Perry believed that Tina was more apt to be on the alert for danger, even where it didn't exist. In the classroom she saw the smallest slights by her teachers and classmates as challenges and often got into fights or acted out sexually.

Almost all of the significant recent research on stress, neglect, and mental health has looked at people like Tina, though it's likely that the effects are similar in many different species. Imagine that an infant girl starts to cry, and instead of tending to her, her mother locks the door of the room and turns out the lights. If this happens once or twice, it will probably not have lasting effects on her development. If it happens every time the baby cries, the parts of the girl's brain that help her attach to people around her, those that release chemicals triggering pleasing feelings when she sees her mother or is held closely, teaching her that attachment to other humans is beneficial, will not be activated. It's possible that when this baby grows up, she will not understand how to have her physical and emotional needs met by other people in a healthy way. This is why comforting crying babies matters; they are learning that crying means help will come. As an older child, adolescent, and adult, she may not trust people to provide for her and she may wind up with an attachment disorder that causes her to attach too strongly and too soon to the wrong people, or not enough to the people who treat her well. Now imagine this infant girl is a gorilla.

Gorilla physiological and emotional development is similar to ours in that it occurs over a long period and bonding with one's mother teaches a young gorilla how to trust. A gorilla who is ignored as an infant may have problems connecting to troop members as an adult in a society that is extremely social. The same may be true for elephants, who also develop over long periods of time and form close relationships with family and herd members. This may, in fact, be the case for any creature whose emotional needs are not met when their brains are developing, or for those animals who are hurt by the ones they are supposed to trust.

Dr. Cynthia Zarling is a psychologist who has worked with troubled children for more than twenty-five years. She also rehabilitates aggressive German Shepherds who have been given up by their families. I told her about Rara the elephant, what I was learning from Perry's research, and what Som Sok had told me about judging animal emotions in Thailand. She wasn't surprised. Instead she told me it made her think of the children she treats. "For kids, the most important relationship is the first relationship," Zarling said, "the mother-infant relationship. This is the base that every single one of the child's future relationships rests on. You get your identity by the mirror that is your mother, and you can end up with a fragmented sense of self if your mother doesn't reflect you well."

Zarling believes something similar happens with the German Shepherds she rehabilitates. Pups who are abused or neglected often fail to become confident adult dogs, and without good dog behavior modeled for them early on—by other dogs or their human companions—they're much more difficult to work with and can be more aggressive, and thus are more likely to end up at shelters. After these dogs are adopted, they're returned when their difficult behaviors surface. E'Lise Christensen calls these canines "recycled dogs" and says that trying to understand why they behave the way they do feels like watching a dog chase his own tail. "Are they problem dogs because they're products of the shelter system, having been adopted only to be returned for their difficult behaviors, often more than once?" she said. "Or are they in the shelter system because they were born difficult dogs?"

Environment, Hot-wired and Otherwise

Where you live and whether your burrow, nest, house, or den stimulates, excites, or calms you matters for your mental health. It's so obvious that it hardly needs stating. Yet many people are still surprised when an animal living in poor conditions or simply in the wrong sort of environment veers into the territory of mental illness and does something spectacular. Whenever an orca at SeaWorld lashes out and kills a trainer, or an elephant tramples her handler, there's a media explosion of surprised accounts by other trainers, park staff, and people in the audience. Of course, PETA and other animal rights groups are never surprised and have press releases ready and waiting for these sorts of events.

Captive animals suffer disproportionately because in many cases their environment has almost nothing to do with the sort of place they would choose to live themselves. These creatures have hours of empty time every day and often a lack of activity to occupy their minds, hands, paws, or jaws. In response, many develop behaviors that are eerily similar to those of emotionally distraught humans. Champions of the animal display industry wave away such criticism, arguing that zoo animals often live longer than their free counterparts and that wildernesses come with their own stressors like hungry predators, no access to veterinary care, and certainly no prearranged meal times. It is also the case that many animals currently living in captivity were born there and may not be able to survive on their own. These points are all trotted out like show ponies anytime the animal display industry comes under attack. But a tally of years lived and calorically balanced meals eaten doesn't account for quality of life or the pleasure that can come from making one's own decisions. It doesn't even account for the kind of suffering that isn't lethal but nonetheless may make an animal unhappy and drive him to gnaw on his toes or swim in endless circles. Just because an animal is born into a certain world doesn't mean that she can't have an opinion about it.

A few creatures, like a few individual people, might prefer life inside of a gilded cage. Somewhere in Kenya or Zimbabwe there may be a lazy giraffe who likes snacking on cut leaves more than stretching his neck out to break them off for himself. I do have a few friends who would jump at the chance to hole up at the Ritz-Carlton and order room service for years on end. But personally, I would get bored of the mimosas, the turn-down service, and the late-night French fries under their silver dome and I'd wonder what lay beyond the carpeted hallway. Unfortunately for most display animals, there's no way to know which giraffe, wallaby, or orangutan might relish hotel life until they've checked in. If they don't enjoy it, if it drives them to distraction and madness, there's no way to check out.

Advocates for zoos and aquariums assert that "good" institutions, meaning well-funded ones, make sure that the animals' needs are met: that they have enough to eat, access to veterinary care, somewhere to sleep, and freedom from predators. As a result, they often choose to reproduce. Many of these animals even have social lives with their exhibit mates or their keepers. But the prevalence of their odd behaviors, obvious to anyone who visits a zoo or aquarium knowing what to look for, is one clue that life in captivity—whether it's a prison or a luxury hotel—isn't the same as being free. Over the past few years I developed a sort of animal mental illness field guide that I share in the form of running commentary before the pacing lions or the compulsively masturbating walruses. I've become a bit depressing to be with at the zoo, and my friends with kids have stopped asking me along. This is just as well.

When I look at a gorilla perched on a fiberglass tree, lovingly hand-painted to look like the most perfect version of itself, inside a faux habitat carefully designed to remind viewers of the environment that she has evolved to live in, I don't marvel at the gorilla in front of me, but instead at the mastery of the exhibit itself. If I were a contemporary zoo animal in America, the exhibit would not remind me of, say, equatorial Africa, because I have probably only ever lived in the

United States. And yet the desire to run fast, swing wildly, roar, fly, or tear another animal limb from limb would exist in me still, just as powerfully, despite having been born in Denver, Cleveland, or Los Angeles.

Admittedly, this is heavy projection. Perhaps the gorilla sitting on the handcrafted stump prefers the feel of fiberglass to bark since that is all she knows. But then why is she rhythmically throwing up inside her mouth, only to swallow it down again, to the exclusion of other activities? Why does she continue to pluck the hair from her shins and forearms and then eat it? Why, of all things, has the zoo called in a human psychiatrist to help? Possibly the fiberglass tree bores her. Or she discovered that the lush plants along the moat are electrified so that the exhibit looks lovely for visitors but cannot be torn off and eaten. Or her favorite young male has been sent elsewhere in order to mate with someone more genetically appropriate. Or possibly the bossiest female in the troop has hoarded the day's grapes. Then again, it could be that her favorite keeper has quit or is out sick, and another, who does not know how she likes her oatmeal, has replaced him. Perhaps one of the troop's babies has died and the body was taken away too soon.

Environment matters. It is the backdrop upon which our lives are lived; we both form and are formed by it. When you are a captive animal living within a circumscribed space, it takes on even more importance. Because of this, understanding animals' responses to their surroundings is fundamental when you attempt to make sense of their behavior.

Some abnormal behaviors are easier to spot than others. The most common are known as stereotypic behaviors or stereotypies. These activities are repetitive, always the same, and seemingly pointless. There are as many types of stereotypies as there are animals busily engaging in them. Certain species have their own preferred types. Human stereotypies include ritualized and repetitive movements that tend to get worse with stress, anxiety, or exhaustion, such as rocking, crossing and uncrossing one's legs, certain sequences of touching oneself, or marching in place. In all animals they tend to be funhouse-mirror versions of normal activities.

Horses may take small, rhythmic gulps of air or endlessly chew on inedible objects like fences or water troughs. Pigs gnaw on one another's tails; caged minks spin in circles like furry dervishes; walruses repeatedly regurgitate and reingest their fish; wombats lie on their backs and wave their paws in the air as if they're doing an odd little back paddle to nowhere. Dogs like Oliver compulsively lick a spot on their paws or flanks, even if there is no irritation or if the irritation that first caused the desire to lick has long disappeared. Whales, seals, otters, or any other creature that swims can develop a pattern swimming stereotypy that is just what it sounds like—swimming in a particular pattern to the exclusion of all other activities. Dolphin stereotypies often include masturbation using their own body or open pipes or hoses in their tanks. Bears and the big cats endlessly pace, wearing dusty trails in their enclosures that read like topographic maps of the compulsive mind. Elephants weave or sway or lift their legs up and down rhythmically, often in a ritualized sequence.

While plenty of domestic creatures develop stereotypic behaviors, they are particularly common in zoos, aquariums, circuses, and large pig, fur, and poultry farms.

In the United States and Europe more than 16 billion farm and lab animals are raised every year, and millions of them exhibit abnormal behaviors. This includes 91.5 percent of pigs, 82.6 percent of poultry, 50 percent of lab mice, 80 percent of minks living on fur farms, and 18.4 percent of horses. A large percentage of the roughly 100 million laboratory mice, rats, monkeys, birds, dogs, and cats used in American labs every year also engage in self-destructive and self-soothing behaviors, from rocking and compulsive masturbation to self-biting and skin picking.

A study published in 2008 found a strong correlation between lab, zoo, and farm animals who were separated from their mothers early and the development of stereotypic behaviors. Early weaning is commonplace on large-scale pig, poultry, dairy, and mink farms. Dairy calves, for example, are often separated from their mothers only a few hours after being born, although cattle do not normally wean their calves until they are nine to eleven months old. Piglets on many hog farms are taken from their mothers at two to six weeks but would otherwise suckle until three or four months, and minks are separated from their mothers at seven weeks, although wild minks stay with their mothers for ten or eleven months. Separating babies from their mothers too early is perhaps most dramatic in the poultry industry. Chicks who would naturally remain with their mothers for five to twelve weeks never even see them; instead the eggs are sent to hatcheries. Those early weaned pigs are more aggressive and more prone to "belly nosing" (rooting at the flanks of other piglets); minks pace and bite their own tails more frequently; foals spend more time eating the wood of their corrals; dairy calves suck on whatever is available to them; mice are more likely to repetitively bite the bars of their cages; and hens born in hatcheries peck one another's feathers out more often and more intensely.

Temple Grandin is a professor of animal science at Colorado State University, a gifted advisor on many things animal, from slaughterhouse design to behavioral training and modification, and the subject of an HBO biopic. She is also autistic and wrote in her book, *Animals Make Us Human*, with coauthor Catherine Johnson, that "really intense stereotypies, stereotypies an animal spends hours a day doing—almost never occur in the wild, and they almost always do occur in humans with disorders such as schizophrenia and autism."

Intense stereotypies can also occur in institutionalized children, like those that Spitz and Bowlby wrote about in the 1950s. Grandin and Johnson point to a study of Romanian orphans adopted in Canada showing that 84 percent of them engaged in stereotypical behaviors when they were in their cribs, repetitively rocking back and forth on their hands and knees; shifting their weight from one foot to another, a bit like circus elephants; and hitting their head against a wall or the bars of their crib like head-banging monkeys and dolphins.

These animal activities remind Grandin of autistic children who sometimes bite their own hands, bang their heads on walls, or slap themselves. She argues that 10 to 15 percent of captive rhesus monkeys housed alone do the exact same thing. She may be right, but Grandin's comparison of autistic children to abnormally behaving animals is controversial. She has categorized autism "as a way station on the road from animals to humans," implying that autistic children may be closer to animals than the rest of us, an assertion that uncomfortably echoes the Victorian idea that certain groups of humans are closer to animals than others. Even if they rock back and forth like upset monkeys do, autistic children aren't more closely related to other animals than nonautistic children.

There is, however, the possibility that other animals can be autistic. And if so, autistic humans and autistic nonhumans may have some things in common. The ethologist Marc Bekoff once observed a wild coyote pup he called Harry. Harry's littermates rolled and tumbled, snarling at one another joyfully, but Harry didn't understand their invitations to tussle and didn't seem to know how to play at all. Despite his best efforts, the pup couldn't read coyote social cues. "For a long time I simply chalked it up to individual variation," wrote Bekoff, "figuring that since behavior among members of the same species can vary, Harry wasn't all that surprising." But a few years later, someone asked him if he thought other animals could be autistic and Bekoff remembered the odd little pup. "Perhaps," he wrote, "Harry suffered from coyote autism."

In 2013, biologists at Caltech took a group of anxious lab mice with poor social skills and stereotypic behaviors and dosed them with a gut microbe, *Bacteroides fragilis.* Their anxiety seemed to lessen, they appeared to communicate better with one another, and they spent less time engaging in odd behaviors. The researchers concluded that the bacteria might help more than mice and suggested that people with developmental disorders like autism should try taking probiotics. This study built on earlier research, also at Caltech, that linked autism spectrum disorders to intestinal problems in both mice and humans. Mice who squeaked at other mice in strange ways, for example, had less *Bacteroides fragilis* in their intestines. So did humans with autism. The lack of this bacteria may not *cause* autism but adding it back in may help animals with their symptoms.

Boredom, a bullying or aggressive exhibit mate, or a disliked keeper or staff member can each send an animal down the road to compulsion. Perhaps the lights are too bright, the darkness too dark; perhaps it is too loud or too quiet, too smelly or not smelly enough.

Many captive gorillas regurgitate their food and eat it again in an endless cycle. This is so common that there is a term for it, *R and R*, for reingestion and regurgitation. Jeannine Jackle, the assistant curator of the tropical forest exhibit at Franklin Park Zoo in Boston, oversees the zoo's troop of eight gorillas and their team of keepers. She has worked with these gorillas for more than twenty years, and her office at the zoo is covered with photos of each of them at various stages of their lives, along with their colorful finger paintings, an inter-ape craft project that involves the keepers sliding paper into the cages and coating the gorillas' fingers in paint. In a yellow tackle box behind Jeannine's desk there is a "primate bite kit" that she has never had to use.

"Each of the gorillas has a particular way of R-and-R-ing," Jeannine told me. "Kiki, a female, keeps it in her mouth or puts it on the glass of the exhibit. She'll also blow it out of her nose and let it dribble down her chin before licking it back up again." Another female, Gigi, who is the eldest in the troop, has perhaps the grossest technique. She spits up all over the floor and then plays with it before eating it.

"They do it more often when they've eaten something sweet,"

Jeannine said. "I think they may like tasting it again, and it gives them something to do. We have a joke among the keepers that if humans did this we'd have '5 R' restaurants."

The behaviorist and wildlife biologist Toni Frohoff specializes in cetacean sociality and communication, has evaluated "swim with dolphin" programs, and served as a consultant on various advocacy campaigns for better treatment of captive dolphins. Toni has seen pattern swimming, masturbation stereotypies, and many instances of what she calls head-ramming, when dolphins repeatedly ram their heads against the sides of pools or tanks.

"Once I was paid to go up to Edmonton, Canada, because there was a shopping mall there with a live dolphin in it," she said. "I had to testify that keeping a lone dolphin in a mall was bad. I got there and saw this dolphin who was obviously exhibiting all sorts of stress behaviors and asked, 'Did you really need a dolphin expert to come up here and tell you that this is a terrible idea?' "

Seals and sea lions also develop weird habits in captivity. In addition to pattern swimming there is "pup sucking," when pups try to nurse from other pups instead of adult females. Captive dolphins and walruses will also throw up and reingest what they've spit up repeatedly, a lot like gorillas. "In the wild it's a normal behavior that marine mammals can use to get rid of squid beaks or rocks they've ingested," the marine mammal veterinarian Bill Van Bonn told me, "but in captivity we see it happen with their actual *food* and we consider it a displacement behavior. People in the captivity industry don't like to talk about it."

They may not want to talk about it because the institutions that depend on captive animals to entertain visitors sell family-friendly experiences or purportedly educational ones, not animal compulsion sideshows. The Association of Zoos and Aquariums (AZA) is a nonprofit accreditation and membership organization made up of animal display facilities invested in convincing the American public that zoos are a series of arklike institutions busily ensuring the planet's biodiversity. According to the AZA, the average zoo and aquarium visitor in the United States is a female between twenty-five and thirty-five years old with children. For the most part, zoos do not want to emphasize the disturbed or disturbing animals in their collections to these visitors, lest they mar the carefully calibrated experience in which everything from the sound track of hissing insects playing from hidden speakers to the hand-painted backdrops inside the exhibits has been designed to promote the zoo's vision of nonhuman nature and family fun.

The differences between display facilities are sometimes as distinct as zebra stripes and sometimes less so. Take, for example, SeaWorld and the Bronx Zoo. The latter is a nonprofit zoo run by an international conservation organization, the Wilderness Conservation Society, while SeaWorld is a for-profit theme park owned and operated until 2009 by the Anheuser-Busch Corporation, and now by the private equity firm Blackstone Group. Many keepers, staff, and veterinarians who work at institutions like the Bronx Zoo have gone to great lengths to convince me that what they do is very different from what goes on in places like SeaWorld. They argue that they educate instead of entertain, or at least *while* they entertain. This debate may not seem to have much to do with mental illness in animals, but it does. These institutions justify keeping their animals on display, and the ensuing mental trouble that the animals may face, with the claim that the creatures inspire visitors to learn about animals and return home more educated about their world and more committed to protecting it. In theory this is a great idea, and if the claim were valid, then perhaps a few compulsive animals would, on balance, be fair collateral. But it simply hasn't worked out that way.

Forty years ago, as the environmental movement began to coalesce into a force that influenced how Americans spent their money and their Saturday afternoons, the country's zoos consisted of barren concrete pit-style enclosures and faced a crisis of diminishing visitorship. They had to become places that no longer depressed people, or they would have to close. The zoos that survived now justify their existence as educational but also as repositories of endangered species and guardians of threatened wildlife. This justification is at best promissory and at worst a smokescreen that allows institutions to remain profitable while the wild counterparts of the animals in their collections go quietly extinct.

In 2007 the AZA published the results of a three-year survey on the educational impact of zoos in an attempt to shore up their argument for their role as environmental stewards and educators. The report argued that zoo visits made people more likely to care about animals and more aware of conservation needs. A follow-up report, however, published by a group of research scientists at Emory University, questioned the validity of the AZA's research methods and argued that the study's educational claims were vastly exaggerated.

Surely some zoo visitors are changed by their experiences watching animals, chatting with the docents, and reading the signage. Facilities like the Bronx Zoo, the Monterey Bay Aquarium, and the San Diego Zoo have environmental education programs, engage in research efforts concerning wild animal populations, and often make significant contributions to conservation. Applying for jobs at these insitutions is a highly competitive process and many new recruits arrive more educated than ever. Yet despite staff education and training, healthy research and outreach budgets, and new exhibits in which the animals are free to munch on native plants or walk on grass instead of concrete, many creatures still wind up with the variety of behaviors that prompt zoo-going children to ask their parents' why, for example, the dolphin won't stop putting his erect penis inside the nozzle of the tank's water filtration system.

When visitors notice and complain about such behaviors, certain display facilities are up front in their efforts to ameliorate them. The Franklin Park Zoo in Boston posts informational signage on the glass of exhibits about why the animals are given things to play with, just in case people wonder why there are blankets or plastic tubs inside the gorilla exhibit. Other zoos and aquariums attempt to deal with the behaviors by removing the most disturbed animals from view.

Ultimately, whether visitors leave motivated, to recycle or donate to conservation causes is not up to the zoo or aquarium but to the visitors themselves.

Contemporary Americans need opportunities to see and interact with other animals, but I do not think zoos are the answer, since the animals are rarely interactive with visitors in the first place. It seems petty, but I also don't like how they smell—a mixture of urine, cleaning solvent, and something else, perhaps the tang of despair, or a bored sort of waiting. The more naturalistic the cages, the more depressing they can be because they are that much more deceptive. To the mandrill on the other side of the glass, the realistic foliage that frames his favorite perch doesn't help him one bit if it has been hot-wired so that he doesn't destroy it. It was conflicts like these that Pavlov so clearly showed cause disordered behavior in dogs. Some of the new natural-looking exhibits may be even worse for their inhabitants than the old cement ones, as the new plants and other features can shrink the animals' usable space. These environments can have drastic effects on the psychological well-being of the animals who spend their entire lives in them.

At the same time, I have met many bighearted, empathetic, and intelligent zookeepers who care about the animals they work with and have made great sacrifices to do what is almost always a thankless job. Keepers work long hours, make little money, are highly replaceable, may frequently find themselves in situations of real danger, perform tasks that are physically demanding, and, perhaps most stressful of all, are not in charge.

A keeper may recognize, for example, that a few of the wild dogs she takes care of are growing compulsive, circling their enclosure in ritualized patterns and no longer stop to play with the pups or curl up to rest. But most keepers don't have the institutional clout to produce the big changes—such as building bigger exhibits or ordering more expensive and varied diets—that ensure the well-being of their charges. These decisions are made by zoo management, who may or may not have firsthand experience working with animals. Their first priority is keeping the zoo profitable and visitorship up, something that can be at odds with animal welfare. It sounds basic, but to be a good exhibit animal a creature must be *visible.* Panda and gorilla births, for example, make for great surges of visitors and lots of *Awwws* on local news programs, but how many mammalian mothers want their newborns on full display under bright lights, often for many hours at a time?

For the zookeepers and trainers, even at amusement parks like SeaWorld or Six Flags, the animals they work with every day become family. These men and women spend more time with whales, dolphins, or wildebeests than their human family members and love them just as much. I have never met a keeper or trainer who doesn't want the best for their charges. Sometimes, when they realize that they can't protect the animals they care for, there's no going back.

Jennifer Hemmett now owns a posh dog boutique, but she used to be a primate keeper who loved working with great apes. After ten years at an East Coast zoo, she reached her breaking point. He was a lowland gorilla named Tom. Because Tom's genetic material would be a good match for another zoo's female gorillas, the AZA ordered him to a zoo hundreds of miles away where he knew no one. He was abused and neglected by the other gorillas, stopped eating, and lost more than a third of his body weight. The transfer was deemed a failure and Tom was sent back home, where Jennifer and the other keepers spent months nursing him back to health. They felt that Tom was in no shape to be moved again. He was emotionally sensitive and didn't do well among other gorillas and human staff that he didn't know. But Tom was sent away again anyway. A few months later, when Jennifer and a few of his other keepers visited him at the new zoo, Tom caught sight of them through the fence of his exhibit and started to cry. This was no quiet whimpering. Tom howled and sobbed and ran toward his old keepers. He continued to follow Jennifer and the others from his side of the fence, step for step, as they circled the exhibit, bawling the whole time. The other zoo visitors complained and told the keepers to "stop hogging the gorilla." Jennifer returned home and gave her notice to the zoo two days later. Administrators at Tom's new zoo informed the keepers that they were barred from visiting the gorilla exhibit again; it was too upsetting for Tom.

Of Mice and Mania

Repetitive hair plucking is known as *trichotillomania*. The disorder affects roughly 1.5 percent of men and 3.5 percent of women in the United States, though this doesn't account for those people who are so embarrassed by the resulting bald spots that they've become quite good at covering them up and may never be diagnosed. Most men and women pluck hairs from their scalp, eyebrows, eyelashes, beards, or pubic area. It's fairly common for someone to start plucking one area, such as their eyebrows, but then over time switch to pulling from another spot. Sufferers say that the plucking is usually preceded by some sort of tension that the pull itself releases, but it can also happen when people are relaxed or distracted, reading a book or watching television. That being said, feeling anxious, angry, or sad often increases the urgency and frequency with which people pluck.

There is a good deal of confusion about how the disorder should be classified. Until 2013 the *DSM* situated hair plucking under impulse-control disorders not elsewhere classified and counseled that it should not be considered a compulsion. Unless plucking is associated with obsessive thoughts, the *DSM* stressed that it was not an obsessive-compulsive disorder, since the plucking isn't usually performed along a rigid framework in the way that obsessive-compulsive hand-washing or lock-checking generally is. The fifth edition of the *DSM* shifted the disorder around and now considers it a form of OCD along with skin-picking.

Whatever its etiology, trichotillomania is in the *DSM* because most people don't do it. We need our hair for all sorts of reasons, some physiological, most not. Bald patches and missing eyebrows can be unattractive and time spent plucking can interfere with daily life. The habit might be a symptom of anxiety or depression, but mostly it makes the sufferer look odd, and this is when it tends to be diagnosed. Some people with trichotillomania, particularly children, may pull hair from other people, or even pets. And it's common for people to play with or sometimes eat their plucked hairs.

In this, like so many of our neuroses, we are not alone. Hair pulling has been reported in six primate species (in addition to humans), as well as mice, rats, guinea pigs, rabbits, sheep, musk oxen, dogs, and cats. Rodents who pluck are called "barbers" because they remove the fur or whiskers of other mice. Like human pluckers, they tend to be female. And these mice, as far as I've been able to ascertain, are found only among captive mouse populations. The online message boards for mouse-rearing humans who raise them as pets or for "fancy" mouse shows are full of hair-removal stories. The owners share photos of mice and rats with little bald spots on their heads, reverse mohawks, or hairless facial patches shaped like tiny *Phantom of the Opera* masks. Perplexed, the rodents' owners look for answers: "Tache seems unable to go more than two weeks in a cage with other mice without beginning to barber again. . . . Today I returned her to the big tank with Pu Manchu and Mrs. Beach . . . but I expect by the end of two weeks she'll be barbering again. How to solve this problem?"

Some fanciers and breeders claim the behavior is about displaying dominance. Others say it happens because of overcrowding or lack of stimulation; a lab mouse, even if he is born and bred to be one, still has sensory, social, and environmental needs that a cage, even a pleasant one filled with exercise wheels and colorful plastic tunnels, doesn't provide. Barbering, like many forms of OCD, including the human varieties, is a normal grooming behavior gone awry. Usually mice groom by scratching themselves with their hind feet, washing their face or fur with their front paws using their own saliva, or smoothing or cleaning their hair with their teeth. Barber mice take these behaviors to an extreme by removing hair on others. These mice don't usually injure the mice whose fur or whiskers they nibble or pluck. In fact it seems that their clients may *enjoy* it. Mice will sometimes follow the barber mouse around until she plucks them, even when the results are the complete loss of whiskers or an unironic mouse mullet.

Because mice are such lab stalwarts, a few researchers have suggested that the barbering rodents might help us better understand overzealous hair-removal in humans. Studies using mice as stand-ins for human pluckers have tried various techniques to make the mice start barbering in the first place (that is, if they haven't already started doing it themselves), then investigated the effects of antidepressants on the behavior.

Barbering may also have a genetic component. An experiment conducted in 2002 demonstrated that mice bred without a group of genes that included the Hoxb8 gene, fundamental in the development of immune cells called microglia found in the brain, became severe self-barbers. The mutant mice not only trimmed hair and plucked their whiskers but also used their paws to scratch bald spots and sores on their rumps. In a later study, published in the journal *Cell*, scientists transplanted bone marrow containing healthy microglia cells from a group of control mice into the population of barbers. A month after the transfer, when the microglia had made it to the mutant mouse brains, many of the barber mice stopped overgrooming. Three months later, their hair had grown back. While no one is suggesting that human trichotillomania sufferers sign up for bone marrow transplants, researchers are now studying the links between the brain's immune system, trichotillomania, and other mental disorders such as OCD, autism, and depression.

Furry animals aren't the only ones to pluck themselves. Avian veterinarians will diagnose feather-picking disorder if the plucking is unrelated to other medical conditions like allergies. Parrot owners, avian vets, and breeders claim that birds pluck when bored, frustrated, and stressed. It can also be related to sexual behavior or premature weaning, a way of seeking attention, a reaction to overcrowding, a sign of separation anxiety, or a reaction to a change in the bird's routine—virtually anything potentially upsetting to parrots.

Phoebe Greene Linden has lived with parrots for more than twenty-five years and is an expert on captive parrot behavior. She told me that the solutions for stopping feather pickers are as individual as strategies tailored for human pluckers. Everyone plucks for different reasons. Phoebe feels that, overall, enriching parrots' environments and helping them learn new behaviors is best. "Making sure that they have opportunities to fly, forage, and socialize is key," she says. And in chronic cases, the class of antidepressants known as SSRIs, such as Prozac can be useful, as well as Xanax and Valium.

At the San Francisco Zoo the lowest-ranking female mandrill started plucking after a male in the troop died. In the wake of his death, leadership in the troop shifted to a mandrill that one docent referred to as "a dictator." The lowest-ranking mandrill, stressed by the new leader, began to pluck so intensely on both sides of her head that she gave herself a mohawk. The zoo put her on Paxil and she stopped plucking as much.

For captive gorillas, the most common sites for hair plucking are forearms and shins, but I've seen plucked patches wherever the apes can reach. Since gorillas have more and thicker body hair than we do, they can pluck it from almost anywhere.

Little Joe, a member of the gorilla troop in Boston's Franklin Park Zoo, is a sixteen-year-old male with ropey muscles and long arms. In

2003 he used those arms to climb out of the tropical forest exhibit and escape the zoo. ("He's like the Michael Jordan of gorillas. His arms are so long and he's so athletic that he challenged the limits of the exhibit in a way that no gorilla had before," said Jeannine Jackle.) Joe roamed around the neighborhood for two hours. When he took a break at a public bus stop, a woman who spotted him there said that at first she thought he was "a guy with a big black jacket and a snorkel on."

When Little Joe isn't planning his next escape, he's often plucking his hair. He pulls out strands on his arms and sometimes eats them, just like a person with trichotillomania. He also picks at his scabs, over and over, leaving tiny wounds on his arms. Jeannine thinks that it gets worse when he's anxious. The keepers haven't been able to stop him from plucking, but they try to keep him busy doing other things, like eating flavored popcorn or taking part in training sessions to present different parts of his body for medical checks.

At the Bronx Zoo in New York a female gorilla named Kiojasha plucked herself so intensely that visitors asked about her. One of the docents said, "They couldn't tell if she was a gorilla or not, she was so hairless. She honestly looked like a wizened old human woman. It was disturbing. After a week of this, the zoo took her off exhibit, and now I think she's in Calgary."

Sufi Bettina, a Bronx gorilla named for the mother of Glenn Close, a zoo supporter, likes to sit on one of the artfully rendered mud banks and stare into the middle distance plucking the hair from her forearms until small bloody scabs form. Then she picks at the scabs.

Primatologists like Frans de Waal and Jane Goodall have observed that nonhuman animals can and do have culture—that is, knowledge passed down from one generation to another or from one group to another. When I heard of the Japanese macaques *(Macaca fuscata)* who teach one another to wash and season their yams in the ocean because the salted yams taste better, I wondered if the same might be true with plucking. The babies of gorillas who pull their hair often turn out to be pluckers themselves, and it's possible they learned this from the apes around them.

Diagnostic Difficulties

Trichotillomania is easy to diagnose because bald spots are often visible even to the untrained eye. But what about other signs of mental unrest that are harder to spot and a bit more subjective to name? Even understanding the causes of certain stereotypic behaviors can be confusing. Temple Grandin has pointed out that just because an animal isn't compulsively pacing, chewing, or repeatedly gnawing someone else's tail doesn't mean that the animal is happy. In the same way that some dogs' separation anxiety can make them more withdrawn than destructive, the catatonic bear in the corner of the exhibit who doesn't trace figure eights in the snow may simply lack the energy to express his frustration. These animals are less likely to be diagnosed even if they're suffering as much as or more than their seemingly more compulsive counterparts.

Even when behaviors are clearly stress-related, they can be difficult to interpret. Mel Richardson was once asked to examine a tree kangaroo at the San Antonio Zoo that the keepers said was acting bizarrely. With the ears of a teddy bear, the rounded chub of a koala, and the tail of a fuzzy monkey, tree kangaroos are very cute. But this female was acting vicious. She was attacking her babies, and the keepers had no idea why. Mel went to check on her. Sure enough, as soon as he approached, the kangaroo ran to her babies and started hitting and clawing at them with her paws. He stepped back, and she stopped. He walked forward, and she ran at the babies again.

"I realized," said Mel, "that she wasn't viciously attacking her babies at all. She was trying to pick them up off the floor, but her little paws weren't meant for that. In her native Australia and Papua New Guinea her babies never would have been on the ground. Her whole family would have been up in the trees." The mother kangaroo wanted to move the babies away from the humans. What looked like abnormal attacks on her young were actually her way of trying to *protect* them. Her behavior wasn't mental illness at all but a response to the stress of being a mother in an unnatural environment.

After the keepers redesigned the kangaroos' cage so that more of it was elevated and farther from the door, she relaxed and stopped hitting her babies.

Mel explained, "As flippant as it might sound, the truth is that in order to know what's abnormal, you must first know what's normal. In this case in order to determine pathology, I had to understand the animal's psychology. It's pretty easy for people to get this wrong."

Years later, Mel was running his own veterinary practice in Chico, California. One of the vet techs came to work complaining about her German Shepherd's anxiety. She said that it started when she had a fight with her boyfriend; he threw something at the wall in anger and knocked a framed picture onto the floor. Ever since, when the German Shepherd entered that room she skulked around the edges, looking fearfully at the walls.

"What do you do when she does that?" Richardson asked the tech.

"I go over and stroke her and talk to her until she calms down," she said.

"Well, you are making this problem worse. You're rewarding her anxious behavior. Ignore her."

The tech started ignoring her dog in their living room, and after two weeks the German Shepherd stopped acting fearful and hugging the walls. Mel believes this sort of thing is common—that expressions of fear and anxiety, or even, perhaps, compulsive behaviors, can be reinforced by well-meaning people who simply don't understand their own role in their animals' seeming madness.

* * *

The Buddhist "elephant monk" Pra Ahjan Harn Panyataro lives in Baan Ta Klang, a village in Thailand's northeastern province of Surin, where women weave silk made by worms raised on mulberry branches that grow in their yards. The community owns more than two hundred elephants, parked like soft gray cars alongside the village houses.

Panyataro, one of the few monks who conducts elephant funerals, built an elephant cemetery where human families leave offerings of fruit, incense, and bottles of water for their deceased elephants at the base of handmade headstones. He also compiles statistics on Asian elephant populations throughout the region and oversees a forest temple in which elephants and mahouts are welcome to traverse the paths among the trees in quiet contemplation. If an elephant kills a man—something that happens a few times per year or so in his community—he leads a discussion between the survivor's family, the elephant's owner, the elephant's mahout, and anyone else affected by the killing. He has been doing this for twenty years.

The morning I met Panyataro he was about to leave for India ("For Buddhism, not elephants," he said). A purring motorbike waited to take him to the airport. He sat down on the front steps of the temple and gestured for me to sit a few steps below him. I wanted to find out if he saw emotional troubles in the many elephants he knows and works with, and if so, how he recognizes their distress.

Nervous, I tripped over my words. "Will you ask him how he knows what the elephants are feeling?" I said to Ann, my friend and translator.

Panyataro looked directly at me and said, "In order to understand other animals, first you have to understand yourself." This seemed both profound and also so commonsensical I wondered if I had needed to come all the way to Thailand to hear it. Then he continued: "Elephants can have mental problems. They are like us. They feel happiness, sadness, hunger, fullness."

I asked how someone should go about making a sad elephant happy.

"First you have to find out what is wrong," he said. "Sometimes this takes a long time. And it is always different."

With that, he gathered up his robes, tucked our donations into the folds, slipped onto the back of the motorbike, and sped off through the holy forest.

Chapter Three

Family Therapy

> I have to listen to Mae Kam Geow's heart. If she doesn't want to go anywhere, I don't make her. She is old. That way she also listens to me.
>
> Dahm, elephant mahout, northern Thailand

> It's up to you to listen to their demands, even when they are unspoken . . . and most of the time . . . they are.
>
> Daniel Quagliozzi, cat behaviorist

Jokia is the only elephant I've ever met who reminds me of the cartoon version. Her legs are stocky, her head and trunk wide and chubby, as if she were a larger elephant stuffed into a smaller, tighter skin. She is also completely blind. Jokia once worked as a logging elephant in northern Thailand. As the story goes, she was pregnant and about to give birth, but her owners refused to give her time off from dragging logs. She birthed her calf while walking up a large mountain, and her baby, still in the amniotic sack, rolled to the bottom of the trail and died. Soon thereafter Jokia began refusing to work. Her mahout shot out one of her eyes with a slingshot, hoping to blind her into obedience. She worked for a few weeks and then refused again. The mahout stabbed her other eye with a knife, blinding her completely, hoping that in total darkness she would be more submissive, more dependent, and more likely to work. Jokia, stubborn and injured, still refused to do what was asked of her. A few years later a Karen woman named Lek Chailert, the founder of Elephant Nature Park, the ecotourism destination in the Mae Tang Valley outside Chiang Mai where I met Rara, heard about the blind elephant. She purchased Jokia for $2,000, and brought her back to the park, a lush swath of land criss-crossed by a winding river.

For an entire year after she arrived, the elephant kept to herself. Then, slowly, she began to make friends with another elephant, a tall, thickly wrinkled, and inquisitive female named Mae Perm. Like Jokia, Mae Perm once worked as a logging elephant. Soon she and Jokia were inseparable, eating and bathing in the river together and grazing shoulder to shoulder through the long afternoons. Thirteen years later Jokia and Mae Perm still spend every waking moment side by side, a tight herd of two. The elephants are rarely out of trunk reach of each other, even during routine vet procedures. Mae Perm walks in front, leading the way, and Jokia follows, stepping slowly and sometimes hesitatingly, using her trunk to feel along the ground. But once a week she makes a two-hour trek to a forest camp, along a road with passing cars and trucks, and then takes a rutted, steep track alongside hikers and dogs through the forest, following Mae Perm every step of the route. When, every once in a while, Mae Perm strays a bit out of reach to graze, pulling long grasses out by the root with her trunk and whacking the dirt off against her knee, Jokia squeaks frantically until Mae Perm rushes back to her side and comforts her by stroking her with her trunk and rumbling deeply. Lek and the rest of the park staff are convinced that if they hadn't found each other, the elephants would not have adapted as well to their new lives or advanced into their matronly old age as smoothly and joyfully as they have.

"Jokia has every right to want to kill people, but she doesn't," Jodi Thomas, a longtime resident and guide at Elephant Nature Park, told me as we stood in the shadow of the two elephants languidly chewing. "She's an easygoing elephant. I think this is because she has so much of what she needs in Mae Perm. Their relationship gives Jokia confidence."

Many of the foreheads of the elephants at Elephant Nature Park are deformed, evidence of past beatings with hooks so severe that they have left permanent grooves and trenches in their skulls. Some elephants arrive with fresh puncture wounds. Many of them have scarred and thickened ankles where spears have been repeatedly struck to urge them to walk faster or stay in line. These wounds are the physical tracings of the logging industry or the performing elephant trade. But there are emotional scars as well. Many of the elephants keep to themselves, as Jokia did, at least for a while, distrustful of new people and elephants. A lot of them have stereotypic behaviors like swaying, head bobbing, or rhythmically lifting their feet in a strange dance routine to which only they know the steps. Some have killed people. Others have killed elephants. In a few cases, like Rara's, they arrive scared of other elephants. The most successful elephant rehabilitations are due, almost always, to the other elephants already living at the park, who often welcome newcomers into their makeshift herds.

Here the elephants don't perform, although they do hew to a daily schedule of interactions with paying visitors, eating fruit at designated platforms and letting themselves be bathed in the shallow river. Lek, who has run the park for seventeen years and built its population of elephants from two to more than thirty-five at last count, told me that the only way she knows to help an elephant recover from a traumatic past is to offer them love, trust, and safety.

"It's pretty simple," she said, as a pack of rescued street dogs, who also live at the park, surged around her, panting and whining for attention. "The elephants also need the companionship of other elephants. Mae Perm and Jokia are a perfect example."

Rabbits for Rabbits, Rats for Rats

Elephants are not the only animals to benefit from healing relationships with their own kind. The rabbit rescuer and rehabilitator Marinell Harriman has worked with hundreds of the creatures over the past twenty-five years. She is one of the founders of the House Rabbit Society and author of the *House Rabbit Handbook: How to Live with an Urban Rabbit.* "In caring for quite a few 'sanctuary' rabbits with long and short-term illnesses," she writes, "we have seen some miracles of motivation. We are convinced that friendship therapy contributes to the recovery or at least stabilization of sick rabbits."

Harriman tells the story of an eight-year-old rabbit named Jefty. After his mate died of cancer, Jefty began to chew his fur in earnest. He soon developed large bald patches, and a vet exam revealed that all the fur that had been once been on the outside of the rabbit was now inside him, in the form of a gigantic hairball lodged in his stomach. The vet thought it was unlikely that the mass would pass on its own and recommended surgery. Harriman started Jefty on a variety of hairball remedies to make him strong for surgery, but she decided to try something else too. She introduced Jefty to a ten-year-old rabbit who had also recently lost her partner. Almost immediately the pair began to treat each other with affection and rabbity care, to the point that Harriman postponed the surgery for a bit, hoping that the new relationship would cheer Jefty.

After a few days with his new rabbit companion Jefty was doing so much better that Harriman canceled the surgery in order to wait and see what happened. An X-ray showed that the fur mass was still in his stomach, but it was shrinking. "I won't try to claim that getting happy cured a furball," she wrote, "but I will claim that it gave Jefty a reason to eat the hay and greens in front of him. He had someone to dine with and to share his pineapple cocktails with."

Over the next few weeks the skinny, bald rabbit regained all of his lost weight and stopped chewing his fur. The massive hairball continued to shrink in size.

Rats too tend to be physically and emotionally healthiest in the company of other rats. American and British rat-fancier discussion forums and Facebook pages—vociferous communities like the National Fancy Rat Society, the Rat Fan Club, and the American Fancy Rat and Mouse Association—are full of dire warnings against keeping rats on their own and success stories of rodents who cheered significantly in the company of new rat and mouse friends. One member of the Rat Fan Club wrote, "Why on earth would anyone settle for just one rat when they could have two?! . . . A single rat lives for his daily freedom, but a pair of rats can entertain themselves for the odd day. . . . Please try to explain to pet stores and any prospective rat owners you know that rats need each other."

My Rat and Me, an earnest, illustrated guide to raising pet rats, notes, "Rats definitely notice when a comrade disappears; they search for the missing rat, and they may grow lethargic and stop eating." The authors suggest giving the survivor a new rat companion or, if that is not possible, extra attention. I often think of this when I'm waiting, late at night, on a New York City subway platform. The Norway rats dash between the tracks, avoiding electrical wires, angling their whiskers into empty potato chip bags and sniffing at crumpled, grease-stained napkins. They are rarely alone.

Phoebe Greene Linden believes that companionship is extremely important for keeping parrots happy as well. One of her birds, a hawk-headed parrot named Hawkeye, has lived with her for more than thirty years. A dedicated feather plucker with a vivacious personality, Hawkeye and her sister Stinker were born to a wild-caught male and female pair who were themselves serious feather pluckers. The female actually plucked herself so severely that she died; the male died within a year. Phoebe believes he died of sadness, never having gotten over the loss of his mate. Years later Stinker died, too, leaving her mate, Henri, on his own.

Henri soon began to falter, becoming withdrawn and silent. Phoebe was extremely worried about him, so she put a little Xanax on walnuts and gave them to the parrot. But the dosed nuts didn't work. He stayed quiet, refusing to eat and puffing up all of his feathers, and couldn't be startled from his funk. "You know," Phoebe said, "we feel a change in the whole orchestra when one voice leaves. It actually sounds different. Henri was quiet for two years after Stinker died. He didn't make a single sound."

Then, for reasons known only to them, Hawkeye started to talk to Henri from across the large room in which their cages were kept. This was something neither bird had done before. Phoebe heard their squawked conversations and started bringing the two birds closer together three or four times a week. Slowly Henri came back to life and is now a chatty bird again.

One afternoon while Phoebe and I spoke in her kitchen, a cozy room with baskets of fruit on the counters, overlooking her sunny Santa Barbara backyard, a parrot sat nearby tearing into a roll of credit card machine paper. "They love how it feels to dig into it with their beaks," she said. "I always try to keep some around." In the living room a flock of two watched a DVD of wild parrots in Brazil. According to Phoebe, they tend to like the segments on their own species best.

She used to raise parrots to sell to other people but decided years ago that they shouldn't be kept in captivity. Parrots live a long time, though, and someone needs to take care of the ones that outlive their owners, or their owners' desire for parrots. So Phoebe has an open-door policy for any of the birds she once sold, and many that she didn't. Over the years she has accepted a number of birds who have been returned for emotional problems and the resulting behavioral issues. These parrots, like Henri, can be difficult to help.

Sometimes people ask Phoebe why she doesn't just release these birds or send them back to their tropical homelands, particularly if they seem sad or upset. She thinks that sending a hand-raised parrot into the wild to cure their troubles is a recipe for disaster. "It would be like taking an anxious three-year-old orphan with freckles who was living in Chicago, and saying, 'Hmm, I hear there are people with freckles in South Carolina, let's send him there.' These birds have different skills than wild parrots and a completely different culture." Instead Phoebe does what she can, keeping them busy with activities they enjoy (like tearing into credit card paper), offering her friendship, and facilitating their relationships with one another. She also tries to build up their confidence; she's convinced that parrots raised in the wild by their own parents are often more confident, and therefore more resilient, than parrots raised in captivity.

"Wild parrots," Phoebe said, "are very good at different things, like yelling loudly, landing on whippy branches, foraging. The whole time I raised baby parrots I was trying to mimic their wild environment. Morphing those skills into ones that fit a captive environment is the goal." If they master these skills they become more sure of their own abilities. This makes them both physically and emotionally healthier and resilient to the challenges they will face over the course of their lives.

The Gorilla and Her Psychiatrist

Of all the surprisingly healing relationships I witnessed, Gigi's, the elderly female gorilla at Franklin Park Zoo, was one of the most interesting. A thirty-six-year-old female gorilla, she likes to watch black-and-white films, especially those with choreographed dance numbers. Gigi also likes to be tickled by men with gray hair and beards. And she loves the Dixie Cups her morning oatmeal comes in. She grunts with joy, a sound like heavy, breathy purring, when she sees either gray-haired men or Dixie Cups. Sometimes she eats the cups. She also has a psychiatrist.

In 1998 a twelve-year-old male gorilla named Kitombe arrived at the zoo. He had been living with his mother and the rest of their troop at the Cleveland Zoo, where he was getting into increasingly violent fights with his father. Because the keepers feared these fights might get worse as Kit became an adult, he was transferred to the Boston troop. The first week there, introductions between Kit and the other gorillas went smoothly. But soon Kit became violent. He also quickly impregnated one of the female gorillas, Kiki. Kit was deeply agitated about the pregnant Kiki and wouldn't let any of the other gorillas in the exhibit near her. His ire was focused on Gigi in particular.

Gigi was the oldest gorilla in the troop and also, perhaps, the oddest. She was born at the Cincinnati Zoo when Ann Southcombe was a brand-new keeper there. Gigi, as was common practice then, was taken from her mother at birth to be raised by the human keepers. Ann was in charge of the "zoo baby nursery," where she cared for Gigi and the zoo's other young gorillas during the day. At night, all of the infant gorillas were shut into lidded boxes, where they stayed, alone and in total darkness, until Ann returned in the morning and let them out. At nineteen years old, Ann had no previous experience with gorillas, let alone a handful of infant ones. She was doing her best.

To make matters worse, the zoo leadership wasn't always supportive of her efforts to give the baby gorillas blankets, toys, and other things for them to use to comfort themselves. Ann was frustrated, and since she followed Dian Fossey's work in Congo closely, she decided to write to her for advice. To Ann's surprise, Fossey wrote back and told her about Penny Patterson's work at Stanford teaching Koko to sign. Ann decided that she'd teach Gigi and the other gorillas a bit of American Sign Language. Gigi picked it up quickly. A few years later, Gigi was transferred from the only home she knew to Stone Zoo in Massachusetts, where she lived in a barren cement cage with a male she didn't particularly like. There Gigi gave birth to two babies. The first she left on the cement floor after giving birth, seemingly ignoring him. The keepers took the baby from her within twenty-four hours. Her response to her second baby, a male named Kubie, was completely different. She picked him up immediately and caressed him.

Paul Luther is one of Gigi's keepers and he's known her since she arrived at Stone Zoo more than thirty years ago. He believes that she learned how to be a mother to Kubie by watching a pair of orangutans, Betty and Stanley, who lived in the exhibit across from her own. "They were really good parents," Paul told me as we walked past the pygmy hippo exhibit at the Franklin Park Zoo. "And the orangs had a baby in those years between Gigi's first and second pregnancies. All she really had to do all day was sit there and watch Betty and Stanley raise their own little orang. Before that, I don't think Gigi had ever seen a mother ape take care of a baby." She certainly hadn't experienced it herself.

As Kit chased Gigi around the exhibit, she screamed and shook. He bit her, tried to drown her in the exhibit's moat, and tore open her scalp from ear to ear. Her injuries required stitches more than once, and Gigi, already an anxiety-prone gorilla given to repeatedly regurgitating and reingesting her food, eating her own feces, and sometimes slamming it on the glass of the exhibit in front of visitors, became a nervous wreck. She barely ate, and anytime she saw Kit she seemed to shut down, rocking and trembling. She also screamed and shook and refused to go back into her off-exhibit holding area at the end of the day with the other gorillas, choosing instead to sleep by herself in the sawdust in the public viewing area.

The keepers were worried and set up cots alongside the exhibit so they could keep her company overnight. After two months of this, Dr. Hayley Murphy, the head veterinarian at the time, reached the limit of her expertise. "It occurred to me that what I was seeing in Gigi was a lot like anxiety and mood disorders in human beings," Murphy said. "I decided to find a human psychiatrist to see if one could help."

This being Boston, she called Harvard Medical School and eventually found her way to Michael Mufson. An assistant professor at Harvard and a staff psychiatrist at Brigham and Women's Hospital, Mufson also has a private practice a few miles from the zoo. "On my first visit," Mufson told me as we watched through the glass as the gorillas shared stalks of celery with one another, "I could see that they were suffering just like people. You don't have to talk to someone to see they're suffering. You see it in their eyes, their physiognomy, their posture."

It wasn't just Gigi's fear and anxiety, or Kit's aggression, that Mufson noticed. He quickly determined that he was seeing mood disorders, probably brought on by anxiety, in the midlevel troop members, like the young male named Okie.

A sweet and slightly dopey five-year-old, Okie hadn't been physically injured by Kit but he was withdrawn, no longer playing with the other gorillas or the keepers like he had before the mayhem started. Mufson prescribed Prozac, and soon Okie seemed calmer, more playful, and "more like himself," according to the zoo staff.

Kit was proving to be a much more difficult patient. Mufson prescribed Prozac and increasing dosages of the antipsychotic Haldol. The drugs gave him diarrhea and slowed him down a bit, but they didn't make Kit less aggressive. The keepers weaned him off the Haldol and Prozac and started him on Zoloft, which didn't work either. They tried one last antipsychotic, risperidone, but after a few months with no change in the frequency of his attacks on Gigi, Kit was separated from the troop and put in a cement and steel holding area by himself. At the end of the day, when the other gorillas returned from the main exhibit, they could see each other through a steel mesh wall. He would often keep a bit of his dinner aside so he could eat with the rest of his troop. Sadly, this isolation period would last more than ten years.

"I tried, but I felt from the beginning that nothing was really going to help Kit," said Mufson. "His aggression came from the fact that he knew Kiki was pregnant and he wanted to protect her. This is a primary biological force. He was upset about anyone else getting close to her. You could sedate him, but the aggression was natural. You can't get rid of that."

Mufson was more hopeful about his ability to help the other troop members. For Gigi, he prescribed a beta-blocker, the same drug that concert pianists and other performers take for nerves. She was on it for three months without much of an effect. Mufson then decided to try a combination of Xanax and Paxil. Gigi soon seemed slightly less anxious, but Kit still intimidated and bullied her. Without a change in Gigi's environment and an escape from her tormentor, he worried that the drugs would only ever be a Band-Aid.

"In general, the Xanax helped relax Gigi, and the Prozac helped Okie get over his depression," Mufson told me, "but the drugs don't stop aggression."

What worked was removing the violent gorilla from the rest of the troop, even if that didn't help him. In the wake of Kit's exile, Gigi was weaned off the drugs.

The zoo staff, to their credit, didn't like the idea of keeping a young male gorilla by himself.

"The gorillas are like your family," said Jeannine Jackle. "You are with them all day almost every day. You wish you could build them a $20 million exhibit. You want to give them the best, but you can't always give them the best. You can only give them what you have."

Seeing Kit all alone haunted Jeannine. She wanted to find a solution. In 2009, twelve years after he was first separated from the troop, Jeannine hoped Kit would now be mature enough to handle himself among the other gorillas without resorting to violence. She had a hunch that Gigi would be more resilient now too. Jeannine pled her case to her supervisors at the zoo, and they agreed to let her try reuniting the gorillas during off hours.

On the day of the reintroduction the staff of the tropical forest and a group of longtime volunteers, curators, and the zoo director showed up early at the gorilla exhibit. The zoo wasn't open yet, and the small crowd of observers thrummed with anticipation.

As the metal doors that separated Kit from the exhibit area slid open, he loped quickly into the enclosure on his knuckles. The rest of the troop was waiting there, and Gigi, catching sight of him, shrieked and ran. Kit made a move to chase after her, but suddenly, the three other females, Kiki, Kira, and Kimani, hurried to Gigi's aid, forming a gorilla barrier between her and Kit. He backed off.

Shaken and scared, Gigi went to her favorite spot along the glass viewing wall, where she curled into a ball and lay down. In a complete reversal from his behavior more than a decade earlier, Kit continued to leave her alone, and Gigi spent the rest of the day looking imploringly at the keeper staff and signing "sex" in American Sign Language to human men as they passed by the exhibit, a vestigial word from her lessons with keeper Ann Southcombe more than thirty years earlier.

"That sign for Gigi means both 'food' and 'sex,'" Jeannine said. "But I think that for Gigi that day it also meant 'help.' She was using sign to try and get our attention. It felt like she was saying 'Get me out of here!'"

Despite Gigi's gorilla SOS, Jeannine thought that the staff needed to let her get through the introduction on her own, provided she wasn't being hurt. If the keepers removed Kit, Gigi would never learn that she could be safe in his presence, gaining the confidence and trust she would need to share the exhibit with him.

When I visited the next day, I found Gigi in her favorite spot along the glass of the exhibit wall, sleeping or pretending to, with a blanket over her head. She wasn't moving around as much as usual, and she gave Kit an extremely wide berth, but she wasn't a terrified mess. Jeannine's plan was working.

Today, more than three years later, Gigi shares the exhibit with

Kit every other day. She is still an easily agitated gorilla, but she and Kit now coexist largely without incident. Jeannine and the other zoo-keepers believe that Gigi's newfound confidence came from learning that Kit no longer posed a danger to her (because of her strong bonds with her female troop members). During the years that Kit had been kept on his own, Gigi helped raise two young females. She groomed them, showered them with affection, played games with them, shared her food, and helped teach them how to behave, day after day, for years. These females had now grown up and defended her. Time and these powerful relationships with the other gorillas ensured Gigi's emotional health. Jeannine noticed these bonds and placed her bets on them. Now all the gorillas are better off for it.

"I'm really proud of that introduction," Jeannine told me, "because it's based on knowing the gorillas individually and on twenty years of experience."

For an entire week after he was reintroduced to the troop, Kit walked around the exhibit smiling, in his gorilla way. "He just looked happy," Jeannine said. "It wasn't a tight-lipped or teeth-baring expression, it's totally different. Also his eyes were sparkling." Now Kit plays with the young gorillas by throwing a sheet over his head and chasing them around. He also shares his celery. Gigi watches from a distance, alert but calm.

Getting Your Goat

Sometimes the best therapist for a distraught animal isn't a member of the same species or even a well-meaning human but another sort of animal entirely.

The practice of giving companions to racehorses in the hope that the animals soothe, comfort, and help the horses run faster is at least a century old. The rationale behind the practice is that horses are prey animals and many are easily startled. Racehorses in particular tend to be high-strung and nervous and prefer not to be alone. Odd collections of animals—goats, rabbits, donkeys, roosters, pigs, cats, and even an occasional monkey—have been used at racetracks and inside stables to calm horses down, a sort of living, breathing security blanket. The expression "getting your goat" may have come from precisely this sort of relationship. Stealing a racehorse's goat companion the night before a big race could make a horse too upset to run well the next day.

Not all horses like goats, however. Before Seabiscuit was a champion racehorse—when he was a promising but pacing, underweight, tired, and fearful colt who broke into a sweat when he saw a saddle and tried to bite the grooms who came too near—his trainer Tom Smith put a nanny goat named Whiskers into his stall. Smith hoped the goat would calm and comfort Seabiscuit. Instead Seabiscuit attacked Whiskers, picking the goat up with his teeth and shaking her violently from side to side, ultimately dropping her outside of his stall. Smith, undaunted, offered Seabiscuit the companionship of a cowpony named Pumpkin. A calm, steady animal who had survived a goring by an angry bull, Pumpkin was, as the author Laura Hillenbrand describes him, "amiable to every horse he met and . . . a surrogate parent to the flighty ones." Seabiscuit didn't attack Pumpkin; they became fast friends and were stabled together for the rest of their lives. Smith was so encouraged by Pumpkin's calming influence on the racehorse that he also adopted a stray little dog with huge ears named Pocatell and a spider monkey named Jo-Jo, all of whom traveled with Seabiscuit to his races. At night the horse slept with Jo-Jo curled on his neck, Pocatell on his belly, and Pumpkin a few feet away. He began to relax and would soon begin breaking records on the racetrack.

At Belmont in 1907 a racehorse named Miss Edna Jackson landed in the newspapers for her interesting transspecies friendships. She supposedly shared her stall with two rabbits and refused to eat unless they were present, until one day she crushed them by accident. Miss Edna then befriended a goat named William. A few years later, the champion of the Kentucky Derby was a horse named Exterminator. He had three successive Shetland pony companions, all of whom were named Peanuts. Exterminator and his ponies lived together for twenty-one years. When the last Peanuts died, Exterminator was said to have mourned.

Giving racehorses animal companions is still relatively common. John Veitch, an American Hall of Fame trainer who has worked with a number of champion horses, believes that since horses live very solitary lives in their stalls, the other animals offer them a great deal of comfort. Another Hall of Famer, Jack Van Berg, became the first trainer to win five thousand races, in 1987. He gives goats to "stall walkers," horses who relentlessly pace whenever they're stabled. "You take a real nervous horse, and sometimes it walks the stall like a damn airplane buzzing around in there," he told a journalist for *Sports Illustrated*. "You get them with a goat, it settles them right down."

Sometimes the horses get so attached to these goats that they have to travel everywhere together. If separated, the horses can pace anxiously in their stalls, refusing to rest. The goats too can become upset when their horse leaves them; one billy goat bellowed every time his horse went off to race. The goats even ride in trailers with the horses when they're being driven from one racetrack to another. And when a horse is sold, their goat goes too. "It's the only humane thing to do," one Chicago trainer observed. "A horse that loses its goat is just bereft."

Recently I visited the Chester Racecourse in England. It was Roman Day, and I watched the race amid a crowd of boozy fans, the men in improvised togas and running shoes and the women in short skirts, wobbly heels, and feathered hats. A man in charge of one of the Jumbotron screens offered me his pass to the course so I could be closer to the track. He attends dozens of races a year and regularly hangs out with the jockeys, so I asked him to corner a few of them and ask if any of the horses I'd watched that day had animal companions. He reported back, excitedly, that many of the stables used sheep and goats to calm the horses, especially when traveling and for horses who haven't traveled much before. Sometimes they use chickens and pigs.

Potbellied pigs may indeed be useful for calming horses, but they can also grow big and stubborn and become too difficult to move around. One of trainer Betty Gabriel's horse-calming pigs once got angry with her, trotted to a neighboring barn, and refused to come back. According to Gabriel, the pig's desertion upset both the horse and a goat that the pig left behind. "The goats have better personalities," she said.

Enrichment

Sometimes getting a pig for your anxious horse, or a dog for your goat, another rat for your depressed rat is just not possible. The enrichment industry intends to help, catering to the minds and downtime of captive and domestic animals by providing things for them to do or play with. The Association of Zoos and Aquariums defines enrichment as "a process for improving or enhancing animal environments and care within the context of their inhabitants' behavioral biology and natural history." Nowhere in the definition does the AZA mention the word *captivity* or *cage*, but the only animals whose environments need to be enriched are captive ones. Wild animals are busy.

When enrichment is done right, it keeps animal minds engaged and stimulated. At the National Zoo in Washington, D.C., the life of the Giant Pacific octopus is a bit more unpredictable because of the various items he receives five days a week—sometimes it's shrimp inside a plastic dog toy; at other times it's a piece of PVC pipe he can stretch his arms through. At the Bronx Zoo cheetahs spend more time exploring their enclosure when it has been sprayed with Calvin Klein's Obsession cologne. In Phoenix, tortoises bob for tasty cactus pads that keepers float in pools. Ocelots and other small cats at the Franklin Park Zoo in Boston receive cardboard paper-towel tubes with frozen pink hairless mice stuffed inside like gory Christmas crackers. The gorillas have blankets, curtains, and towels that they use to make soft nests, or to cover their heads and run around looking like kids pretending to be ghosts pretending to be gorillas.

There are even consultants who specialize in helping zoos, sanctuaries, and laboratories become more playful places for animals and keep them from stereotypical behaviors. The Shape of Enrichment is one such company. Their website lists their library of instructional videos, such as *Bungee Jumping Monkeys, The Bear Necessities, Fruit Bat Enrichment*, and *Tree Kangaroo Pouch Check Training*. The latter could be to see if they're hiding contraband, but more than likely it's to teach the kangaroos to present their babies for veterinary checks on command.

Some of this animal brain teasing is actually legislated, a sort of sideways acknowledgment of the prevalence of animal mental illness in certain groups of captive creatures. In 1985 the USDA began to require enrichment for various laboratory animals. That year's amendments to the Animal Welfare Act obligated labs to provide primates with perches, swings, mirrors, or other forms of environmental or social enrichment in their cages and to provide dogs with some form of exercise. Sadly, a single mirror does not ensure a happy primate, but it's a start.

Enrichment isn't new, even if the term, the regulations, and the industry are. Zookeepers have given their charges things to do for a long time. Sometimes these activities included ape tea parties, roller-skating or bicycling chimp shows, elephants water-skiing, or high-diving horses. At their best the activities occupied the animals' time without causing them undue stress; at their worst they were fear-inducing and dangerous and may have caused the early deaths of the animal performers.

Today's enrichment programs are, in a way, similar, even if they're far less deadly or dangerous. Offering polar bears giant plastic puzzles or lions cardboard zebras to savage is still making the animals do things that the humans in charge think is good for them and their audiences. The keepers, trainers, and zoo directors who oversaw chimp tea parties and kangaroo boxing matches also believed that what they were doing was appropriate. It's simply that what we think is good entertainment for the animals, and for us, has changed.

This isn't to say that enrichment is bad. It's not. It's simply that enrichment programs, alongside an increased focus on zoos as biological banks that ensure the survival of endangered wildlife, are part of a long-standing effort to make watching caged animals palatable to a new generation of Americans who are uncomfortable seeing metal cages or seemingly neurotic animals. The plastic toys in the arms of the octopus or in the toothy grip of the aquarium seal are there not only to occupy the animals' minds, but also to make us feel better about ourselves watching them.

Jeannine Jackle says that for enrichment to really work it must be personalized. "Take Gigi, for example," she told me. "She really likes to see human feet. I'm not sure why, but maybe it's interesting for her when we take off our shoes and she sees that human feet are like hers. And yet I can't exactly ask a zoo visitor to show Gigi their feet even if it would keep her entertained." Occasionally, though, a zoo volunteer, Gail O'Malley, who has visited the Boston gorillas at least three times a week for twenty-seven years, will take off her shoes and wiggle her toes in front of the exhibit's glass windows.

Gigi also enjoys watching the black-and-white films that keeper Paul Luther tapes for her off the American Movie Channel and plays behind the scenes from a TV/VCR on a rolling cart. The Franklin Park Zoo's now-deceased mandrill Ushindi also liked to watch television, especially Disney feature films. His shelf of DVDs at the zoo included *Free Willy* and *Free Willy II*, *101 Dalmatians*, and, oddly enough, *National Geographic African Wildlife*.

Recently, the Wilhelma Zoo in Stuttgart, Germany, installed a flat-screen TV inside the bonobo exhibit. The apes can toggle between different channels streaming short clips of bonobo life (taken from a documentary originally produced for children in the Congo): bonobos eating and searching for food, a female being tender with infants, two males fighting, play scenes, and bonobos mating. Surprisingly, the apes have not seemed very interested in their custom porn channel. A zoo employee told the NBC evening news that "maybe they are not so interested as bonobo apes very often have sex anyway."

Sometimes what stimulates captive animals the most are the humans staring back at them. James Breheny, the director of the Bronx Zoo, said, of the multimillion-dollar Congo gorilla forest exhibit, "We thought we were building this great exhibit for people to look at gorillas. We found we were building a great exhibit for gorillas to look at the people."

Kate Brown has been a docent in this exhibit for more than ten years. She is convinced that the gorillas' favorite time of year is Halloween, when children and adults come to the zoo on two different weekends in October dressed in costumes. "The gorillas get really interested in all the funny hats and colors," Brown told me one day as we watched dozens of visitors filter past the viewing windows and tap on the glass near a small female picking her teeth with a twig. "The gorillas come right up to the glass and look. It's something different for them."

Unfortunately the rest of the year humans tend to be extremely boring. Great groups of us come into the observation areas and excitedly point to the gorillas closest to the glass. We say things like "Look—a gorilla!" or "Wow." We pull out phones and digital cameras to take photos, squinting at the tiny screens. We wave—always a jaunty, open-palm human hello to the nonplussed gorillas. We comment to one another on how similar the apes are to us. And then we move on to the Asian monorail, the face-painting booth, the lemurs, or the gift shop to purchase plush giraffes or meerkat-shaped erasers. The gorillas have seen it all before, most of them for years.

Gail O'Malley says that there are some surefire ways to engage the apes on the other side of the glass but that few people do it. One of her favorite games is something she calls "playing purse." "Every zoo I've gone to, it has worked," she said. "Gorillas always want to know what's in your bag. . . . But you have to make a game out of it, you can't just dump everything onto the floor. Take one item out at a time—it doesn't matter what it is, it could be your sunglasses or keys. But you have to do it slowly, with anticipation and a flourish."

She demonstrated for me by pulling her wallet from her purse with a dramatic wrist twist. A burlesque routine came to mind, albeit a clothed, interspecies one. O'Malley also swears that most gorillas like human babies, something that has been confirmed for me by a number of keepers. She used to take her friends' babies to the zoo to entertain the gorillas since she didn't have children of her own. She says it worked every time. They always came up to the glass to get a good look, especially the females.

Kate Brown used to show one of the gorillas at the Bronx Zoo a picture book. "She really liked to look at the images," Brown said. "She was actually somewhat bipedal too. That is, she would stand up on her hind legs and walk around like a human, which is not something they are supposed to be able to do for long. But she would see me, and I would come over to the glass and open a book so that she could see it. She would be standing up on her back legs and then cross her arms over her head and lean in against the glass. I would show her the book, and when she wanted me to turn the page, she would tap on the glass."

Enriching the Rest of Us

Enrichment isn't just for captive wildlife or lab animals. The American pet products industry is banking on the tendency of Americans to buy things for their pets—playing to our potential guilt, love of shopping, and interest in helping the creatures that we live with. As of 2010, the industry was the fastest growing retail sector in North

America, with over $53 billion in sales per year. The majority of dog, cat, and parrot toys that are for sale at Petco, PetSmart, and independent pet shops—chewy puzzles that release morsels of kibble or elaborately knotted ropes around bits of birdseed and suet—are intended to stimulate pet minds, paws, jaws, and beaks. There are also the nontoy, nonpharmaceutical products intended to soothe animals that I was first introduced to during those early, anxious dog park conversations about Oliver. These include CDs like *Music to a Dog's Ear,* liver-flavored Happy Traveler chews meant to ease travel anxiety, the health-food-store staple Bach Flower Rescue Remedy for pets, Tranquility Jerky, lavender-scented Chillout Biscuits, pheromone diffusers that plug into the wall like the oddest of air fresheners, chamomile- and lemon-scented Relaxation Gel to rub on paws, Serene Drops for rabbits, and a range of meal supplements like Avi-calm for birds and Grand Calm pellets, whose upbeat packaging promises to cure anxiety and give focus to scatterbrained equines.

One of the most popular mental health products for dogs is a snug doggie jacket with Velcro tabs called a ThunderShirt. Its manufacturer claims that the shirt eases thunder and firework phobias, separation anxiety, problem barking and jumping, and travel stress. Recently the company released the ThunderCap, an elasticized hood that goes over the eyes and around the muzzle, which, if it was white and a tad more pointed, would be uncomfortably Klanesque. Instead it looks more like a shower cap for the face.

There is also a ThunderLeash, a ThunderToy that dispenses treats, and a ThunderShirt for cats. The only studies of the ThunderShirts' efficacy on treating anxiety have been done by the manufacturer itself. Still, Donna Haraway, the philosopher of science and technology who has written extensively about her own dogs, believes that the ThunderShirt helped her Australian shepherd named Cayenne cope, not with thunder (which didn't bother her at all) but with gunshots and fireworks.

There is also the Anxiety Wrap, another tight doggie jacket

created by trainer Susan Sharpe and her business partner, and their Quiet Dog Face Wrap, which looks like a soft rubber band for the snout. The most colorful option is the Storm Defender, a bright red cape-apron hybrid with a built-in metallic lining meant to shield dogs from static electricity.

When I asked the behaviorist E'Lise Christensen if these things actually work, she said that if they don't, they can't do any harm. Unless, that is, a dog is deathly afraid of being clothed or handled. Most veterinary behaviorists can't say definitively whether such products are effective because the animals who wind up as their patients didn't respond to the over-the-counter pellets, botanical paw rubs, or canine straitjackets. If they had, they wouldn't be there.

It seems to me that there are as many pet owners who swear by these products as there are those who think they're worthless. My suspicion is that their utility is as individual as the dog (or, unrealistically, cat) who is being shirted, capped, wrapped, or defended. I never tried to put Oliver into any of the calming outfits, though perhaps I should have. And I didn't give him lavender biscuits or paw drops. I did try plenty of other products, spending a small fortune on Rescue Remedy drops, a slew of new puzzle toys, and music I put on the stereo when I left the house. None of it really helped him, but it made me feel as if I was doing something.

Heavy Petting

Mac, the twenty-three-year-old Sardinian miniature donkey who lives on the ranch where I grew up, is, as I mentioned in the introduction, adorable, ferocious, and unstable. In the wake of his mother's death, when he was given to me to care for, I had excellent intentions but knew nothing about raising a baby donkey. I fed him bottles of formula, and, for a time anyway, Mac had the run of our house. The few rules he did have, like no biting, I didn't always enforce. His long furry ears and soft nose obliterated my better judgment, and that of my parents'. I also weaned him too early. Mac moved to his corral having bonded only with people. He had no equine social skills, but he did have an attitude. In this way, he was a bit like my own personal hotel elephant, preferring the company of humans to that of his own kind and expressing his displeasure dramatically when things did not work out as he wished. Mac became a tiny but potent hazard to our other donkeys, the pony in the adjacent corral, and a pair of goats. He attacked all of them with a savagery that belied his small size. When he was isolated from the other animals, he turned on himself, biting his legs until they bled, pulling out his fur with his teeth, and gnawing on the metal bars of his corral. He stopped only when someone was with him or when there was some sort of human activity unfolding nearby: he watched all of our goings-on with interest.

As I grew up and developed more of a conscience, Mac's behavior made me feel guilty and sad. To keep him from biting himself, I hung expensive banana- and cherry-flavored Lick-its in his corral (a horse popsicle that takes effort to lick), rubbed him with lavender-scented Equine Calm Balm that I tested on both of us, and gave him a horse ball covered in molasses that interested him for about thirty seconds—enough time for him to realize that it was just a plastic ball covered in molasses. What Mac *actually* relishes is chasing chickens out of his corral when they wander in, sussing out pomegranates that roll under the fence from a nearby tree, menacing the ranch dogs when they get too close, and occasionally escaping from his corral only to show up in our garage or to look balefully into the living-room window of a neighbor's house. He also likes to eat avocado leaves and strip the bark from newly planted trees. More than anything, however, Mac enjoys being firmly massaged. His eyes roll back and he relaxes, swaying on his feet. This liquid state lasts for a long while, then he will suddenly change his mind, and you have to jerk your hand out of the way before he bites down hard.

When I began to talk to other people about how much Mac liked massage, I learned that plenty of other creatures enjoy a good rubdown. Many trainers I spoke to gave me suggestions for how to touch Mac more effectively—most involved patience and quick reflexes. Then they mentioned Linda Tellington-Jones.

Tellington-Jones is something of a horse-therapy rock star and a prophet for people who do things like massage their hoofstock. In 1994 she was named "Horsewoman of the Year" by the North American Horsemen's Association. She has been inducted into the Massage Therapy Hall of Fame and has written fifteen books and many articles on what she calls the Tellington, or TTouch, method. She teaches massage for not only horses but also dogs, cats, and llamas, and the latest of her publications, *TTouch for Healthcare,* focuses on humans. Her inspiration came in the 1970s, when she was studying with Moshe Feldenkrais, whose technique for working with people uses nonhabitual movements like twisting, swiveling, and stretching the body in various ways to lessen physical pain, increase flexibility, and even, Feldenkrais devotees argue, enhance creativity. Tellington-Jones was curious to find out whether Feldenkrais-like movements might help other animals, and she began experimenting successfully with horses.

The cover of her 1995 book, *Getting in TTouch: Understand and Influence Your Horse's Personality,* is reminiscent of a soft-focus cover on an '80s Crystal Gale album, only Tellington-Jones, wearing a turquoise mohair sweater, is embracing a tall white horse. Her patented TTouches have names like the "clouded leopard," the "python lift," "tarantulas pulling the plow," and the "flick of the bear's paw." According to her own promotional materials, TTouch helps dogs suffering from everything from nervousness and aggression to car sickness and much in between. Her tagline is "Change your mind & change your animal."

The touches she uses are different from typical massage. They are nonintuitive, things I likely would not have thought to do with my own dog: sliding fingers lightly back and forth horizontally on their ears, for example, or gently tugging on the base of their tail.

As wacky as it sounds, people throng to her classes and workshops and talk about Tellington-Jones's methods in reverent tones that sometimes veer into the cultish. She now works with all sorts of animals, from nervous camels to asthma-suffering people, but no one, including Tellington-Jones herself, can explain why her touches are successful. A variety of studies on humans have demonstrated the power of massage to enhance emotional well-being and lessen anxiety, but studies on other animals have concentrated solely on physiological benefits. Research on the effects of equine massage—not TTouch in particular—has documented how much it helps racehorses, for example, recovering from injuries. Massage has also been used on dressage horses and even pet ponies, and there is now a thriving community of therapists with their own membership organization, the Equine Sports Massage Association. The website features photos of men and women in barn jackets rubbing glossy equines to a high polish, alongside video clips of racehorses galloping on treadmills while women in velvet riding helmets look on encouragingly.

The California-based dog trainer and wildlife photographer Jodi Frediani discovered Tellington-Jones when her daughter's horse started refusing to do what she was asked. "She'd lay her ears back," said Frediani, "and threaten to bite. Her fight response was well-developed, though she was quite friendly and would come running to the gate when called. In retrospect, I believe she learned her bad habits in self-defense from the man who bred her. He kicked his horses and used other dominance tactics to get them to do his bidding."

Frediani hired a local TTouch practitioner to help with the horse's aggression and was so impressed by her progress that Jodi enrolled in TTouch training herself. When, during the course, she saw an anxious horse instantly melt into a calm and relaxed state as Tellington-Jones massaged his gums, Frediani decided to incorporate TTouch into her own work as a trainer. She believes it's helpful because it surprises the animal but not in a frightening way. The touches are different from what the animal is used to and the disconnect between what he or she is expecting and what happens makes the animal "stop just using their fight-or-flight response," Frediani said. Calm animals have an easier time learning what's expected of them, and this leads to less fear, confusion, and stress. Based on observations with her clients and with her own dogs and horses, Frediani believes that TTouch eases muscle tension and may lower heart rate and blood pressure. She finds this especially helpful for animals with emotional problems, like the dogs she treats for separation anxiety.

If TTouch specifically, and massage in general, has significant effects on animal well-being, it may not be because of the "clouded leopard" or the "raccoon touch" but because of the calm, confident presence of a trustworthy human. I believe this was the case for Oliver. After he jumped out of our window he was so sore he could barely move. And when he wasn't on Valium, he was a knot of anxiety. Our dog walker, a kind fellow named Kelly Marshall, was Oliver's favorite person after Jude and me. One afternoon shortly after Oliver's jump, Kelly stopped by to check on him and told us that he had just begun a course on dog massage. He wanted to know if he could practice on Oliver. Jude and I looked at the Beast splayed on his bed, his body uncomfortably folded in on itself, and said yes. The next afternoon Kelly came by to work on Oliver. The results were immediate. Oliver relaxed, and his stiffness began to ease. He began to walk again, for the first time, a few minutes after his second session with Kelly.

Healing Humans

The Three-Legged Elephant and Her Family

Unfortunately there is no single pill, calming balm, masseuse, or magical therapeutic product that works for all disturbed animals, just as there isn't a single one for disturbed humans. Relief most often flows from an individualized cornucopia that may include exercise, behavior therapy, pharmaceuticals, and healthy new relationships. Sometimes these relationships are with people.

Tall, with wide cheeks and a skeptical manner that softens after a few glasses of Chang beer, Preecha Phuangkum is one of the most experienced and respected elephant veterinarians in Thailand. He started working with the animals thirty-two years ago; for fifteen of those years, he was in charge of the Thai government's logging elephants. The two hundred elephants and their four hundred mahouts—one man riding the elephant's neck and the other on the ground with the logs—lived in forest camps far from villages and towns. When they had finished logging an area, they would pack up and walk to a new site. Preecha went from camp to camp to oversee these elephants and their mahouts and make sure that the right elephant was matched with the right two men. If the match wasn't successful, the team would not be good at logging—dangerous work that demands elephant and mahouts respect and listen to one another.

After logging was outlawed in the mid-1990s, Preecha became the director of the Thai government's Mahout Training School. There, he trained hundreds of new mahouts and oversaw the health and well-being of the state-owned elephants, including the king of Thailand's royal elephants, chosen for auspicious traits like perfect toenail shape, their skin color, and, oddly enough, the sound of their snores.

"Lots of people believe that being with an elephant is a one-way relationship, that the elephant is to be controlled," Preecha told me. "But this is wrong. It should be a long-term relationship of love. If a mahout is cruel, the elephant will be cruel right back. If the keeper or mahout is upset or sad, the elephant will be concerned. They will get uncomfortable."

In an echo of many of the other mahouts, veterinarians, and elephant dealers that I spent time with in Thailand, Preecha believes the match between man and elephant is the most important part of an elephant's mental health: "When mahouts and elephants worked together for a long time in the camps, you would see the elephants taking care of the men. Like carrying their mahout back to camp after he got too drunk to walk. Now things have changed. Being a mahout is not a very respectable job in northern Thailand, and young men dream of buying things and moving to the cities. This has major effects on elephant emotional well-being."

Today elephants tend to have the same mahout for only a few years before the man (it is almost always a man) moves on. This is hard on the elephants, who often think of their mahout as family. According to Preecha, continuity and the right relationship is especially important when an elephant is very young. Of the more than thirty calves whose training Preecha oversaw, only a few grew up to kill people. He says that now, looking back, he can see that they chose the wrong keeper or mahout for the young elephant.

And yet, some elephants, no matter how kind or compassionate their mahout or keepers are, may still be emotionally unstable, violent, or aggressive. Preecha remembers one wild elephant matriarch in particular who, he believes, was simply born angry. Since the matriarch guides the behavior of the entire herd, if she is not a good leader, there may be problems with the whole group. This elephant was very aggressive and a crop raider of nearby villages. Preecha is convinced that her example led the entire herd to be more aggressive and more likely to raid village gardens and orchards, even when there was plenty to eat in the forest.

Most captive working elephants kill for the first time by accident. Then, realizing their power, they may do it again. The most common explanation I heard for this had to do with the emotional bonds elephants have with one another and with humans. "When an elephant is in love, they are the most dangerous," said Preecha. "They will stop at nothing to be with the elephant that they love."

He believes that this accounts for 80 to 90 percent of human deaths caused by elephants, the dramatic tramplings or targeted gorings that tend to remind human onlookers of psychotic behavior.

During the logging years more killings happened when the mahouts set up camp near villages. This wasn't because there were more strangers around the elephants or because they didn't like their new surroundings. Rather, it was because the mahouts found girlfriends. The elephants, used to being with their men around the clock in the forest camps, became jealous and sometimes violent.

Even today many mahouts told me that no matter how much they bathe after returning from visits to their girlfriends, their elephants can be sullen and standoffish, sometimes even aggressive. It can take days, and many proffered treats of sugarcane, bananas, pineapple tops, and affectionate ear scratches, to rouse them from their funks.

Preecha worked hard to instill in his students the belief that a good mahout isn't intimidated by the vicissitudes of elephant emotional life, isn't afraid of the elephant, is brave, and has self-control. He is convinced that elephants are generally quite reasonable and that the best way for a mahout to build a good relationship with his elephant is to expect the best of him or her. "If the mahout believes that the elephant is crazy for no reason, then the mahout will be much harder on the elephant and not treat them as well."

In 2007 Preecha retired from the mahout training school and became the head veterinarian for the nonprofit Friends of the Asian Elephant Hospital. It's a peaceful place. Unlike many other Thai elephant organizations, the FAE Hospital doesn't exist to serve ecotourists. There are no shows or demonstrations, and you can't interact with the elephants. Everything is quiet, and the only movements are the switching tails of the elephant patients and the hurried steps of the staff, in matching orange uniforms, as they change IV bags or rake out stalls.

Mosha, a nine-year-old female, is a permanent resident of the hospital. When she was seven months old, she was walking through the forest along the Burmese border behind her mother, a logging elephant. Mosha triggered a land mine, planted by the Myanmar Army against the Karen National Liberation Army and Shan rebels fighting for independence from the ruling military dictatorship. Mosha's front left leg shattered in the blast. Her mother was uninjured. Both elephants arrived at the hospital nervous and frightened. The veterinarians amputated Mosha's leg from the knee down. She stayed with her mother at the hospital for eight months, until the family who owned her mother demanded her back; they needed the income the older elephant generated from logging. Mosha, a newly weaned calf who would never be able to work, stayed behind. She was clearly upset by the loss of her mother, as well as the shock and pain caused by the land mine and her subsequent surgery. But she was still a curious, playful elephant calf, even with three legs, and she had a yawning, nearly bottomless need for affection. Preecha and Soraida Salwala, the founder of the hospital, were determined to find the right keeper for her.

His name was Paradee, Ladee to his friends. The first time I met him it was through the bars of Mosha's corral. A shy, gentle Karen man in his mid-twenties, Ladee was arranging thick blue gym mats of the sort that pole-vaulters use. Mosha followed his every move like an elephant-shaped shadow, hopping along behind him on her three good legs, stroking his shoulder with her trunk, and squeaking to him. Finished with the mat arranging, Ladee moved on to oiling Mosha's prosthetic leg, custom-built by a team of human-prostheses makers. Unfortunately for the young elephant, her problem isn't merely that the land mine blasted away her lower leg; it also embedded shrapnel in what was left of it. Nine years later she can stand for only a few hours at a time. The staff try to make Mosha wear her prosthesis a bit every day to relieve pressure on her other legs. She hates it and struggles to undo the buckles as soon as they are latched.

During my first visit to the hospital the heat was unbearable—even the birds quieted in the hottest part of the afternoon—and Ladee announced it was naptime. He walked over to the mats and motioned for Mosha to lie down. She did so, and then, in one practiced motion, raised her trunk and her front right leg in invitation.

Ladee crawled between her front legs, each as thick as his torso, and played idly with the tip of her trunk as she wrapped it around him. Mosha then laid her head down on the mats and lowered her eyelids halfway like a child pretending to sleep. "She won't fall asleep if I'm not here," Ladee called to my translator and me, as we stood alongside the corral, incredulous at the sight of the large elephant and the young man curled up together on gym mats as if it were the most normal thing in the world. "And she gets excited by visitors, so she may not sleep if *you're* here."

Mosha also won't sleep at night if she thinks Ladee is gone. His room is only about twenty feet from her corral, but sometimes she has nightmares and wakes up in the middle of the night agitated and scared and throws her mats around. "She will scream and scream until I wake up," said Ladee. "And then, if I come to the door and call to her, she'll calm down and go back to sleep."

She also hates it when he goes home to his village to see his family, 185 miles away, a trip he makes only two or three times per a year for a few days at a time. "The staff tell me that she calls for me the entire time I'm gone. And then, when she hears my motorbike on the road, she gets really excited. She knows I'm coming back."

Ladee used to go out and see friends nearby, but he doesn't do that anymore. It was too hard on Mosha. Before they found him, Preecha and Soraida tried three different keepers with the young elephant. Ladee began to work with her when she was two years old, and from the first days, it was obvious to hospital staff that he was the one. "The other keepers worked for themselves first, not Mosha. Ladee is different. He's kind. And he's also single. Mosha is the most important thing to him," said Preecha.

They tested Ladee for three months before they agreed to let him be a mahout, and then Preecha tested him out for another two years before he gave him to Mosha. Ladee had arrived at FAE as a nineteen-year-old assistant mahout to a sick elephant who had come from an elephant circus camp outside Chiang Mai. The hospital accepts and treats elephants without charge, as long as the owners provide transportation and send along the elephant's mahout. The hospital pays the mahout for the duration of his stay. FAE adopted this policy in the hope that visiting mahouts would learn more about basic elephant health care and take the knowledge with them when they left.

Soon after Ladee arrived, Preecha noticed that he liked feeding and washing the elephants and offered him a job on the condition that he live at the hospital and refrain from drinking alcohol. In exchange, Ladee would earn double what his salary would be in a camp, plus three meals a day and free housing. Ladee sends all his earnings home and is saving to buy land. Because he works so well with Mosha, Ladee got a raise in 2011. He now earns 10,000 baht a month (roughly $325). He's a wealthy man in his village.

A year and a half after Ladee and I first met, I returned to the hospital to see him and Mosha. They had moved to a slightly bigger enclosure. Mosha had grown feet taller. She was filling out, too, but she still squeaked like a young elephant and flapped her ears contentedly while nibbling on sugarcane stalks. Ladee greeted me with a smile and a ring woven from elephant tail hair. "Not Mosha's," he said, "I would never cut *her* tail hair." The two of them no longer slept together in the afternoon since Mosha was now so heavy she could crush him by accident. Ladee still put her to sleep at naptime and at night, with a sort of elephantine-tucking-in ritual that involved stroking the young elephant until she dozed off. Only when he was sure she was asleep did he move to his own bed.

As Ladee swept the corral, Mosha hopped along behind him, taking occasional breaks to come over to me and extend her trunk through the bars to touch my hands, my camera, the tops of my shoes, and to sniff my head. When Ladee had accumulated a big pile of sweepings and left to get a dustpan, Mosha plunked her entire body down on top of the pile, covering it entirely, so that when Ladee returned with the pan, there was no way he could finish.

"Mosha, Mosha, Mosha," Ladee cooed, laughing along with me and, it seemed, Mosha. She flopped onto her side and he came over and stroked her flank, smiling. When she'd had enough of his attention, she rolled off the pile so that he could finish cleaning.

The next day I asked Preecha if he thought Mosha was traumatized at all by her experience with the land mine or her odd life at the hospital without her mother and the other older females who would have been her aunties and helped raise her. "Especially with female elephants," Preecha said, "the mahout is like family. I am sure that Ladee thinks about getting married and leaving one day, of having his own human family. But Mosha does not think about him leaving. She does not think about him having another family. I think she thinks about him as another elephant, just like her." For now, this is enough for both of them.

The Bonobo and His Therapists

One afternoon, in a lull between patients, Harry Prosen, chair of the Department of Psychiatry at the Medical College of Wisconsin, received a call from the college president, who asked Prosen if he would consider treating Brian, a young male bonobo at the Milwaukee County Zoo. Prosen offered the ape a three o'clock appointment for the following Wednesday at his office. Prosen was only partly kidding. He had more than fifty years of experience with human patients, treating everything from paranoid schizophrenia to severe depression and psychosis, but he had never met a bonobo.

If there were such a thing as an undercover great ape, bonobos would be it. *Pan paniscus* have the long limbs of pro basketball players and brows so furrowed they seem to be perpetually struggling to remember something. The apes have never had the kind of famous human champion that chimps have in Jane Goodall, gorillas in Dian Fossey, or orangs in Biruté Galdikas. The most recognized bonobo researcher is the Dutch primatologist and ethologist Frans de Waal, whose work is primarily about empathy and morality in the primates, but no Hollywood hunk has ever played him in a film.

If bonobos are known for anything, it's sex. The apes use frequent and hearty sexual activity to express affection, settle or avoid disagreements, and ease all manner of other social interactions. Their repertoire of sexual behavior is extensive and gender-nonspecific, including oral sex and tongue kissing, females rubbing their genitals together until they orgasm, and males penis fencing while hanging from trees, which is just what it sounds like. They are the only other apes, besides humans, to have sex in the missionary position. All of this sex is part of the reason that bonobos tend to be absent from American zoos, lest parents should find themselves awkwardly explaining bonoboning to their children.

Bonobos have their aggressive moments too, but in general they are peaceable apes. De Waal believes that we owe them some recognition for their equanimity. "Everything we [humans] do negatively," he told a PBS interviewer, "is associated with our biology. . . . And everything we do that is nice, or when we're altruistic and empathetic, and so on . . . we claim that as our own unique human nature. So the stories out of Gombe on the chimpanzees [who waged war on one another] confirmed that negative biological view that we have of ourselves as purely competitive, purely aggressive. And when the bonobos came along later, they didn't fit that view."

Brian, the young male bonobo at the Milwaukee Zoo, was a special case, however. He was neither sexually fluent nor peaceable. "He was my first patient," said Prosen, "who tried to throw feces [and] spit and urinate on me."

Brian arrived in Milwaukee in July 1997, and the staff quickly realized that his psychological needs were beyond anything they had seen before. Barbara Bell, the head bonobo keeper at the zoo, has worked with bonobos for more than twenty years. Brian "would vomit thirty, forty, fifty times a day, pace in circles all day long," she said. "I never saw him sleep. He was not able to eat a meal in the group. He had no social culture to blend in. He lived in fear of having the snot kicked out of him by all the other animals because they did not see him as worthy of being in their troop."

He also tore off his own fingernails, repeatedly thrust his fist into his rectum so hard that it bled, rubbed his genitals on sharp objects, stared blankly at the walls, and was extremely aggressive toward the keepers. He was also scared of objects that he hadn't seen before, so giving him toys or puzzles to distract him from mutilating himself only served to make Brian more upset. Undaunted, the zoo staff tried to reward him anytime he wasn't hurting himself. But after six weeks they felt completely defeated. Brian still couldn't eat in the presence of other bonobos; he didn't know how to play or how to relate to the adult females, and he was scared of the adult males. When he became even slightly stressed, he curled into the fetal position and screamed. "When an animal is self-destructing to this extent," Bell said, "it has to be stopped or that animal won't survive at all."

On Prosen's first visit to the zoo, he was overwhelmed by the sad state of Brian, who was constantly clapping his hands and spinning in circles in a holding area behind the scenes, by himself. "I have had some difficult interview situations," said Prosen, "but establishing communication looked at first glance to be very difficult indeed."

As he does with his human patients, the psychiatrist's first step was to put together a complete history for the bonobo. He called the first of many case conferences with the keepers and zoo veterinarians in the basement kitchen of the primate building and tried to gather as much information as he could on Brian and his experiences before he arrived in Milwaukee. While keepers chopped bananas and watermelons for bonobo meals, Prosen and the staff discussed the ape's history.

Brian's past, they discovered, was as abnormal as his behavior. Brian was born at the Yerkes National Primate Research Center at Emory University in Atlanta, and lived there alone, except for his father, who sodomized and intimidated him, for the first seven years of his life. Sodomy may be the one sexual act that bonobos don't engage in and sexual violence is rare. Brian's father, a research subject, doubtlessly had his own emotional problems.

Bonobo society is matriarchal. Mothers and older females are extremely important to the development of young bonobos. There is group mothering, and male infants stay with their mothers twice as long as females do, learning how to communicate, share food, resolve disputes, and express themselves sexually. In the wild, the males stay in close contact with their mothers for fourteen or fifteen years. As a baby, Brian was taken from his mother and housed alone with his father inside the laboratory. He didn't receive any mothering and had no opportunities to bond with older females who would have taught him how to trust others, or relate to bonobos in general. His environment was entirely unnatural and his early sexual experiences—of his father violently attempting to mount him—were likely traumatizing.

Brian's fisting habit developed sometime during his stint at Yerkes, where, by the end of his tenure there, he was doing it so often and so intensely that he was losing large amounts of blood. At this point, Yerkes researchers feared for his survival, so they removed him from his father and kept him on his own for eight months. MRIs showed no physical problems, except thickened rectal and colonic tissue due to the chronic fisting, so Brian was given Prozac and Valium. His fisting continued though, so the researchers decided to send him away for help.

The Milwaukee County Zoo has a reputation for healing distressed bonobos. This is due in part to Barbara Bell's decades of experience with the apes, but also because of the kind and stable pair of bonobos who led the zoo's troop for many years, Lody and Maringa. Captured in Congo, the young male and female were sold to sailors bound for Amsterdam when they were two years old. They arrived at the Milwaukee zoo in 1986 and, for more than twenty-five years, they, together with Bell, helped manage the largest group of captive bonobos in the United States. Because of Bell's and the troop's reputation for healing disturbed apes, troubled bonobos like Brian continue to arrive.

Not all of the bonobos get along with one another, and managing their interpersonal dramas can be difficult. Sometimes the bonobos have their own ideas about what should happen. The zoo's troop is so large and the apes' personalities and preferences are so various that they are not all kept together on exhibit. Some of them stay behind the scenes in a play area while others go into the public viewing area, depending on the day. If the bonobos do not like the groupings the keepers choose for them, they form their own and refuse to go anywhere until they are allowed to stay with their friends. Companions change frequently. Bell says that this part of her job is a bit like "mixing volatile chemicals."

The only bonobos who would put up with Brian's volatility were a forty-nine-year-old blind and deaf female named Kitty and Lody, the twenty-seven-year-old male from Congo. Brian often allowed Kitty to groom him, and then he would help the elderly female find her way to the outdoor area. Lody took Brian by the hand when he was too panicked to move and led him around, to their playroom or outdoors. Lody would even postpone eating his own meals to sit with Brian and comfort him. Once, when a younger male stole a mailing tube of treats the keepers had prepared for Brian, Lody put his own treats back in his tube and gave it to the anxious ape.

These small kindnesses weren't enough to make Brian feel better. He still made himself vomit for hours on end and continued his fisting. He was also very attached to his OCD rituals and refused to eat a meal until he'd gone through a whole sequence of them.

"I began to see Brian's self-destructive behavior as an attempt to soothe himself in extreme anxiety situations," Prosen said. By touching himself, even in a hurtful way, Brian was trying to make himself feel better in a world in which he had no other means of comfort and no control over his own life. "He had what in humans would be called a social phobia of mass proportions, as well as a complete inability to understand his environment or to accurately interpret attempts to relate to him as helpful, not dangerous."

Prosen's first course of action was to prescribe a low dose of antidepressants for Brian to help him deal with his fear and anxiety, as well as his obsessions. Since Prozac had not worked in the past, he prescribed Paxil, but only in the hope that Brian might relax long enough for a therapeutic program to begin. Occasionally the keeper staff supplemented Brian's Paxil with Valium, but only for short periods of time, on bad days when his panic and anxiety were dire. The Paxil took away his underlying anxiety and "once that was gone," Bell said, "he was able to stop some of the obsessive compulsive behaviors he had," like the long rituals before mealtimes.

"But the beauty of the drug therapy was that the other bonobos could start to see him for who he really was, which was really a cool little dude. Once you got rid of all the clutter in his world and he started to learn a few social behaviors, then ever so slowly his life started to pick up."

Brian's therapy began in earnest. Prosen, Bell, and the keeper staff set to work making Brian's world safe and predictable. All his meals were served at the same time, in the same place. He was given quiet time every day after lunch. The keepers also kept their voices low and tried their best to use consistent mannerisms and words of praise. Every new object was introduced to his environment slowly so that he could look at it, touch it, smell it, and get used to it at his own pace. Daily training sessions were short, and the keepers made sure to end them on a positive note. Prosen said that because Brian was a captive animal, therapy was in some ways easier than treating his human patients because he and the keepers could control Brian's environment and his daily life so completely. Brian's daily schedule of activities, for example, never deviated because he was so fearful of change. The other bonobos at the zoo were quite flexible, easygoing, and open to all sorts of new experiences. For Brian that was unthinkable.

"One thing we did," Bell said, "was team him up with other bonobo kids who were much younger than him and could teach him play behavior. We paired him with two- and three-year-olds so he could learn. We all know that's why you go to kindergarten. You are learning social skills. Brian had to go all the way back and learn proper play behaviors in order to grow."

Watching over Brian's therapy, Prosen became fascinated with the similarity between the bonobo and his human patients, especially those who suffered from developmental deficits.

"A very successful businessman I once worked with lost his father when he was twelve," said Prosen, "and literally grew up from twelve to twenty overnight. This occurred not through a normal kind of developmental process, as being patiently taught, but rather a mimicking behavior. Very rapidly, the man appeared to be effectively doing things that he really had not learned in the usual way of growing up—that is, from a mentoring father. The results of this only appeared later, when, as he developed his own business, he began to have great difficulty with employees as they reached the adolescence of their employment, and with his own children when they reached their own adolescence. He had what I'd describe as an acute developmental deficit."

In Brian, Prosen saw a hairy mirror of this man. He was convinced that Brian had serious developmental deficits and that this was why he was acting at any one time as if he were three or four different ages. Brian did quite well in certain training situations, but as soon as he found himself in a new and different environment that required more mature behavior, he fell to pieces. Interacting with adult females, to whom he'd had no exposure as a youngster, caused him all sorts of anxiety. This was confusing to the rest of the troop because

Brian *looked* like an eight- or nine-year-old young male, but developmentally he acted like a five- or six-year-old. And even this wasn't stable. One minute Brian was a confident young fellow, and the next he'd suddenly try to nurse from one of the females, which annoyed and confused them. He often had his toes bitten in retaliation. Only Lody reliably came to his rescue.

Prosen believes that the older male's kindness and encouragement were the reason Brian survived his own self-defeating behavior. But he also thinks that bonobos may be more resilient than people when it comes to overcoming developmental deficits. Another psychiatrist, who works with chimpanzees, believes he knows why.

Dr. Martin Brüne treated ten traumatized chimps who had once been used for research and then retired to a Dutch sanctuary. He was impressed by their ability to recover from years of mistreatment, abandoning their rocking, self-mutilation, and R-and-R-ing habits relatively quickly by spending time with other chimps, eating healthier food, throwing themselves into enrichment activities, and being given antidepressants. Brüne believes that a human who grew up in an environment as abnormal as a laboratory wouldn't be able to recover as quickly as chimps do. He posits that the human ability to adapt easily and fluidly to new situations, social groups, and environments may not always be a good thing. "That's why we populated the globe, not chimps," he said. "But on the other hand, this could perhaps come at a cost. And the cost is perhaps increased susceptibility to psychological disturbances."

Bell and Prosen share a slightly different belief about bonobos: that they're flexible in their way and that humans might be able to learn something from them. In a joint paper about the process of helping Brian, they wrote, "Brian, who was initially a very 'weird' bonobo, began to view the world more calmly. . . . He grins and glows. I think we have to recognize that perhaps in contrast to humans, development can be opened up and restarted in bonobos and other primates. If this is the case, then the study of the treatment of developmental deficits in bonobos might contribute to the literature of working with human primates as well."

By 2001, four years after Brian arrived as an uncertain, self-destructive, and undeveloped creature, he was correctly reading social cues in a group overseen by two dominant females and politely following bonobo custom. A new mother in the troop even trusted him to gently stroke her ten-day-old infant with his finger. A year later, he was comfortable with much larger groups of bonobos. Lody even let him take charge of the troop from time to time. By Brian's sixteenth birthday, in 2006, he was finally acting closer to his age. In a remarkable reversal of power, as Lody aged and grew more frail, Brian became a troop leader.

"They still get along fine," says Bell, "but their roles are reversed. Brian eats first and Lody second. Frankly, I think Lody just doesn't want to be the leader anymore."

Brian now also enjoys the interest and affection of the ruling females. Bell says that what makes him the happiest is when he's allowed to carry the troop's babies, and in the last few years he has even fathered a few of his own. She doesn't remember how many years ago he was taken off Paxil but she knows that they stopped giving it to him when he started to share it with the other bonobos (a phenomenon Prosen has observed among other great apes he's written prescriptions for). "When he became a drug runner we had to stop," Bell said with a laugh. "Periodicially he regresses, but rarely. He's social and wonderful with females and babies."

The staff continues to carefully manage Brian's social groupings, making sure that he is always with the calmest members of the troop. His days remain predictable, and he is given a lot of time to adapt to new situations. Prosen occasionally wonders whether Brian feels true empathy for the apes around him or has simply learned to artfully mimic Lody. "Brian really may have an empathy deficit due to his extremely abnormal upbringing," he told me, "and he could be somewhat of a psychopath. Only bonobo psychopaths apparently aren't violent like human ones can be."

In the meantime, Brian has grown up, and is now a well-muscled young male. The females have noticed and the same ones who used to violently rebuff him now show interest in him. After Lody died of an enlarged heart in 2012, Brian's role as troop leader was cemented. Brian has also formed new alliances in the troop and Bell says that he has "abandoned the road rage approach."

Over the last fifteen years both Prosen and Bell have received many requests for psychiatric consultations from other zoos. "I will get calls from other keepers saying that they want to get rid of their eleven- or twelve-year-old male because 'he's ripping girls to shreds or doing this or that . . . he's crazy!' That's the time you need to hang on to them. Those are the young males that need nurturing, and more guidance, not less. We've been really blessed by Brian. In a weird way he was a gift to our zoo. He and I have a loving relationship that's very guarded. He is still a pistol, and I think he could hurt me, but we are still learning from Brian, we will always be learning from him."

Brian's therapy seems to have been a resounding success, and yet Prosen refuses to take credit for the bonobo's recovery. He believes that Bell's and the other keepers' efforts were heroic and that it is actually the zoo's bonobo troop that is responsible for Brian's transformation, that Lody and Kitty were Brian's real therapists.

"Empathy knows no country, no species, is universal and has always been available," Prosen said. "I discovered after arriving at the zoo that it belonged to the bonobos long before us."

Chapter Four

Proxies and Mirrors

The maddest thing I ever did was done under orders.

Pat Barker, *Regeneration*

Every creature in the world
is like a book and a picture,
to us, and a mirror.

Alain de Lille, c. 1200

What if Oliver had lived in the late nineteenth century instead of the early twenty-first? Victorian bystanders may have caught sight of him in full frothy panic at our bedroom window, confused him for a mad dog and shot him on the spot. If he'd been born a few decades later, just after the turn of the twentieth century, newspaper reporters, dog fanciers, and sidewalk observers who noticed his jump from our apartment window might have chalked up his behavior to mortal homesickness or heartbreak.

The labels we have used for oddly behaving animals over the last hundred and firty years have often corresponded to the ones we've used for humans. Like our human diagnoses, they have never been stable. Veterinarians, zookeepers, natural historians, farmers, pet owners, and physicians have applied terms as old as *hysteria* and *melancholia* and as recent as *OCD* and *mood disorders* to other creatures.

Diagnoses have come and gone like whalebone corsets and Elizabethan ruffs. That is, men, women, and other animals were stuffed into them, somewhat awkwardly, until another diagnosis that was more fitting or fashionable came along, one that people or their physicians felt was a better fit.

Turn-of-the-twentieth-century cases of nostalgia and heartbreak, for example, unfolded alongside an increasing tendency to medicalize and treat mental health. As the century wore on, physicians who treated various forms of insanity became specialists and the process of therapy became more rooted in individual patient-physician relationships. By early midcentury, these physicians were known as "psychiatrists."

Efforts to make sense of other animal minds often reflected these shifting ideas about human mental health. People use the concepts, language, and reasoning they have at hand to understand puzzling animal behavior. Disorders such as mortal heartbreak and nostalgia may sound quaint or old-fashioned today but contemporary Internet addictions and attention deficit disorders may, by the twenty-second or twenty-third centuries, seem antiquated. In this way, looking at instances of animal madness in history and how we've mapped ailments such as nostalgia, mortal heartbreak, melancholia, hysteria, and madness onto other creatures is like holding up a mirror to the history of human mental illness. The reflection isn't always flattering.

Mad Elephants, Mad Dogs, Mad Men

For centuries, the genesis of madness in animals was confusing and hard to pin down. Even the word *madness* has meant many different things. By the sixteenth century, *mad* was a common word for "insane," and by the eighteenth it became a standard term for "anger" in Great Britain and, later, in North America. In the second half of the nineteenth century and well into the early twentieth, any animal acting strangely or aggressively could be deemed mad whether or not he or she was rabid. It wasn't until the late nineteenth century, and in some cases later, that mad animals were seen as victims of physical rather than mental disease.

Rabid dogs were especially terrifying because the disease was at first silent, sometimes incubating for many months in an exposed person, until it bloomed into excruciating pain and certain death. The disease was also scary because the main carrier, or at least what people thought was the main carrier, was man's best friend. It is difficult to imagine today the fear of contagion present among city dwellers in the late nineteenth century. Dogs were not yet uniformly pets. Some were certainly closer to the coiffed and dominated denizens of our contemporary urban dog parks, creatures whose lupine tendencies have largely been bred into floppy-eared, doe-eyed obsolescence, but in the late nineteenth and early twentieth century, dogs were much freer to roam and pursue their own interests, even if this came at the cost of mange, early death, or hunger. Their potential rabidity was disconcertingly present and more difficult to contain. A mad dog could be anywhere, and even though the fear was often disproportionate to the actual public health risk, it was real and paralyzing fear all the same.

The public's anxieties about mad dogs were evident in inflammatory newspaper headlines: "Mad Dogs Running Amuck: A Hydrophobia Panic Prevails in Connecticut," "Mad Dog Owned the House," "Lynn in Terror," "Suburbs Demand Death to Canines . . . Members of the Household Lock Themselves in Their Rooms While the Raving Beast Roams about the Halls."

It wasn't until Louis Pasteur successfully inoculated the first person against rabies in 1885 that widespread understanding of the disease morphed into biological narratives of contagion. Before Pasteur, rabies symptoms were often referred to as "insanity" instead of signs of infection. The historian Harriet Ritvo has argued that contracting rabies was not only thought to be bad luck, but was also considered a punishment that the infected animal somehow deserved and had brought upon itself by being unclean, engaging in sinful behavior, being overly lustful, or having too many unsatisfied sexual urges. In Britain, pets of the poor were considered especially at risk of madness, but so too were the pampered and seemingly corrupted pets of the upper classes.

Infection was also thought to jump from dogs to other animals or from other animals to dogs. Horses in particular were frequently bitten by vicious canines and then quarantined to await signs of hydrophobia. Sometimes they were simply shot. One early twentieth-century burro, bitten by a mad coyote, went on to kill a mastiff, bite a chunk out of the neck of a horse, and attack a party of miners in Death Valley. In 1890, a few hundred miles away, a mad lynx attacked a horse, killed one dog and beat up another, wounded several pigs, chased a herd of cattle, and was finally shot by a woman with a musket. Other cases concerned circus animals. In Chicago a little girl named Mabel Hogle was bitten by a monkey while visiting a museum of curios with her father. The monkey, reportedly foaming at the mouth, was assumed to be suffering from hydrophobia and killed.

Not all of these animals were rabid, however. Since so many people used the term *mad* to describe both rabidity and insanity, it wasn't always easy to tell the difference. As early as 1760, when Oliver Goldsmith published the poem "An Elegy on the Death of a Mad Dog," there was little distinction between being mad with rabies and other forms of insanity. The poem included the lines "This dog and man at first were friends / But when some pique began / The dog, to gain some private ends / Went mad, and bit the man." This dog, fictional or not, was not rabid. He bit his human companion because he "lost his wits."

Labeling an animal mad was not only a way of explaining irrational anger, it also described creatures' strange behavior, aggression, or some other form of insanity, such as hysteria, melancholia, depression, or nostalgia. One small dog, for example, discovered with a pig on a shipwreck afloat in midocean in 1890 was said to have gone mad with loss. Animals could also go mad from a lifetime of abuse, such as

Smiles, the Central Park rhinoceros, who reportedly did so in 1903. Maddened horses, as they were known, could simply take off running through Central Park or Williamsburg, Virginia, or anywhere at all, still attached to their carriages or dragging their riders behind, often with fatal consequences. Other horses, suffering from "equine insanity," could, in a flash, turn on their grooms or riders and stomp them to death. Madness was also used to explain other seemingly bizarre animal actions. In 1909 Henry, the monkey mascot of a New Orleans baseball team, supposedly went mad when fans of the opposing team taunted him past his breaking point. He broke free from his cage at the stadium and climbed into the grandstand, creating a stampede and causing the game to be called in the seventh inning. As late as the 1920s and 1930s there were mad cats yowling in "madder orgies," cows gone mad on the way to slaughter, at least one mad parrot, and some unruly Hollywood primates. In 1937, just a few months before forging an alliance with Hitler, Mussolini made international news when he was attacked by a mad ox during a parade to welcome him to Libya. He escaped unharmed and commended Libyans for their support of fascist Italy.

The attribution of madness to a variety of animals was widespread but many of the most enduring stories concern elephants. One early article typical of the mad elephant genre, published in the *New York Times* in 1880, told the story of an Indian elephant who one day began terrorizing nearby villages. Police who followed him found a trail of smashed buildings, trampled corpses, and a creature who doubled back to attack his pursuers. "[The elephant] was not merely wild—it was 'mad,' and as cunning and as cruel as a mad man," relayed the reporter. "But insanity itself is a tribute to the animal's intelligence, for sudden downright madness presumes strong brain power. Owls never go mad. They may go 'silly,' or they may be born idiots; but as Oliver Wendell Holmes says, a weak mind does not accumulate force enough to hurt itself."

While an elephant could theoretically contract rabies, most were not physically ill but more likely reacting against poor treatment and abuse. These mad elephants were newsworthy, not simply because they smashed buildings or cars or trampled people but because they expressed themselves in often spectacular ways—choosing particular individuals on whom to vent their anger or exact revenge, biding their time until they found the right, most devastating moment to act. Captive elephants have been known to suddenly explode into violence, going after their handlers, grooms, or trainers. This is so common that, since the nineteenth century, expressions like *running amok* came to characterize just this type of event. These accounts were commonplace in the nineteenth and twentieth centuries and still appear in the twenty-first.

On August 20, 1994, in front of thousands of people eating cotton candy and peanuts, a twenty-year-old female African elephant named Tyke entered Honolulu's Blaisdell arena, part of the Circus International show. She was wearing a headdress of golden five-pointed stars. Her trainer, Allen Campbell, wore a sparkly blue jump suit. Even on a shaky home video recording, Tyke seemed agitated. She began to turn in quick circles at the edge of the brightly lit ring. Campbell was frustrated and pushed and prodded her, trying to gain some control over the whirling elephant. She trumpeted loudly and knocked her groom, who'd been standing nearby, to the ground. Quickly, she bent down on her front knees and pressed him against the floor with the full weight of her body. Then she rolled and kicked him, like the lightest of logs, along the ground. This is when Campbell went after Tyke, trying to stop her. But she knocked him to the ground too and began to kick him more forcefully than she had her groom, pausing to sink down on her knees and smash him against the floor. As she stood back up, Campbell flopped to the side, limp.

"It looked like the elephant had a rag doll tied to her leg—the way the man's head was moving," said a woman who'd brought her daughter to the circus, in an interview for a special episode of the television show *When Animals Attack*. "Then people started panicking. The people closest to the ring began to realize that this was not part of the show. That something was wrong."

After Allen stopped moving, Tyke turned back to her groom, kicking and rolling him across the floor one last time. At this point, Allen looked dead or unconscious. The crowd screamed and panicked. People began to run and push toward the exits of the arena. Tyke burst out of the building, throwing one of the heavy wooden doors off of its hinges and sending it sailing twenty feet. She headed into the adjoining parking lot, followed by a police car, and then into a nearby street, stopping traffic. More cops sped to the scene, dozens of cars converging on the streets around the arena, training their guns on Tyke.

Tyler Ralston was driving down Waimanu Street when he saw Tyke running toward his car. "Initially, I was confused," Ralston told a reporter for the *Honolulu Advertiser*. "The elephant was coming at me and the police were behind it."

He swerved out of the way in time to see Tyke chase a circus clown into a vacant lot while another circus employee tried to lock her in by closing a pair of chain-link gates. She charged through the flimsy barrier and into him, shattering his leg. Then the police started shooting. "That's when I was, like . . . 'I don't want to see an elephant get killed.' And the next thing I knew, it was running by me, bloody."

The police shot Tyke more than eighty times. Of the men she went after, only her trainer, Allen Campbell, was killed. After news of Campbell's and Tyke's deaths began to spread, more of the elephant's story came to light. According to USDA and Canadian law enforcement records, years earlier Tyke had been performing with another circus when her trainer was seen beating her in public, to the point where she screamed and bent down on three legs to avoid being hit.

Whenever the trainer walked by Tyke afterward, she would scream and veer away from him. He claimed he was punishing her for trying to gore his brother. She had also escaped twice before. In April of 1993, Tyke charged through a door of the Jaffa Shrine in Pennsylvania during a Great American Circus performance—ripping off part of the wall (causing more than $10,000 in damage) and running onto an upstairs balcony. She was later coaxed back by her trainers. And in July of the same year, during a performance at the North Dakota State Fair, Tyke escaped from her trainer again, trampling an elephant show worker and breaking two of his ribs. Tyke belonged to the Hawthorn Corporation. The company, managed and owned by John Cuneo Jr., leased animals to circuses and other entertainment ventures around the world for more than thirty years—including Circus Vargas and Walker Brothers Circus. The company had a terrible track record of Animal Welfare Act violations. In 2003, the USDA seized an elephant from Cuneo. It was their first elephant seizure in history. Her name was Delhi and she was suffering from skin abscesses, lesions, and severe chemical burns; a trainer had soaked her feet in undiluted formaldeyde. A year later the USDA charged Cuneo with nineteen more counts of abuse, neglect and, mistreatment, and he was forced to relinquish his entire herd of sixteen elephants.

Outside of mistreatment, bouts of madness in male elephants may be, at least partially, explained by musth, a hormone-fueled period that can last weeks to months. A male in musth is considered more aggressive and stubborn, their penises may be erect, and a sticky substance leaks from the glands in their temples. Sometimes these males become violent; musth periods have been described as passing bouts of erotic madness.

Chunee, a once docile Asian elephant who lived at Exeter Change in London in the mid-nineteenth century, was killed when his annual attacks of "sexual excitability" made him too violent for the comfort of his keepers. His execution in March 1826 was gory and went on far too long. Chunee refused arsenic, three rifle shots only made him more upset, and repeated volleys of military muskets by a group of soldiers called in at the last minute couldn't finish the job. Eventually a keeper delivered the final blow with a sword.

Gunda too was once an approachable star elephant at the Bronx Zoo just after the turn of the twentieth century. But he became, upon sexual maturity, "most troublesome and dangerous," according to William Hornaday of the New York Zoological Society. His repeated, six-month "bouts of erotic frenzy" made him so violent that he was put under extreme restraints for half of every year. Debates over what to do with him captivated New Yorkers, and articles and editorials about his fate, the ethics of chaining him in place, and his possible execution peppered the New York press on the eve of World War I. In the end Gunda was shot point-blank in the elephant house by the famed elephant hunter and taxidermist Carl Akeley. His folded, dehydrated hide was taken to the American Museum of Natural History in New York, where it remains today, stored on a large metal shelf underneath the Planetarium. Gunda's execution for mad behavior was representative of many other elephants' experiences, their rights to life hinging on how their sanity was perceived by the humans charged with caring for and confining them.

Tip: Reform or Die

On January 1, 1889, an eighteen-year-old Asian elephant named Tip walked off the Pavonia Ferry onto Twenty-third Street in New York City. He was a New Year's gift to the people of New York from Adam Forepaugh, a circus-owning competitor of P. T. Barnum and Ringling Brothers who'd made his fortune by selling horses to the U.S. government during the Civil War. Forepaugh's shows included Russian acrobats and Wyoming cowboys, "comedian pigs, donkeys and canines," bicycle battles, a Museum of Savage People and Living Freaks, a boxing kangaroo named Jack, and a white elephant known as "The Light of Asia." The shows advertised elephants riding velocipedes on wires strung high in the air, walking tightropes, and knocking down human boxers. The shows also included Tip, but for some reason (Forepaugh called it generosity) the elephant was now being given to the city of New York to be its first publicly owned pachyderm.

Over the next few years Tip would become first a lovable celebrity, then an example of violent animal madness, and finally an allegedly unrepentant criminal who divided natural historians, big game hunters, and animal collectors, as well as thousands of New Yorkers, into vocal and impassioned camps. But on that New Year's afternoon Tip seemed to be a peaceful elephant whose presence was about to turn the animal pens inside Central Park into a proper zoo. According to the man who was giving him away, Tip was "docile as a lamb." He was also worth $8,000 and had been a star of Forepaugh's elephant show. What the first few newspaper articles didn't question was why Forepaugh, a shifty businessman who hired pickpockets to work the crowds at his own shows, would want to give up a healthy trained elephant worth so much money, even if it meant scads of good publicity. It is more likely that Tip wasn't docile at all.

Forepaugh had purchased the elephant from the legendary animal collector and zoo man Carl Hagenbeck, who had himself purchased Tip from King Umberto of Italy. Tip was probably captured like other Asian elephants of the time, by being forcibly separated from his mother in the forest where he was born. Or perhaps he was born to a captive female and taken from her just after weaning. Either way, his early years and his long voyage first to Italy, then to Germany, and finally to the United States would have been difficult. He would have been continually separated from the people and other elephants he was familiar with. His diet would have consisted of hay, bran mash, or sometimes wine, not the grasses he was born to favor. He wouldn't have been able to roll in mud and swim in rivers but would have drunk from a bucket or hose and spent long hours chained in place on hard-packed ground, perhaps swaying to relieve the pressure on his knees and ankles. He was trained under threat of beating, and the tricks, such as riding velocipedes, weren't easy for an elephant to perform. When the adolescent hormones started flowing through his temples, stoking his desire for female companionship, Tip likely grew even more frustrated by his strict confinement.

For his first few years inside the Central Park elephant house, Tip's life as an attraction was rather uneventful. But in 1894 the *New York Times* announced that Tip "must reform or die." The article claimed that unless the elephant controlled his temper he would be killed and his bones sent uptown to the American Museum of Natural History. His keeper, William Snyder, was clamoring the loudest for Tip to be done away with, convinced that the elephant was mad and planning to kill him and that it was only a matter of time until he did so. Snyder was right. One morning, as the keeper went to feed Tip his breakfast, the elephant snapped the chains that held his tusks to the floor and hit Snyder hard with his trunk, knocking him to the ground, and then tried to stomp the life out of him. Snyder screamed, and a park policeman came running, dragging him away from Tip just in time.

The elephant waited three years to try to hurt Snyder again. One afternoon, before the keeper finished his work for the day, he went into Tip's pen to add an extra chain to his already heavy manacles. Snyder immediately sensed that Tip was ready to attack but before he could jump out of the way, Tip swiped him with his tusks. The blow sent Snyder flying into the wall and the elephant quickly moved to gore the keeper while he lay prone on the floor. But Tip missed, hitting the wall of the pen so hard that the building shook. Snyder crawled to safety, and his hatred of the elephant hardened into resolve to see him dead.

The Central Park commissioners began a week of deliberation over what to do with the elephant. Daily newspaper articles covered his plight and the pros and cons of keeping him alive or killing him. Attendance at the zoo, concentrated in front of Tip's pen, increased with the news coverage. The animal dealer Hagenbeck, who sold Tip to Forepaugh, was in favor of death. The commissioners weighed the loss of a popular zoo attraction against the gain of a great exhibit for the American Museum of Natural History. One commissioner noted that chaining Tip in place for five years might be the reason for his vicious temper but that he was so dangerous now there was no way he could be unchained. The debates revolved around two major questions: There was no record of Tip going after anyone besides Snyder, but would he? And could Tip be held accountable for wanting to kill his keeper in the first place?

Despite the newspaper claims that increasingly called Tip mad, he was probably more frustrated than insane. He almost certainly wasn't rabid. He also may have been going through musth. Perhaps Tip grew so frustrated that he tried to change his situation. The most logical way for him to do this, he may have thought, was to kill the keeper responsible for his extreme restraints.

As the Central Park commissioners argued Tip's fate, the public did as well. Echoing the reporter who claimed that elephants were intelligent enough to go crazy, people who saw Tip as smart and calculating called the loudest for his death. These death-for-Tip advocates clearly felt that in order to want to kill Keeper Snyder, that is, to plan it and wait for the perfect opportunity, Tip had to be self-aware and capable of reason. By calling for him to reform or die, they demonstrated their belief that Tip was intelligent and sane enough to be culpable for his actions. On the other side were newly formed animal rights groups and activists who urged the Park Commission to see Tip as a creature worthy of pity who should not be blamed for his behavior. This view is, in some sense, similar to today's plea of insanity.

The late nineteenth and early twentieth centuries saw a new wave of animal advocates establishing societies and agitating for more humane treatment of certain kinds of creatures, including captive wildlife and domestic animals. Books like *Black Beauty*, first published in 1877, reflected these shifting attitudes about animal protection. In

Tip's case, the people who wanted him spared may also have believed he was too dumb to have gone insane in the first place.

On May 10, 1894, the Central Park commissioners unanimously decided that the mad elephant should die. They claimed that Tip had killed four men while he was with Forepaugh's circus and had gone after at least four more in Central Park with the intention of murdering them. It also cited an escape attempt, his great strength, the flimsiness of the elephant house, and testimony from an employee of Barnum's circus who said he'd always thought Tip was a danger to the public.

The park flooded with visitors, all of them assembled in front of Tip's pen, paying their respects or perhaps hoping to catch a glimpse of his death. Photos taken outside the elephant barn that week show a sea of men in bowler hats and fedoras, dressed in dark jackets against the spring chill, looking expectant. The first execution attempt was a hollowed apple filled with cyanide. He refused it. He also refused cyanide-laced carrots and bread. Meanwhile thousands more people thronged the edges of the pen, waiting for something dramatic to happen. Representatives from the Natural History Museum who had brought rifles wanted to shoot Tip then and there, but the head of the Society for the Prevention of Cruelty to Animals wouldn't allow it. It wasn't until Keeper Snyder showed up with a big pan full of wet bran that Tip succumbed. Snyder mixed capsules of potassium cyanide into the bran and rolled it up into a big ball. Tip ate quickly; within minutes he seemed agitated, and droplets of blood dripped from his mouth. He made a final, powerful attempt to escape through the back of his pen toward the green lawn of the park, snapping all of his chains except the one around his ankle. This last chain tripped him and he fell to the ground, trumpeting faintly as he died.

I went to the American Museum of Natural History 117 years after Tip's death to look for him. After scouring the specimen acquisition books, heavy rectangular volumes with flaking red spines that document every animal, plant, mineral, and artifact donated to the museum since its inception in 1869, I found the entry for Tip. He arrived the day after his death in 1894 and became specimen number 3891. The official record says that 3891 consists of a skull and mandible. His tusks are stored inside the museum's ivory vault. His skeleton is at the museum, too, though its arrival isn't noted in the books.

A few days later I followed the curator of mammals up a slim metal staircase to the first floor of a storage space tucked under the eaves. "The Africans are here," he said, "the Asians are upstairs." He was referring to the museum's collection of elephant skulls. Large and hulking, the skulls sit in trays along the floor, covered with plastic sheeting to protect them from roof leaks. The first two skulls whose tags I inspected belonged to a mother and calf killed by Teddy Roosevelt and his son Kermit in 1909.

On the second level, a single bulb hung from the ceiling. Every surface was covered with dust, so thick it looked like gray snow and had a similar muffling effect. A long row of skulls stretched the length of the floor, and more than a century's worth of accumulated bone fragments rested between them, small shards of ossified jaw and eye socket. The tallest skulls reached almost to my waist. At the very end of the row, flush against the sloping roof, was Tip. His skull had gone bronze with age, and the cavities where his tusks had been were gaping open, as if in surprise. He had been there ever since a team of men dragged his body by horse cart to a nearby barn and, working by lamplight, finished skinning his body and cleaning his skeleton for display. As I looked at Tip's skull, I thought about his trial in the park and then his long, strange tenure after death as a specimen. Tip was an example not only of *Elephas maximus* but also of the frustrated mind. He was deemed mad not because he was rabid or demonstrably insane but because he acted violently toward the men who sought to control him, keep him in chains, and diminish his sensory, social, physical, and emotional world to a small barn. His badness caused his madness, and his madness cemented his badness. Tip was a victim of the human tendency to punish what we misunderstand or fear. New York of the 1890s was a world in which elephants killed men out of vengeance and spite, and insanity could leap from animal to human. How Tip was treated for his behavior, his increasingly restrained world, and his eventual execution reflected the anxieties of the people around him who fretted about the causes of madness and just who was susceptible.

Mortal Homesickness in Gorillas, Geisha Girls, and Everyone Else

Other forms of infectious insanity plagued nonhumans too. Certain of these illnesses have gone extinct, the diagnostic equivalents of passenger pigeons or dodos. Two diseases in particular, homesickness and nostalgia, felled men, women, and a fair share of other animals, from aquarium-dwelling sea lions to pet ducks. Throughout the nineteenth century and well into the twentieth, homesickness was considered a physical illness like tuberculosis or scarlet fever. It was thought to weaken, kill, or even inspire suicide. The ways the disease manifested reflected the era's fears about increasing urbanization and the isolation that recent city transplants, far from their families, often felt, the psychological trauma of war, and the new, widespread immigration made possible by train travel and steamships. The term *nostalgia* could be used interchangeably with *homesickness*, and both afflictions were considered potentially deadly ailments. During the Civil War, for example, Union doctors diagnosed five thousand men with homesickness, seventy-four of whom died of the disease. In some cases army bands were prohibited from playing "Home Sweet Home" out of fear that the song might inspire mortal cases of homesickness or nostalgia in the soldiers who heard it. After the war the diseases grew more commonplace as Americans moved from farms to cities and millions of immigrants from around the world flooded into the United States, many of them pining for home.

Some groups of people, such as African Americans, Native Americans, and women of all races, were thought to be more susceptible to homesickness than white men, and many psychologists and social commentators argued that this must be proof of Darwinian evolutionary theory at work (that is, people who succumbed to homesickness were culturally underdeveloped and unfit for American society, which favored the adaptable and sturdy). One charity worker observed in 1906, "Nostalgia . . . is the first and most effective aid to the natural selection of desirable immigrants."

Other creatures were caught up in ideas about loss, longing, physical deterioration, and evolutionary fitness too. Animals served as convenient mirrors for these sorts of preoccupations since many exotic species were also far from home for the first time. The same forms of transportation that made human travel possible on a previously unheard-of scale in the late nineteenth century also did the same for nonhumans. These animals' behavior, once they arrived in their new homes, often reminded people of themselves.

The Homesick Gorilla

A few floors below Tip's final resting place at the American Museum of Natural History is the mammalogy collection. Its hallways look a lot like typical high school corridors lined with lockers. Instead of textbooks and algebra binders, however, they are stocked with gorilla skulls, orangutan skins, and tiny cardboard boxes of teeth organized by species. The smell of formalin billows out in faintly sweet clouds when you ease open the doors.

At the end of one row in the great ape section is a locker labeled "G. gorilla. Casts, Zoo. No data." This is where John Daniel is. Or at least this is where the parts of him are that aren't on display in the Hall of Primates, where his mounted and stuffed skin and glass eyes have stared out at visitors, in a kind of simian thinker pose, since 1921. The odd and surprising life of John Daniel, a western lowland gorilla captured in the forests of Gabon in 1917 and taken to London to live in the front window of a department store, is a true-life parable of the way that labels like homesickness and nostalgia were applied to other animals, and why, in cases like John's, it may have made perfect sense.

John Daniel was a superstar. No one remembers him anymore other than a few circus historians and die-hard gorilla fans. (The latter have a name for themselves, gorillaphiles, and they sometimes vacation together, traveling around the United States to see zoo gorillas.) In the 1920s John Daniel was famous for his surprising mind and for upending what it meant to be an ape, a scientific object, and a circus attraction. His short, curious existence suggested to the Western world for the first time on a grand scale that gorillas weren't bloodthirsty brutes but affectionate and intelligent creatures who thrived when treated with kindness and love and who were susceptible to the same emotional stresses suffered by humans when they were denied it.

John was captured in Gabon after his mother was shot by a French army officer sometime between 1915 and 1916. When he was roughly two years old, he was shipped to England in the company of a group of monkeys ordered by the British government for experimental purposes, and purchased by the animal dealer John Daniel Hamlyn. Working out of his shop in London's East End, Hamlyn bought and sold exotic animals captured throughout the British Empire and is credited with inventing the chimpanzee tea party. These staged shows of chimps dressed in pants and shirts and drinking tea out of cups while sitting on chairs, were a fixture of Western zoos well into the twentieth century. Hamlyn was also said to have kept chimpanzees as children in his house, wearing clothes and eating at the dinner table with him and his wife. One of these chimps reportedly answered the door at Hamlyn's shop and then trundled off to find a human to help the waiting customers. When the young gorilla arrived from Gabon, Hamlyn promptly named John after himself and sold him to the department store Derry & Tom's with the idea that in a few months' time he would make an excellent Christmas attraction.

A young woman named Alyse Cunningham and her nephew, Major Rupert Penny, spotted the gorilla in the department store's front window. They were intrigued, purchased him soon after, and brought him back to their house in central London. John had a bad case of the flu, and Cunningham described him as rickety and underweight. She also said that he'd been lonely. "We soon found it impossible to leave him alone at night because he shrieked every night, nearly all night, from loneliness and fear!" wrote Alice.

She was convinced that his fears stemmed from his long nights at the department store, when he was left all alone after the salespeople went home. The staff told Alyse that he'd cry and cry when they started to pack up at the end of the day. Alyse and Rupert felt that John's night terrors were contributing to his slow weight gain and sickly demeanor. They decided to build him a bed in the room adjoining Rupert's. He loved this new sleeping spot and his nightly shrieking ended. He began to grow and put on weight.

Alyse's goal for John was that he become a member of their family, as if he were a human child, and so she started teaching him how to brush his hair, handle a fork, drink out of a glass, turn the taps on and off, and open and close doors. It took him only six weeks to learn all of these things, and then he was free to roam about the house as he chose.

John was a picky eater. Alyse didn't know this, but had he been with his mother, it's likely he still would have been breastfeeding. Gorillas generally nurse until they are about three years old. John always wanted milk, a lot of it, and warmed on the stove. He was also quite fond of jelly, especially fresh lemon jelly. The gorilla wouldn't touch anything if it had been sitting out a few hours, though he always had room for roses. "The more beautiful they were, the more he liked them," wrote Alyse, but he would never eat faded ones.

John also loved to have guests and would become so excited

around new people that he'd show off like a young child, greeting them at the door and taking them by the hand, leading them around the room in circles. He was fond of shutting his eyes and then running around, knocking into tables and chairs. According to Alyse, he relished taking everything out of the wastebaskets and strewing the contents around. When asked, he would pick everything up and put it back, looking bored.

One afternoon Alyse put on a light-colored dress to go out. John went to hop up into her lap as he often did, but she pushed him away and said, "No," as she didn't want him to make her dress dirty. Offended, he lay on the floor and cried for about a minute, then stood, looked around the room, and picked up a newspaper, spread it on her lap, and hopped up. The newsprint dirtied her dress, too, but Alyse was too impressed to care.

Stories of John's exploits were covered in newspapers in England and the United States, and accounts of his humanlike nature intrigued famous naturalists like William Hornaday. As the director of the New York Zoological Society and the Bronx Zoo, he had been trying to secure a gorilla for New York City since 1905, when he received a letter from a young daughter of a member of the Zoological Society. "My father says I may give a gorilla," she wrote to the director. "Please place an order for him. I would like to name him 'Cheese.'"

Unfortunately for Hornaday, it wasn't easy to purchase a gorilla that would live long enough to become a zoo exhibit. Before John Daniel, captive gorilla deaths were considered inevitable and chalked up to homesickness, nostalgia, or mortal melancholy. One of the few gorillas to live more than a few months was Dinah, a young female captured by Professor R. L. Garner, a popular naturalist and animal collector who was convinced gorillas could speak. On a trip to Gabon in 1893 he decided to test his theory. Garner set up a forest cage he called "Fort Gorilla" and moved in, waiting to be approached by a talking ape. When this did not happen, Garner befriended a chimp whom he named Moses and then attempted to teach him to speak

English. This, too, did not go as planned; Moses did not speak. On a subsequent trip, in 1914, Garner came across a baby gorilla he named Dinah and brought her back to New York. She was sickly but survived eleven months, long enough to be taken around the Bronx Zoo regularly in a pram, wearing a frilly white cap and red mittens. She supposedly liked to watch the buffalo.

John Daniel was the first gorilla that seemed to thrive among people and it surprised many naturalists that his fine health didn't seem to be due to his diet, the temperature of his quarters, or any other physical aspect of his environment. Instead it appeared to be a result of his affectionate family life. This shocked Western scientists and zoo men in particular. Just three years earlier Hornaday had proclaimed that there was no reason to hope that a gorilla would ever survive in captivity. He believed that when gorillas were captured as adults, their "savage and implacable nature" made them impossible to keep and that even if babies could be "captured and civilized," they were liable to die soon thereafter. John contradicted all of this by flourishing.

For more than two years Alyse and Rupert encouraged John Daniel's development, challenging and stimulating his mind without teaching him any tricks. "He simply acquired knowledge himself," Alyse said. They would take him as an ordinary train passenger, without a cage, chain, or leash, to their cottage in the country. He liked the garden and woods but was fearful in open pastures. He was also frightened of full-grown cows and sheep but was fascinated by calves and lambs. Occasionally, they brought him to the London Zoo to see the animals and to be marveled at by the other visitors.

John Daniel was growing and would soon become a large male gorilla, or silverback. Alyse and Rupert felt that a free-roaming three-hundred-pound adult gorilla would no longer be considered acceptable in public. John also couldn't be left alone as he became an anxious mess, howling until his family returned. Alyse and Rupert tried to find someone to help look after him, but that proved to be impossible, as most people tried to physically discipline the young gorilla. According to Alyse, they never hit John: "The only way to deal with him was to tell him he was very naughty, and push him away from us; then he would roll on the floor and cry and be very repentant, holding one's ankles and putting his head on our feet."

Alyse and Rupert decided that they needed to find him a new home. It's not clear why they weren't able to locate a suitable spot for him in England, but what is certain is that a man appeared with an offer to buy the young gorilla, saying that he represented a private park in Florida where John would have his every need met and be able to live out his life in a garden. This was not to be. The buyer, they discovered too late, was a representative for Ringling Brothers Circus. In March 1921, John was put on a ship to New York City, where he was housed in the cold and drafty tower of the old Madison Square Garden building and put on display.

The first reports of John Daniel's mental and physical deterioration appeared almost immediately upon his arrival in the United States. The *New York Times* reported that he was homesick and spending most of his time "sitting quietly in one corner steadily watching for some familiar face in the crowds that come to see him." "It is only when Mr. Benson [the agent who had traveled with him from England] arrives that he shows any animation and then he reaches his fingers through the bars to shake hands with his friend."

The loneliness and isolation John must have felt inside his cage at the Garden was probably crushing. First he'd been separated from his gorilla mother, then he had been raised like a hairy human child and, at four years of age, would have been developmentally like one. What John Daniel felt when he was taken from Alyse and Rupert is likely similar to what a human child of the same age would feel upon being separated from his parents and the only home he knew, to sit in a cold room with only the gaze of strangers to keep him company. John responded to English. He had culture. He knew a gorilla version of love and affection. He also knew a gorilla form of sadness.

Soon both circus-goers and the press reported that the young gorilla was literally dying of loneliness. Alyse set out for New York by steamship as soon as she realized what had actually happened to John, but she didn't make it in time. Three weeks after John Daniel arrived in New York, he was dead. Reporters for the *Times* claimed that homesickness, confinement, and improper care did him in. At least one argued that he had actually died of pneumonia. Both things may be true, as John's immune system was likely weakened by his loneliness and isolation. In the weeks before his death he had refused food and would crouch on his iron bed, covering himself with a blanket, facing away from the front of his cage and the crowds who came to see him. By the time the wife of one of the circus performers began to spend time with him, putting warm compresses on his forehead and giving him the attention he craved, it was too late. A Ringling employee who knew John said that he had been treated like any ordinary museum specimen and this was the problem: "I think myself that he might have lived if allowed to stick to his former habits."

There was a reason that he'd been barred from his normal habits, and it was financial. In the three weeks that Ringling Brothers displayed John in New York, even considering that he was listless, sad, and staring at the wall—not the most cheery of circus exhibits—the company earned back the $32,000 it paid for him. Had he lived and continued to attract the same number of people, John would have earned the circus roughly $500,000 per year in the 1920s, more than $5.6 million today.

Alyse must have been devastated by the death of her beloved gorilla. And yet, her interest in the apes was undiminished. Shortly after John Daniel died, she purchased another baby gorilla, whom she named John Sultan. He too moved into her London apartment and country house, but this time she did not let him out of her sight. She signed a contract with Ringling Bros. and Barnum & Bailey Circus to display the gorilla under the name John Daniel II, but she retained ownership of him and stipulated that she was always to be at his side. She also required that they stay together in hotels and that he travel by car, train, and ship, not in a crate like the other circus animals but next to her, in a passenger seat.

John Daniel II and Alyse arrived in New York City on April 24, 1924. He was three years old. Unlike the first John Daniel, who crossed the Atlantic in a cage in the ship's hold, John Daniel II shared a stateroom with Alyse. Once they arrived, they stayed together at the luxurious McAlpin Hotel on West 34th Street and Broadway, where he was allowed to play on the rooftop for exercise. The gorilla was displayed at the circus like his predecessor, but this time Alyse was nearby and they went home together at the end of every day, in a taxi. He and Alyse also visited the American Museum of Natural History, where the young gorilla was, morbidly, shown the mounted body of the first John Daniel. The famous primatologist Robert Yerkes came to the museum that day to meet them, along with physicians from Columbia University and the famed big-game hunter and taxidermist Carl Akeley, who was responsible for the museum's dramatic dioramas, including a mountain gorilla family, frozen before a misty painted volcano in the Hall of African Mammals. A *New York Times* reporter who visited John Daniel II at the McAlpin mused that "William Jennings Bryan has not called on the visitor but John certainly would have been a valued acquaintance for Mr. Darwin. He offers . . . ocular proof of everything Mr. Darwin has affirmed and Mr. Bryan denied." The legendary Scopes Monkey Trial, in which Jennings Bryan would forcefully argue against the teaching of evolution in American public schools, would happen only months after John returned to England.

Throughout his tenure with the circus, traveling across the United States and then Europe, and his years with Alyse in England, John Daniel II was a playful, wide-eyed little fellow prone to "nervous tension" around too many people. He relaxed with the clowns in between shows, was gentle with young children, and only occasionally bit his mistress. Despite the attentions of a tropical disease specialist who served as his private physician in London, he died in 1927. I do not know what became of his body or if Alyse, well aware that her first gorilla had been mounted and studied, kept him for herself, buried in the fields near their country home in Gloucestershire.

Almost a hundred years after his death, New Yorkers can still visit the first John Daniel. His taxidermied body is inside a glass cabinet on the third floor of the American Museum of Natural History, next to Meshie the chimpanzee, another human-raised ape-child, whose "father" was the famous comparative anatomist and gorilla hunter Harry Raven. John has been there since his death and his tag, which reads "*Gorilla gorilla*," says nothing of his remarkable life. Upstairs in the metal locker, where his skull and bones are stored, is a tiny orange tin with a handwritten tag. It holds his milk teeth. I believe this is the same little tin that Alyse must have stored them in while he was teething, and gave to the museum upon his death. The cursive handwriting is beautiful, painstaking. His teeth are small and only slightly colored with age.

John's life in England and his subsequent trip to the United States unfolded in the aftermath of World War I. The psychological effects of the fighting existed on a previously unthinkable scale: 3.9 million American men were in the armed services, and 72 percent of them had been drafted. Many soldiers grew homesick and had to be treated at the front for emotional problems. Persistent homesickness was not only dangerous on its own but thought to be an indicator of imminent nervous collapse or "emotional shell shock." During the war era and for some time afterward, newspapers ran stories on homesickness or tried to raise funds to buy soldiers at the front musical instruments to allay their emotional pain. Homesickness and nostalgia were also used to explain desertion in terms other than cowardice. Away from the front, war brides suffering from homesickness gassed themselves or threw themselves and their babies into San Francisco Bay.

Homesickness was not a new affliction. The *Oxford English Dictionary* lists the first mention in 1748, but a flurry of deaths by homesickness was reported around the turn of the century, and increasing numbers of cases were diagnosed during and after World War I. Country boys new to cities were thought to be particularly susceptible, but homesickness plagued many other people as well, from Geisha girls brought from Japan to the 1904 World's Fair in Chicago to a man who was so homesick that he stole a parrot who spoke to him warmly.

Other animals were dying of homesickness and nostalgia too. One case in 1892 concerned a mule shipped by railroad to a farm near Independence, Louisiana. Three weeks later the homesick creature reportedly showed up back in Tennessee, having reportedly walked four hundred miles to return home. Dogs who whined with nostalgia made the papers in Chicago, and just after the turn of the century, Jocko, a monkey mascot of a U.S. Navy ship acquired during the Spanish-American War, was reportedly so homesick for his original Spanish crew that he tried to swallow poison. His death was attributed to fatal melancholia aboard the ship. That same year an African elephant named Jingo, who was being shipped in a wooden box from England to New York to replace Jumbo in Barnum's circus, refused to eat and died on the boat. His body was heaved overboard and his death reported as a possible case of homesickness. And soon after John Daniel's death, a young female mountain gorilla named Congo, who had been studied at great length by the primatologist Robert Yerkes, died at John Ringling's estate in southern Florida. The *New York Times* mused, "What caused her death is unknown but it may well have been loneliness, a broken heart and the wistful desire for companionship of her own kind."

Even birds were considered susceptible. Toward the end of World War I, the family of a young San Francisco boy moved from a house into an apartment, and he was forced to give up his pet duck, named Waddles. The boy took her to Golden Gate Park and left her there. After a few days of endless honking and searching, the *San Francisco Chronicle* claimed Waddles died of nostalgia for her lost companion. The story ran next to another: "Despondent over Wounds Soldier Jumps out of Window."

John Daniel was afloat on an ark crowded with nostalgic and homesick animals who, like humans around the turn of the twentieth century, were finding themselves far from home and not adjusting easily. The fact that animals could suffer from these diseases seemed to imply widespread acceptance of the idea that a mule, a gorilla, or another creature could have a sense of self and understand their displacement.

The reports of homesickness- and nostalgia-related deaths were also sometimes used to justify racial hierarchies that placed white men above everyone else—patronizing and unfair attempts to make some humans and other animals seem more emotionally fragile. Ota Benga, an African pygmy man brought to the United States in the early twentieth century and displayed for a time inside the ape house at the Bronx Zoo, was one example. He shot himself in 1916, in Lynchburg, Virginia. His death was held up as evidence of his inability to adjust to life in the United States, a result of mortal nostalgia and homesickness.

Heartbroken Bears, Men, and Mothers

Throughout the late nineteenth and early twentieth centuries, another popular catchall diagnosis for puzzling behaviors and untimely deaths was heartbreak. Like homesickness or nostalgia, brokenheartedness was considered a potentially lethal medical problem that affected both humans and other animals. Furthermore, a broken heart was not merely bad on its own but was also thought to be a gateway problem that could lead to melancholia and other forms of insanity. One 1888 treatise on the subject proclaimed, "The asylums of this and every country are full of these mental wrecks occasioned by these emotional cyclones."

Evaluated from a twenty-first-century perspective, many deaths attributed to heartbreak might have been suicides, but well into the twentieth century it was far more socially acceptable to attribute self-killings to heartbreak and, incidentally, much easier to collect life insurance on. Despite some skeptics' pooh-poohing heartbreak postmortems, the press breathlessly reported the stories. There were lovers' hearts that gave out at the exact same moment, or after one ran off with a younger paramour; bankers' hearts that ruptured when markets collapsed or their investments tanked; mothers and fathers whose hearts broke when their children fell through the ice while skating or were hit by a train or were kidnapped. Even one of Brigham Young's wives supposedly fell victim after he accused her of sleeping with another man. Sad veterans and vanquished generals succumbed, as did immigrants stuck in limbo at Ellis Island. Women whose husbands were serving time in Sing Sing and at least one Indian princess were considered heartbreak victims as well.

The diagnosis grew out of a complex moralizing, often about the risks of bad behavior or the cost of love. It was also a convenient way to explain strange behavior, a physiological rationale for emotional pain that hadn't yet been medicalized into depression or suicidal impulses in quite the same way it would be later. And perhaps most critically, the stories were entertaining.

As with homesickness, the popular and sometimes the scientific press covered a menagerie of brokenhearted animals, many of whom were dogs. Canine death by broken heart wasn't a new phenomenon. Loyal hounds who died of heartbreak and sadness when their owners or companions passed away have been celebrated and held up as ideals for human beings since antiquity. Greyfriars Bobby, a Skye Terrier, supposedly spent fourteen years camped out on the grave of his deceased master in Victorian Edinburgh until he died too. Other dogs reportedly passed away because an animal friend died first. In 1937, a German Shepherd named Teddy stopped eating when his horse companion died; he stayed in the horse's stall for three days until he died himself. Horses were also supposedly done in by heartbreak; mules perhaps not. According to one World War I source, a horse stuck in a shell hole full of water "will strive and struggle to get out, until he actually dies of a broken heart. Not so the mule. He has no imagination and no such outlook on life. He calmly and philosophically lies in the shell hole until someone comes along and digs him out."

Besides faithful dogs and other companion animals, late nineteenth- and early twentieth-century stories of brokenheartedness centered around zoo and circus animals, possibly because these creatures also lived closely with people and weren't destined for the dinner table. There may have been less incentive to recognize humanlike brokenheartedness in a future steak or chicken breast. Between the late 1880s and the mid-1930s, animals like Bomby, a taciturn rhino that lived in Central Park, a blind sea lion named Trudy, and an Emperor Penguin who refused to be force-fed after his mate died in Washington, D.C., were each said to have died of heartbreak. Wild animals were occasionally thought to suffer from mortal heartbreak too, but mostly in association with their capture. And the inability to keep many animals, from lions to songbirds, alive in captivity was ascribed to heartbreak well into the twentieth century. In 1966 a killer whale named Namu was the second orca to be caught alive. He was brought to Seattle's Marine Aquarium, where observers watched him ramming his head against the side of his tank and screaming loudly, his calls sometimes returned by passing orcas in Puget Sound. He drowned after becoming tangled in a net, a death the *New York Times* chalked up to heartbreak. A calf that had just been captured in order to keep Namu company was sent to SeaWorld in San Diego instead. That calf became the first Shamu.

Plenty of twentieth-century zookeepers discussed the risks of loneliness, grief, and heartbreak in their charges and the physiological problems they believed came with it. Belle Benchley, director of the San Diego Zoo from 1927 to 1953, once said, "Solitude brings melancholia to the majority of animals. They pine away and die from sheer loneliness, which explains many of their strange friendships."

One of these friendships, at the Berlin Zoo in 1924, cheered a monkey from his melancholy. The keepers gave him a porcupine.

Monarch

Until the exhibits were shifted around for renovations in 2012, the large mounted body of a male grizzly bear stood just outside the cafeteria at the California Academy of Sciences in San Francisco. Visitors wandered past his glass vitrine on their way to buy soup and pizza slices, unaware they were passing a legend. Much cuddlier in death than he was in life, the bear's taxidermied form is disturbing. It's as if every ounce of ferocity was drained out and replaced with soda pop. His face is mounted in a loopy, awkward smile, the kind of strained grimace you manage when chatting with someone you'd rather be walking away from. It's even worse, however, because bears don't actually smile. The only honest parts of his mount are his claws, overgrown and curling under. He's obviously a bear who didn't walk much.

His name is Monarch and he has been on display at the academy ever since he died inside his cage in Golden Gate Park in 1911. In the 1950s Monarch's mounted body served as one of the models for the redesign of the California flag. State legislators decided the bear on the existing one looked far too much like a hairy pig than a majestic figurehead. Since then, Monarch's likeness has been duplicated millions of times, printed on everything from boxer shorts and bank logos to travel mugs and tattoos. Yet few people know the bear was an actual living, breathing, scratching creature; fewer still know he was once said to be suffering from the deepest ennui, at risk of dying from a broken heart.

The only known example of a California grizzly mounted for exhibit, Monarch is a shaggy metonym for the drastic ecological and social transformations that took place in the state before and after the gold rush and a bear-shaped metaphor for the changing attitudes of San Franciscans toward the wilderness that surrounded them until the mid-nineteenth century. Americans who shuffled past Monarch's cage or read about him in the paper made sense of his behavior in ways that reflected the times in which they lived, just as they had with Tip and John Daniel. Monarch was an icon of a recently neutered wilderness, and the worries over his emotional health reflected society's newly romantic attitude toward the nation's wildlands—which had been increasingly "tamed" by the slaughter and erasure of Native Americans and fearsome predators like Monarch from the landscape.

Until the latter half of the nineteenth century, the forests, meadows, and riverbanks of California were thick with grizzly bears. If you knew what you were doing, it was fairly easy to capture one. In 1858 a sheriff in Sacramento sold a wild grizzly for $15.50; a trained one went for $20.50. When the trapper George Yount arrived in California in 1831 and settled in Napa Valley, he said that the bears "were everywhere—upon the plains, in the valleys and on the mountains, venturing even within the camping-grounds, so that I have often killed as many as five or six in one day, and it was not unusual to see fifty or sixty within twenty-four hours."

In the 1850s, Grizzly Adams, the famous bear hunter and showman, traveled with two trained bears, Lady Washington and Ben Franklin, and exhibited dozens more in a menagerie in San Francisco. Ben Franklin had been captured as a still-nursing cub, so Adams gave him to a greyhound dog who had recently had a litter of pups and made buckskin mittens for the bear's paws so he wouldn't hurt the dog. Benjamin nursed from the greyhound for weeks, until Adams started feeding him meat. Both bears traveled hundreds of miles with Adams, sometimes chained to the wagon, other times walking freely alongside, and occasionally inside the wagon with him and the dog. Lady Washington also carried a pack, dragged a sled, and moved timber, and both bears helped Adams hunt grizzlies and other game that they shared at mealtime.

Well into the 1860s captive bears could be found chained or

caged at train stations, where they performed tricks or ate sweets and cakes fed to them by waiting passengers. One bear was said to have played the flute. People bought tickets to watch bears fight with bulls. Some Californians even kept them as pets. The actress and dancer Lola Montez chained two large grizzlies by the front door of her cottage in Grass Valley. By the late nineteenth century, however, the bears were few and far between. Those that hadn't been killed had become more reclusive, and captive bears were harder to buy. The animals who, only a few years before, had been everywhere were now hunted almost out of existence.

William Randolph Hearst, the eccentric California newspaper magnate, watched shrewdly as the bears became ever rarer. He decided that he could exploit his readership's interest in the impending extinction of such a charismatic animal. In 1889 he hired Allen Kelly, a newspaper reporter with a bit of hunting and trapping experience, to capture a grizzly bear as a mascot for one of his papers, the *San Francisco Examiner*, known as the "Monarch of the Dailies." Hearst hoped that the tale of capturing one of the state's last grizzlies would boost readership. He would name him Monarch, after the paper.

Kelly began his hunt in the hills behind Santa Paula in Ventura County, but the bears avoided his traps. A few weeks became months and still he had nothing. Kelly's editor at the paper fired him, but he continued undeterred. A few months later a Mexican man trapped a large grizzly in the San Gabriel Mountains of Los Angeles County and offered to sell it to Kelly. The bear furiously tried to escape his wooden trap, biting and tearing at the logs, hurling his body against the walls. For a full week he raged and refused to touch food. It took an entire day just to chain one of his legs. Finally, the bear was hauled onto a rough sled to be pulled by a team of skittish horses. The rest of the long trip to San Francisco was made by wagon and then railroad.

Egged on by wildly embellished tales of his capture in the *Examiner*, twenty thousand people came to see Monarch on his first day at Woodward's Gardens, an amusement park in the city's Mission District. He was kept there, in a steel cell, for five or six years, until visitors lost interest in watching him. Hearst gifted the bear to the new Golden Gate Park in 1895. Shortly after Monarch arrived, the park commissioners were preoccupied with a number of more pressing issues than the bear, such as bicycles, new machines that the park leadership worried would frighten horses or lead to violent collisions. Monarch's arrival took up a mere two sentences in the commissioners' annual report. The large gift from the *Examiner* first "objected to his strange surroundings and tried to make his escape but now seems reconciled to his fate, and is a very popular attraction."

Over time, however, Monarch was becoming depressing to look at. By 1903 he had taken to spending all day inside a hole he dug out in the center of his pit between two boulders. Lying inside this hole, he placed his immense head on his paws and stared blankly through the bars of his cage. He may have liked the cover the boulders provided from the eyes of visitors, or perhaps he enjoyed the coolness of the exposed dirt, but February in San Francisco is hardly hot, and it seemed to be evidence of a long, slow change in his behavior. The park commissioners declared that Monarch had not "been himself" for some time and seemed to be suffering from an extreme case of ennui. They also claimed that he might be grieving his old life as a free bear, living among other bears, and suggested that he might be at risk of dying of a broken heart.

Indeed, by 1903 this adult grizzly bear, whose natural range would have been dozens, if not hundreds, of square miles and whose diet would have included a variety of grasses, berries, rodents, grubs, fish, and occasionally big game, had lived for fourteen years inside a small metal cage and then a slightly larger but barren and barred one. The extreme change in his daily life from that of a free-living bear who hunts and forages for his meals to one of total confinement, a completely different diet, no exercise, an environment full of noisy humans, and only the occasional whiff on the wind of the park's bison herd was probably enough to change his behavior. How passersby interpreted his behavior, however, often had more to do with themselves than with the bear.

San Franciscans wandering past Monarch's cage in the park, reading about him in the paper, and attempting to make sense of his blank stare were also changing, or at least the world around them was. In the years leading up to Monarch's capture and throughout his early years in captivity, vast numbers of new roads, canals, railroads, and steamships had been built. Whitney's cotton gin and other recent inventions revolutionized agriculture. More Americans lived in cities than in rural areas for the first time in U.S. history, and those cities were proving to be disturbing places, brought to life in books like Upton Sinclair's *The Jungle*, published in 1906. Western wilderness had given way to pasture, farms, grazing land, and bigger towns and cities; the buffalo had disappeared, packs of wolves thinned, and animals like the California grizzly went extinct. The nation's native peoples, their populations decimated, were being moved off their remaining lands and could no longer stop development, mining, logging, agriculture, or the acquisition of grazing country.

These vast changes, along with intensive mining and logging and the growth of manufacturing, helped make the United States an economic powerhouse by the end of the nineteenth century. In 1896, seven years into Monarch's stint in San Francisco, the historian Frederick Jackson Turner announced the closing of the American frontier. He argued that the frontier had not only made America different, it had made it better.

The country's wildlands and wildlife had long been a badge of nationalistic pride to people like Thomas Jefferson, but it wasn't until the 1880s and 1890s that a growing number of Americans began to feel that this source of pride might be in need of protection. While Monarch sat with his head on his paws in Golden Gate Park, John Muir traveled the state's mountainous spine and founded the Sierra

Club. Many people joined the newly formed Audubon Society, and Turner, Roosevelt, Muir, Gifford Pinchot and others lamented the loss of the country's wild places and the possible effect this loss would have on the national character, a largely white and masculine affair. A flurry of new national parks were established, from Glacier National Park to Yosemite.

Nostalgia for the passing frontier inspired men who could afford it to turn to recreational camping, hunting, and other outdoor pursuits on estates in the Adirondacks or with guides on the Great Plains. In 1910, Monarch's last full year in the park, Ernest Thompson Seton helped found the Boy Scouts of America in order to teach young boys frontiersmen skills and keep them from becoming too citified. Wealthy urban tourists and sportsmen visited the new national parks, complete with fancy lodges, resort towns, game wardens, and park rangers. Ideas of the American West became increasingly idyllic.

In order for this to happen, however, its history was sanitized. Places like Yosemite and Yellowstone could now be seen as antidotes to increasingly unhealthy and dirty cities only because the wilderness was no longer a place of war or suffused with animal predators. Wildlands could now be a place of renewal, at least for people of means. Efforts to protect and celebrate these places were an attempt, in some sense, to protect the origin myth of the United States and the individualistic frontiersman who brought it into being. There was no better figurehead for the contradictions inherent in this new idea of wilderness than Monarch. A once ferocious creature who could easily have eaten the average human visitor to Golden Gate Park, Monarch was now caged. His great bulk was reduced, his claws curled with disuse. Visiting him was entertainment, cheaper than a trip to a resort lodge in Montana. Since grizzlies were no longer a threat to Californians, Monarch could be a nostalgic, last-of-his-kind figure. The grizzly could be pitied. Noting his apathy, listlessness, and seeming sadness, the park commissioners ordered the capture of a female bear as a mate for the old mascot.

Unfortunately there were no grizzlies left to capture in California, so a female bear was trapped in Idaho. When her shipping crate was unloaded into the enclosure next to Monarch's, he stood up, tore at the ground, and sniffed at the air. One observer said that the new female "typified the saying 'cross as a bear.' She was ferocious and objected to photographers. . . . Perhaps old Monarch would better have gone on suffering from ennui after all." Eventually Monarch and the Idahoan grizzly got along quite well. They mated, and just before Christmas in 1904, two cubs were born.

And yet the tinge of mental ill health associated with Monarch did not quite dissipate with the jubilant arrival of his "wife," Montana, as she was called in the press, or the successful birth of their cubs. Just as people had seen humanlike ennui in Monarch, when one of the cubs died after only three days, the *Chronicle* now saw abusive behavior in his slatternly mate: "The poor little cub died from a combination of neglect and disgust for life." It was a melodramatic story of an unnatural, selfish, and negligent mother who refused the baby any care or comfort. In actuality, however, the park officials who tried to rouse Montana to a sense of responsibility for her tiny, furry offspring had taken the cub from her. When he sickened and died, they reported that he had "fallen out of humor with existence."

Monarch garnered less and less attention as the years wore on, with one exception. In the aftermath of the 1906 earthquake and the raging fires that reduced the city to smoldering rubble, a poster with an artist's rendering of the bear, perched Godzilla-like atop the ruins of San Francisco with an arrow in his back and a snarl on his lips, urged residents to stay strong and rebuild their ravaged city.

Four years later Monarch was confirmed as the sole California grizzly still alive, captive or free, though he wouldn't be for long. In 1911, after twenty-two years in captivity, Monarch was deemed decrepit by park officials and euthanized. It was announced that his skin would be on display at the park museum in time for Labor Day. His skeleton, minus much of his skull, which was included in his mount, was buried nearby. Later it was dug up, cleaned, and given to the Museum of Vertebrate Zoology at the University of California, Berkeley, where it remains today. Like John Daniel, Tip, and countless other animals before him, Monarch had become a specimen. Unlike John Daniel and Tip, however, Monarch had lived so long due to his own fortitude, good health, determination, and luck. He hadn't thrived, exactly, but he had survived.

One of my favorite books as a child was E. B. White's *Charlotte's Web*. The ranch where I grew up was home to Mac, the surly miniature donkey, but also barn cats, chickens, the occasional goat, a few rabbits, regular-size donkeys, and a pony named Midnight. I was convinced for much longer than I probably should have been that the animals gossiped and argued with one another whenever I was out of earshot. If I could appear slowly and silently enough alongside the donkey corral or behind the chicken coop, I thought I might catch them at it. I never did. But I was sure that if I had, I would have heard something like Wilbur, the talking pig, discussing heartbreak.

In a dramatic scene toward the end of the book, Wilbur is attempting to save the eggs of his dying friend Charlotte, a spider. He implores Templeton, the selfish rat antihero, to scurry up to the roof and rescue her egg sack.

> "Templeton," said Wilbur in desperation, "if you don't stop talking and get busy all will be lost, and I will die of a broken heart. Please climb up!"
>
> Templeton lay back in the straw. Lazily he placed his forepaws behind his head and crossed his knees, in an attitude of complete relaxation.
>
> "Die of a broken heart," he mimicked. "How touching!"

Even fictional animals had begun to get a bit snarky about the idea.

I can't imagine what the veterinary behaviorist would have said if, over Oliver's sore and bruised body after his jump from our apartment, I had asked how to treat his broken heart. The possible headline: "Rejected Lovesick Dog Jumps from Building to Search for Family; Pines Away for Months; Barely Survives 50-Foot Fall; Owners Despair over Vet Bills."

Strangely enough, long after the first animals were said to be dying of heartbreak, the idea has stubbornly refused to go away. It still pops up from time to time, as a way of explaining mysterious animal deaths. And while most veterinarians would perhaps not choose to write "heartbreak" in their patients' charts, tales of animals dying of exactly that exist alongside accounts of animals suffering from more modern afflictions like depression or generalized mood disorders.

In 2010, two elderly male otters who had been inseparable for fifteen years died within an hour of each other at a New Zealand zoo. Only one had been ill. Their keepers believed that the second otter died of a broken heart. The ethologist Marc Bekoff writes of animal heartbreak too. In *The Emotional Lives of Animals* he tells the story of a Miniature Schnauzer named Pepsi that a veterinarian gave his father as a gift. The dog and the elderly man became extremely close, sharing the same food, chair, and bed for years. When he was eighty years old, the father committed suicide. Pepsi grew weak and withdrawn; he never recovered after the death of his companion and eventually died. The veterinarian was convinced it was due to a broken heart; that is, the dog had lost the will to live after his human was gone.

In March 2011 another heartbreak story went pinging about the Internet. A British soldier with the Royal Army Veterinary Corps, Lance Corporal Tasker, was killed in a firefight in Helmand, Afghanistan. His dog, Theo, a Springer Spaniel mix who was trained to sniff out explosives, watched the whole thing. Theo wasn't injured in the firefight, but hours after Tasker died, she suffered a fatal seizure, brought on, according to witnesses, by stress and grief over the loss of her companion.

These contemporary stories, like the earlier tales of animal heartbreak, are as much about the humans telling them as the animals themselves. We imagine ourselves inside the dog's or otter's head and heart. We make sense of their behavior by seeing our feelings and fears reflected in *them.* This is surely a kind of anthropomorphism, but it can also be a valid one. As humans, we can imagine fading and dying after we lose someone dear to us. Most of us know someone this has happened to.

In the early 2000s, the UCLA cardiologist Barbara Natterson-Horowitz encountered her first cases of Takotsubo cardiomyopathy, a newly identified syndrome characterized by crushing chest pain and a severely abnormal EKG. She rushed her patients to the operating suite for an angiogram expecting to see blood clots or signs of heart disease, but there was nothing obstructing the coronary arteries. These men and women weren't having heart attacks. The only abnormalities in their hearts were odd, lightbulb-shaped bulges in the left ventricle that kept the organs from making strong contractions.

Japanese cardiologists named the syndrome in the mid-1990s; the bulbous tissue reminded them not of lightbulbs but of octopus pots, the round ceramic *takotsubo* that Japanese fishermen use to catch the cephalopods. The flabby, ballooning area of heart muscle makes the ventricle contract unpredictably and weakly, pumping blood in intermittent spasms. This is the source of the chest pain in people who arrive in emergency rooms after coming down with sudden heart trouble. What surprised Natterson-Horowitz, however, was that Takotsubo wasn't brought on by cardiac disease or congenital defects but by acute stress and emotional pain. Patients appeared at the hospital suffering from weak contractions after seeing a loved one die, before being sent off to prison, after losing a lifetime of savings, or having survived an earthquake. In *Zoobiquity,* the book she coauthored with journalist Kathryn Bowers, they write that this new diagnosis was proof of the powerful connection between mind and heart health, confirming a causal relationship that many doctors thought to be "more metaphoric than diagnostic." She and Bowers point to a few fascinating public health statistics, such as the increased rates of heart failure among Israelis fearful of Scud missile strikes during the 1991 Gulf War—statistics that suggest that panic and dread of the strikes may have killed more people than the missiles themselves.

In the days surrounding the Al-Qaeda attacks of 9/11 there was a 200 percent increase in the number of life-threatening heart rhythms in American patients with implanted tracking devices. And in 1998, when England lost the World Cup to Argentina in a suspenseful last-minute penalty kick, heart attacks across the United Kingdom spiked by 25 percent in a single day. Since then, a number of other European studies have corroborated the relationship between spectactor stress and heart health. Ironically, games that end in "sudden death" shootouts seem to be particularly dangerous for fans.

In the spring of 2005, Natterson-Horowitz was called to the Los Angeles Zoo by the chief veterinarian to consult on an emperor tamarin named Spitzbuben with heart failure. The tiny monkeys all have dramatic, white Fu Manchu mustaches that make even the young females look like wise old men. Natterson-Horowitz was excited to meet one and made eye contact, trying to calm the primate as she would a human patient. The veterinarian stopped her, warning that she might give the tamarin "capture myopathy," killing her before they even had time to intubate. When animals, particularly high-strung prey animals such as deer, rodents, birds, and small primates like Spitzbuben find themselves caught in the teeth of a predator, tangled in a hunter's trap, or eye to eye with a veterinarian, for that matter, they're flooded with adrenaline and other stress hormones. The overflow of these hormones is so powerful it can injure the pumping chambers of the heart; the contractions become so weak that blood stops circulating, and the animal can die. Capture myopathy was first identified by hunters more than a century ago. Big game such as zebras or moose sometimes died after a long chase even though the hunters hadn't actually hit their targets. Since then, sudden death among terrified animals has been observed in every corner of kingdom Animalia—from Norway lobsters trawled from the ocean floor to wild mustangs terrorized by the Bureau of Land Management's helicopter roundups, to a mid-1990s concert by the Royal Danish Orchestra, whose version of Wagner's *Tännhauser* in a Copenhagen park caused a captive six-year-old okapi within hearing distance to anxiously pace, try to escape her pen, and then die. Her veterinarians cited capture myopathy.

Looking at the different ways people have described emotional well-being and illness over the years offers something of a parallel history of how we've made sense of our own minds and hearts. They not only expose the futility of attempting to separate emotional trauma from physiology but also the impossibility of dissociating disease from history. Where earlier generations saw madness, homesickness, nostalgia, and heartbreak, veterinarians and physicians now see anxiety and mood disorders, obsessive compulsivity, depression, and capture myopathy. Similarly, debilitating fears of horse-drawn fire trucks or flickering gas lamps don't scare many people or their pets today, but perhaps they once did.

Chapter Five

Animal Pharm

> Life is no way to treat an animal.
>
> Kurt Vonnegut

When the reality-TV star Anna Nicole Smith overdosed, in 2007, on a mix of prescription drugs that included Prozac, her dog Sugarpie was taking Prozac too. So was the former French president Jacques Chirac's normally friendly Maltese and Bichon Frise mix, Sumo. The small, white dog accompanied him everywhere when he was in office and was often spotted on Chirac's lap as they were chauffeured through Paris in the back of a shiny presidential Citroën. But after Chirac lost the presidency to Nicolas Sarkozy and the family was forced to leave the palace, Sumo lost his appetite. He was also lethargic and not acting like himself. Chirac's wife, Bernadette, believed that Sumo, used to the large gardens of the Élysée, couldn't adjust to their postpresidential life in a spacious apartment and seemed anxious and depressed. Their vet put the dog on Prozac, but he twice bit the ex-president, hard enough to send Chirac to the doctor. Finally, Sumo was sent away to live on a farm in the countryside with family friends; they say he hasn't bitten anyone since.

Sugarpie and Sumo are not anomalies. Prozac Nation has been offering citizenship to nonhumans for decades. Fluoxetine, or generic Prozac, is available in a dizzying array of forms and flavors that call to mind a snow-cone stand at a carnival for animals. Pet owners and veterinarians can choose fluoxetine flavored like anchovy, apple and molasses, banana and marshmallow, beef, bubble gum, butterscotch, cherry-vanilla, chicken, chocolate mint, double beef, double chicken, double fish, double grape, double liver, double marshmallow, double molasses, orange, peanut butter, peppermint, piña colada, raspberry, strawberry, tutti frutti, watermelon, wintergreen, and—as if double fish isn't enough—triple fish. It's deliverable in equally bizarre forms, from chewable pills marketed as Gourmeds to injections, drops, and transdermal gels for animals who can't or won't swallow pills.

What is surprising about all of this is not that we are giving animals psychoactive compounds, it's that we are doing it to help them cope with *us*, closing a loop of pharmaceutical development that began in nonhuman animals in the 1950s, transitioned to millions of people around the world, and has now returned to certain animal species. The dosing of other creatures with psychopharmaceuticals also serves as a sort of tacit acknowledgment of emotional (and neurochemical) parallels between humans and other animals. You could argue, and people like the behaviorist Nick Dodman have, that this is not the story of animals taking human drugs but of humans taking *animal* drugs. Almost all of the contemporary psychopharmaceuticals—from antipsychotics like Thorazine to minor tranquilizers like Valium to the antidepressants—were developed in the mid-twentieth century, and animals were test subjects from the very beginning.

Executive Monkeys and the Making of Miltown, Xanax, Valium, and the Antipsychotics

Most small-animal doctoring was, until the turn of the twentieth century, performed by the owners of pet shops. People tended to give their pets over-the-counter medications and the same invalid folk remedies (rice gruels, beef stews) that they used to treat themselves. Milk of magnesia and castor oil, for example, were used for both canine and human constipation. Cough syrup and chamomile steam baths were used to treat respiratory infections in people and other creatures. By 1910 dog-specific medicines for mange or worms could be bought at feed stores and neighborhood drug stores. Cats simply received smaller amounts of the same compounds. Some pet owners concocted medicines at home with the help of recipe books that included things like asthma treatments for canaries.

It was not until the 1950s, when a new, transformative class of drugs became available, that nonhuman animals were ushered into the psychopharmaceutical age along with, and in some cases before, humans. Throughout the 1950s and 1960s experimental animals played integral roles in the making of the new psychopharmaceuticals. Monkeys, rats, mice, and cats were important human proxies in the search for nonsedative solutions to anxiety, psychosis, and other problems of the mind. The animals' emotional and behavorial responses to the new drugs also helped to define the disorders themselves.

In May 1950 Henry Hoyt and Frank Berger, researchers at a small pharmaceutical company in New Jersey, submitted a patent application for a substance called meprobamate. They were impressed with the way the drug relaxed muscles in mice and calmed their notoriously testy lab monkeys: "We had about twenty rhesus and java monkeys. They're vicious, and you've got to wear thick gloves and a face guard when you handle them. After they were injected with meprobamate though, they became very nice monkeys—friendly and alert. Where they wouldn't previously eat in the presence of human beings, they now took grapes from your bare hand. It was quite impressive." The drug caused such relaxation in the monkeys that it prompted researchers to wonder if meprobamate might be a productive complement to psychoanalysis in people.

At the same time, another company was relaxing rats. In the late forties and fifties, pharmacists working for the French company

Rhône-Poulenc were developing antihistamines. In 1951, a company pharmacist screened one of these new drugs, called *chlorpromazine*, for behavioral affects. (Previously compounds were screened for toxicity, not necessarily for how they made people or other animals *behave.)* Rats were given the antihistamine and put into a cage with a platform that had food on it. To reach the platform the rats simply had to climb a rope. If they didn't climb the rope, they would receive a shock. The drugged rats didn't climb the rope, even when they learned that the shock was coming. What was interesting to the Rhône-Poulenc researchers was that the rats seemed totally indifferent: they weren't concerned with the shock *or* the food. And it wasn't because they were sedated or uncoordinated; they were wide awake and physically unimpaired.

The rats' indifference piqued the curiosity of other researchers in Switzerland, Canada, and the United States, and soon the new drug was being tested in cardiac surgery, on battlefields, and in psychiatric practices as a light sedative for humans. It was back in France, however, that the drug was poised to drastically change psychiatry. At Sainte-Anne Hospital in Paris in the early 1950s, doctors began giving chlorpromazine to patients with delirium, mania, confusion, and psychosis. The drug didn't sedate these people or put them to sleep as other sedatives had done. Instead, patients on chlorpromazine were aware and, like the rats, indifferent to the outside world but could engage with it when needed. The drug also caused certain patients at Sainte-Anne, and soon many other mental hospitals, to awaken from catatonic states that they had been lost in for years.

A barber from Lyon, France, was a typical case. He had been hospitalized for psychosis for years, and was completely unresponsive to people and activity around him. After a few doses of chlorpromazine he woke from his stupor and told his doctor that he knew where he was, who he was, and that he wanted to go home. Then he asked for a straight razor, water, and towels and shaved himself perfectly. Another patient, who had been frozen in a series of odd postures for years, responded to the drug in a single day. He greeted the hospital staff and asked them for billiard balls, which he then juggled. It turned out that he had worked as a juggler before being institutionalized.

In 1954 Rhône-Poulenc sold the U.S. chlorpromazine license to Smith Kline, which named the drug Thorazine. It was marketed as an antinausea agent, but by then, everyone knew of its remarkable results with psychotic patients. The market for the new drug was mind-boggling, generating $75 million in sales in its first year. At many state asylums every single patient received it, and a slew of outpatient clinics were established to care for all of the newly released mental patients now able to live on their own.

Veterinarians and researchers soon began giving antipsychotics to a few of their animal patients. One 1968 journal article summed up the veterinary use of the new psychopharmacology as helpful for treating nervousness, anxiety, fear, fright, conflict, panic, viciousness, agitation, excitement, and also as a relaxation agent before the collection of semen for breeding purposes. "Litter-savaging" pigs—pigs who cannibalized piglets—also reportedly responded well to chlorpromazine. Soon other new antipsychotic drugs, such as reserpine, were being used to treat cannibalistic chickens and pheasants, along with view-restricting plastic spectacles (which ostensibly were to be strapped onto the chickens' small heads to disorient them or make the rest of the flock look less appetizing).

A year after Smith Kline bought the chlorpromazine license, Hoyt and Berger made a film about their own new drug, now called Miltown. The film, *The Effect of Meprobamate (Miltown) on Animal Behavior,* featured rhesus monkeys in three different states: their vicious, sober selves; totally unconscious on barbiturates; and calm but awake on meprobamate. They screened it in April 1955 at a meeting of the Federation of American Societies for Experimental Biology in San Francisco. It excited the audience as well as executives at Wyeth Laboratories, who offered to purchase the drug from Hoyt and

Berger. They still wanted a name for this new class of compounds, and at dinner one evening Berger complained to two of his friends that he didn't know what to call this new kind of drug. "The world doesn't need sedatives," one of his friends said. "The world needs tranquility. Why don't you call this a tranquilizer?"

Meanwhile scientists at Walter Reed Army Institute's Neuropsychiatric Division were conducting a second set of animal experiments that would truly usher in the tranquilizer era. Researchers restrained two monkeys on opposite sides of the same cage and shocked their feet every twenty seconds. One was deemed the "executive monkey" and could protect both monkeys from the shocks if he pressed a lever at his side every twenty seconds. The monkey with the responsibility to protect himself and his cage mate quickly learned that he was capable of saving them both from the shocks, but over the course of repeated experiments, he became more and more agitated and finally died. When the scientists gave the executive monkey tranquilizers, he pressed the lever more calmly and with more success. The researchers came to the conclusion that what worked for responsibility-laden monkeys might also work for responsibility-laden men (though they argued about who was more stressed: male newspaper reporters, for example, or businessmen). Drug companies jumped on their study results, and soon Roche Laboratories was even suggesting, via promotional films like *The Relaxed Wife,* that tranquilizers would help breadwinners and their families enjoy one another more after a long workday.

The 1950s was a key decade in the forging of new links between men, women, and the psychopharmaceutical industry, a circle that would eventually widen to include pets and zoo animals. Articles about pharmaceutical miracle cures dotted the pages of *Newsweek, Time, Cosmopolitan,* and *Ladies' Home Journal.* The authors suggested that women's infidelity, frigidity, or uncertainty could be fixed by taking a pill. The historian Jonathan Metzl has argued that the popularity of psychoanalysis throughout the 1950s contributed to popular notions that women's mental health directly affected men, particularly that psychiatric symptoms were the result of early life experiences people had with their mothers. Pills could help women be kinder to their sons and husbands. As a result, the people most often implicated as needing tranquilizers in the popular press were unmarried women, loose women, women who wished to keep their wartime jobs, and women who rejected their husbands' sexual advances.

Pharmaceutical marketing pitches built on earlier notions that single women, lesbians, and inconveniently opinionated women were pathological. Hysteria had been the most common diagnosis, but by the 1940s, American single women, working women, and those who chose not to be mothers were pathologized in books like *Modern Women: The Lost Sex*, which argued that women who wanted to leave the home were deeply ill. When tranquilizers arrived, these dangerous states could be resolved with a pill. The drugs promised to be a new form of behavioral control, and would soon come to be prescribed back to the animals the drugs were first tested on.

Before the mid-1950s, talk therapy, not drugs, was the standard for helping emotionally distraught patients cope with anxiety. With the creation and rapid diffusion of meprobamate, this approach shifted slightly. Many psychoanalysts had already been trying to understand the possible biological causes of psychiatric problems (something that interested Freud as well). When Miltown proved to facilitate the talking cure instead of replacing it, the drug only spurred interest in the new biological psychiatry. Even the *Physicians' Desk Reference* stated that the drug could make a patient alert and more amenable to psychotherapy. Miltown went to market in 1955 and became the fastest-selling drug in U.S. history. Physicians' reference manuals published by pharmaceutical companies emphasized the risks of untreated anxiety among businessmen. Roche Laboratories' manual *Aspects of Anxiety* gravely cautioned that male breadwinners suffered the effects of unchecked anxiety, arguing that workplace stressors were very hard to change: "The physician must attempt to change the patient's outlook on life and attitudes to his worth through 'pharmacotherapy.'"

A mere two years after Miltown was released, one survey of American executives found that a third of respondents were using tranquilizers. Half were habitual users and almost three-fourths claimed that the drugs helped their job performance. Americans in the 1950s greeted the pill with curiosity and excitement. Patients who heard about Miltown, now nicknamed "emotional aspirins" and "peace pills," asked doctors for the drug directly. The sensation that began with monkeys and executives soon drifted to Hollywood. Lucille Ball took it in her coffee on the set after a fight with Desi Arnaz. Lauren Bacall was given a prescription after the death of her husband, Humphrey Bogart. Tennessee Williams took it while working on *Night of the Iguana.* Soon everyone else wanted it too. Even the military: between 1958 and 1960 the U.S. armed forces spent millions of dollars on the drug, and dosed everybody from air force pilots to patients at VA hospitals. Psychiatrists self-prescribed it. Athletes took it. President John F. Kennedy took it, too, for his anxiety and colitis. Children were given the drug for agitation. A few dogs were also given the drug to treat car sickness, hyperexcitability, viciousness, and shyness. By 1957 more than 36 million Miltown prescriptions had been filled and a billion tablets manufactured. Tranquilizers accounted for one-third of all prescriptions in the United States. As Metzl argues, the drug was active in redefining the very idea of what anxiety was and who could suffer from it.

Another interesting result of the Miltown craze was that, at least among scientists like Frank Berger who were developing the drugs, blame for mental distress was shifting away from Freudian theories of mother-child relationships, repressed subconscious conflicts, and flawed interpersonal relations toward biological explanations, such as problems with the limbic system. If a drug could cure anxiety, Berger later argued, it was more likely a result of physiological problems rather than past experiences someone had with their mother. The fact that these drugs seemed to also be effective in nonhuman animals may have only served to make the biological claims more valid.

By the mid-1960s, however, the addictive nature of Miltown came to light and fascination with the drug dwindled. In 1967 it was placed under abuse control amendments to the Food, Drug, and Cosmetic Act. Its impact on the pharmaceutical landscape, though, was long-lasting. The industry's success with its first lifestyle drug paved the way for benzodiazepines like Librium, Valium, and Xanax, as well as the broad range of antidepressants. As the historian Andrea Tone has argued, the popularity of Miltown demonstrated that a drug could be fashionable, and the fact that, for a while anyway, seemingly everyone took it made certain kinds of drug taking socially acceptable.

Other psychopharmaceutical drugs were also in development in nonhuman animals at the same time. In 1957 a chemist working for Hoffmann–La Roche discovered a new compound, Ro 5–0690, that caused mice to behave in surprising ways in the "inclined screen test." Mice were given an experimental drug and then placed on an angled screen with their heads facing down. Undrugged mice turned around and ran to the top of the screen; tranquilized mice tended to slide slowly to the bottom. The mice given Ro 5–0690 slid to the bottom of the screen with their muscles relaxed, but unlike tranquilized mice, they were alert and active the whole time. These mice also had no problem walking when they were prodded.

The new drug also passed the industry-wide "cat test," which involved giving a group of cats a drug and then holding them by the scruff of the neck to see what they did. Cats given Ro 5–0690 hung limply without struggling, even those who'd been specially selected for their meanness. Remarkably, these extra-difficult cats were, according to the researchers, transformed into content, sociable, and playful felines after taking Ro 5–0690. Researchers compared the cats' reactions to the same cats on Thorazine, Miltown, and phenobarbital. The new drug was comparable to Miltown, only it was even more potent. It was also less toxic and less sedating than anything else on the market. Clinical trials in people, including ethically problematic studies on prisoners, showed that Ro 5–0690 was good at reducing anxiety, agitation, and aggression. Roche named the drug Librium, from the word *equilibrium*, and released it in 1960.

This cat-mouse-human relaxant quickly became the top-selling drug in the United States, that is until Hoffman–La Roche released Valium, its second blockbuster benzodiazepine, in 1963. Valium went on to become the first $100 million pharmaceutical brand; between 1968 and 1981 it was the most widely prescribed medication in the Western world. Benzodiazepines like Valium and Librium, and the animals who first demonstrated their usefulness, made Hoffmann–La Roche into one of the most profitable businesses on earth.

Early Adopters in the Cuckoo's Nest

One of the first nonhumans to be given psychopharmaceuticals not as a test subject but as a *patient* was a gorilla named Willie B. Like the midcentury women given tranquilizers to ease their anxiety and encourage their compliance with the status quo, Willie was given the drug to ensure his good behavior and to keep him from expressing his displeasure over the strict limitations placed on his daily life.

Willie B. was a western lowland gorilla, famous in Atlanta, Georgia. Sometime in the 1960s he was captured in Congo as an infant and sent to Zoo Atlanta, where he lived for thirty-nine years, twenty-seven of them alone in an indoor cage with a tire swing and a television. Named after the city's mayor, William Berry Hartsfield, Willie was the subject of countless newspaper articles and television programs, and was the inspiration for the city's soccer team, the Atlanta Silverbacks. When he died in February of 2000, eight thousand people attended his memorial service.

According to Mel Richardson, who was working as a veterinarian at Zoo Atlanta at the time, Willie broke a glass window in his enclosure in the winter of 1970–71 and had to be transferred to a much smaller cage for six months while the glass was replaced with heavy metal bars. "He weighed around four hundred pounds, and the cage was way too small for him," said Mel. "If he stood up and stretched each arm all the way out he could almost touch both sides of the cage at once." The vet staff decided to medicate him so that the six months would be more bearable. They put Thorazine in the Coca-Cola he drank in the morning. According to Mel, Willie responded to the drug as many institutionalized humans do: he shuffled back and forth across his cage with dulled eyes. "It was a little like watching the men in *One Flew Over the Cuckoo's Nest*," Mel said, "except Willie was a gorilla."

Since then, human antipsychotics like Haldol (haloperidol) have been given to animals in zoos and aquariums throughout the world. The drugs are used to overcome phobias in birds, such as one yellow-naped Amazon parrot's fear of being held. Haldol was given to red-necked wallabies to ease their transition to captivity. A captive black bear cub was given Haldol to treat her separation anxiety after she was moved to a cage by herself. SeaWorld has dosed their performing California sea lions. Six Flags Marine World gave antipsychotics to a young female walrus who was compulsively regurgitating her food. At the Toledo Zoo Haldol was used to calm anxious Grant's zebras, a group of wildebeests, a pair of ostriches, and a swamp monkey named Maxine. The keepers hoped Maxine would get along better with her daughter. She didn't. According to the vet staff, antipsychotics did help Trouble and her sister, Double Trouble, two of the Toledo Zoo's birds of paradise. They were serious feather pluckers, but after three days on Haldol, they stopped. The drugs are "definitely a wonderful management tool," the mammal curator told a Toledo newspaper reporter. "And that's how we look at them. To be able to just take the edge off puts us a little more at ease."

The fact that these antipsychotics are often used to make captive animals more "manageable" is reminiscent of the debates about antipsychotic drugs given to the institutionalized, or the tranquilizers prescribed for 1950s housewives. When the antipsychiatry movement began to unfold in the 1960s, in the wake of newspaper exposés of abuse in insane asylums that positioned psychiatry as an exercise in "chemical straightjacketing," antipsychotics began to be seen as the root of mental illness rather than its solution. Ken Kesey portrayed the psych ward as an oppressive place where inmates shuffled about, their minds dulled by antipsychotics.

From roughly 1965 to 1975 the discipline of psychiatry changed in response to the public's growing concern about the treatment, and confinement of the mentally ill. As the historian David Healy points out, psychiatry saw many of its practices, such as electroshock therapy, demonized. Pharmaceutical companies, however, continued to advocate the use of drugs to eliminate unwanted behavior. In the 1970s and 1980s their doctor-focused ad campaigns urged the use of antipsychotics as "behavioral control" in young men to limit antisocial and violent acts. More than thirty years later, psychopharmaceuticals are still being used to control the behavior of human prisoners, to manage patients held for compulsory psychiatric treatment, and by people who feel pressure to take their medications in order to limit their own outbursts.

Behavior control, perhaps a slightly less draconian term than chemical straitjacketing, is also alive and well in the care and treatment of nonhuman animals. This doesn't mean that it's not often helpful. Antipsychotics, antidepressants, and antianxiety medications have, for example, been used to treat macaques and other primates used in research, who cannot be released from their labs for one reason or another. These monkeys, distressed beyond measure, biting themselves and feeling despondent, are given antipsychotic or antianxiety drugs because it's more compassionate than not drugging them. In another case, an easily agitated male gorilla in Ohio, who became upset when any gorilla in his troop was tranquilized for surgery or other medical treatment, was given Valium beforehand; it sedated him but didn't stop his nervous diarrhea. These animals' caretakers determined that prescribing rather than denying psychopharmaceuticals was their only option to alleviate the animals' suffering.

At the Guadalajara Zoo in Mexico a sixteen-year-old female gorilla stopped eating, started vomiting, and developed a bad case of diarrhea. According to the veterinary staff, she also seemed depressed. After a fecal sample showed a salmonella infection, the keepers separated her from her baby and the troop so that she could be treated and not infect the others. For ten days she was kept by herself while the salmonella infection cleared up. On the tenth day she began biting her fingers and toes for hours at a time, until they bled. Even after she had been reunited with her troop and baby, she continued to gnaw on herself until she'd made deep wounds. The vet staff started her on Haldol. According to her keepers, the gorilla eventually stopped the biting, and the staff began to taper her antipsychotic dose. Six months later she was off Haldol entirely and still wasn't hurting herself.

After their experiences treating Gigi the gorilla at the zoo in Boston, the veterianarian Hayley Murphy and the pyschiatrist Mufson were curious about the use of psychopharmaceuticals in other captive gorillas, so they surveyed all U.S. and Canadian zoos with gorillas in their collections. Nearly half of the thirty-one institutions that responded had given psychopharmaceutical drugs to their gorillas. The most frequently prescribed were Haldol (haloperidol) and Valium (diazepam), though Klonopin, Zoloft, Paxil, Xanax, Buspar, Prozac, Ativan, Versed, and Mellaril had all been tried.

Mufson keeps photos of the Boston gorilla troop on his desk alongside pictures of his wife and children and, every year, he brings medical students on psychiatry rotations to the zoo to see the apes. Since he first began working with Gigi, Mufson has treated a number of gorillas in other American zoos for problems like trichotillomania and coprophagia (feces eating). For the plucking he prescribes Luvox or Celexa, as he does for his human patients, and at the same dosage, milligram for kilogram.

Sometimes apes are dosed so they can deal with the stress of travel. In 1996, a male Western lowland gorilla named Vip was being flown from Boston, where the females found him sexually unappealling, to the Woodland Park Zoo in Seattle, part of an AZA program to manage genetic diversity in the captive gorilla population. Vip was tranquilized and his keepers began to load him into a crate. Vip wanted nothing to do with the crate and the process took a lot longer than the staff had anticipated. Eventually, he was loaded onto a passenger jet at Boston's Logan Airport and his keeper, Shanna Abeles, took her seat in the cabin. A few hours later, somewhere above the Midwest, Vip's sedatives wore off. He woke up in the cold dark of the cargo hold, terrified and confused, without any idea where he was. Abeles could hear him beating his chest and bellowing inside the cabin. The pilot could too and it terrified him. By the time the plane was over Utah, the pilot, nervous the gorilla may have gotten free, made an unscheduled landing in Salt Lake City. After a short layover to check on Vip, the plane taxied back out onto the runway, but when the pilot heard Vip's loud banging in the cargo hold, yet again he brought the plane back to the terminal. He would fly no farther with Vip. The gorilla and his keeper were unloaded on the tarmac. To calm Vip down, Abeles fed him a banana laced with crushed Valium and they waited for a truck to take them the rest of the way to Seattle. Vip was the last gorilla to be allowed on a commercial airplane. Ever since, they are shipped by FedEx.

Dolphins, whales, sea lions, walruses, and other marine creatures in parks like SeaWorld have also been given psychotropic drugs for what their vets see as depression, anxiety, compulsive regurgitation, flank sucking, or other distressing behaviors. The drugs have been given to dolphins, too, so that they can be flown, like Vip, to new aquariums or amusement parks without becoming upset.

There are incentives at these facilities to keep such information quiet, particularly in the wake of tragedies like the 2010 killing of the trainer Dawn Brancheau at SeaWorld Orlando by Tilikum the orca. The whales' motivation for the grisly attack has been attributed to the extreme stressors inherent in keeping such an immense and social predator in crushing confines, away from his family. His offspring, products of captive breeding, also seem disturbed. SeaWorld treated two of Tilikum's sons with antianxiety medications after they exhibited aggression themselves. One tried to mate with a nine-day-old calf. The calf's mother was also given diazepam while nursing, a contraindicated use. Other mother orcas may be given antianxiety medications after the forced removal of their calves.

Giving animals psychotropic drugs to treat signs of mental illness, even if it has become common practice among humans, may invite unwanted criticism of the industry. Many marine mammal trainers and zookeepers have signed nondisclosure agreements with their employers, and a thicket of public relations protocols cocoon animal-care staff from the public. In many cases I wasn't able to secure permission from public relations departments to talk about the use of antidepressants, antipsychotics, and antianxiety and antiobsessional medications in their animal collections. Finding out that the gorillas, badgers, giraffes, belugas, or wallabies on the other side of the glass are taking Valium, Prozac, or antipsychotics to deal with their lives as display animals is not exactly heart-warming news for most people who go to zoos, theme parks, and aquariums. Two marine mammal veterinarians who have spent decades on staff or consulting for American animal-display facilities, the military's marine mammal program, and doing research told me that antidepressants and antipsychotics are commonly used but that "no one was going to talk to [me] about it." Even they wouldn't speak about the subject on the record.

A few of the published cases include a four-year-old beluga whale at Shedd Aquarium who was given antidepressants to treat her compulsive regurgitation. She began to throw up whole fish after her training sessions, to the point that she was losing a dangerous amount of weight. The aquarium veterinarian prescribed an antidepressant that seemed to reduce the frequency of her regurgitation. The young whale was kept on a maintenance dose even after she gained weight.

The word *antidepressant* was coined in 1952 by the psychologist Max Lurie, but both the term and the substances it referred to took time to catch on. From roughly 1900 through 1980, depression was considered a rare disorder, as opposed to nervousness or anxiety. In Europe before the 1950s depressive disorders were understood as melancholias. People with severe depressive personality disorders were admitted to hospitals at a rate of 50 to 100 cases per million people. In 2002 depressive disorders affected 100,000 per million people; 250,000 more reported depressive symptoms. The historian Edward Shorter has argued that the cause of the thousandfold increase in cases are the antidepressant drugs themselves. That is, the idea of depression didn't become common until there were antidepressant drugs that seemed to treat it.

Antidepressants, particularly Prozac, appeared in 1990s popular culture and were often positioned as making those who took them better-than-well. Suddenly serotonin was a conversation topic and the benefits of the new drugs were debated in books like *Listening to Prozac.* Other animals were part of the conversations about the new drugs too and, as it happened with antipsychotics and antianxiety meds, many of the creatures frequently given antidepressants were primates. A human psychiatrist prescribed Remeron for Minyak, a male orangutan at the Los Angeles Zoo, in the wake of a respiratory infection that made him too listless and lethargic to mate. The chief veterinarian at the zoo thought Minyak was depressed. The drug stimulated his appetite for both sex and food. Two years after Minyak fathered a baby, however, he was still being weaned off the drug.

Johari, a female gorilla at the Toledo Zoo, was prescribed fluoxetine, or generic Prozac, for what keepers understood as her premenstrual symptoms. Her keepers tracked the number of injuries Johari inflicted on other troop members against her menstrual cycle and found that she was most likely to be aggressive the week before her period. After a month on antidepressants her violent episodes ceased. When Johari later became pregnant the zoo staff hoped that the hormonal changes associated with pregnancy and nursing would reduce her PMS. They took her off Prozac, and according to her keeper, "She got kind of psychotic on us." So she was put back on the meds.

When tabloids broke the story in the mid-1990s that Gus, one of the polar bears in the Central Park Zoo, was compulsively swimming figure eights in his pool for up to twelve hours a day, every day, for months, and the zoo had paid a behaviorist $25,000 to help him, something of a Gus moment took hold of the city. The bear was on the cover of *Newsday*, Letterman cracked jokes about him, the *New York Times* ran an editorial, political cartoons featuring "bipolar" bears appeared in American papers, and the Canadian band The Tragically Hip wrote a song called "What's Troubling Gus?" While much of this was tongue-in-cheek, Gus was also a symbol of his times. Bipolar disorder came into vogue in the 1990s. The frequency of cases went up, and the threshold for presumed age of onset dropped to such an extent that one- and two-year-old children were suddenly being diagnosed with bipolar disorders and treated with mood stabilizing drugs.

The zoo's public affairs manager said that Gus's story was so captivating because "it's like Woody Allen always being in therapy—the idea that all New Yorkers are neurotic." In the wake of the news coverage, people called in from around the country to ask how the bear was doing. The answer was complicated. Gus lived in a 5,000-square-foot enclosure—less than .00009 percent of what his range in the Arctic would be. He was also a major predator who, despite being born in captivity, no doubt still felt predatory impulses. In fact when Gus first arrived from an Ohio zoo in 1988, his favorite game was stalking children from the underwater window in his pool. "He liked to see them scream and run in terror—it was a game," the zoo's animal supervisor told a reporter. But the zoo staff didn't want Gus to scare children or their parents, so they put up barriers to keep visitors farther away from the window. Gus soon started to swim in endless figure eights.

Hoping to curb the neurotic behavior, the zoo hired Tim Desmond, an animal trainer who had trained the orca who played Willy in the film *Free Willy*. Desmond was able to reduce Gus's compulsions by giving him new things to do, such as bear food puzzles or snacks that took him longer to eat: mackerel frozen in blocks of ice or chicken wrapped in rawhide. The zoo redesigned his exhibit and installed a play area stocked with rubber trash cans and traffic cones that Gus could pretend-maul. They also reportedly put him on Prozac. I do not know how long he was on the drug, or even if it was as effective as his new exhibit and entertainment schedule, but eventually Gus's compulsive swimming tapered off, though it never went away entirely.

In August 2013, Gus was euthanized. At twenty-seven years old, he was an old bear with an inoperable tumor. On my last visit to see him, he hadn't been neurotically swimming. Instead he was tearing open a brown paper bag of meat as if it were bear takeout.

Because it's impossible to replicate even a slim fraction of the kind of life polar bears have in the wild, the animals may be among the most common zoo dwellers to be given antidepressants. Other bears, though, have taken them too.

Abdi is a male brown bear *(Ursus arctos)* who was born in the Kure Mountains of Turkey in the middle of winter in 1992. Hunters shot his mother when he was a small cub at her side and kept Abdi as a pet. For two years he was left outside on a short chain, with no shelter from the sun, rain, or cold of winter. Eventually Abdi was moved to a concrete-floored cage inside a hut where, for the next eight years, he saw light only through cracks in the roof. Villagers threw food to him from a dark hole but never cleaned his cage or let him out; Abdi was thin and infected with parasites, and his coat was dull and balding in patches. After spending more than a decade under these conditions, the Karacabey Bear Sanctuary rescued Abdi and moved him to an indoor-outdoor area at their facility. When he'd lived at the sanctuary for a month, the staff began to urge him to spend time with the other bears in an attempt to socialize him, but the sight of them terrified Abdi so much that he wouldn't even leave his den. The staff moved Abdi to a smaller enclosure where he could see the other bears but have no physical contact with them. After six months he had gained weight and his coat was thick and bearlike again, but Abdi was still desperately frightened of other bears. Even more worrisome to his keepers, he paced incessantly. From time to time the staff would bring another bear over to Abdi to introduce them, but he wouldn't stop his pacing, not even to acknowledge the other creature. Over time Abdi's pacing slackened a bit, but he was still spending most of his day walking in tight circles. The staff decided to give him fluoxetine, hoping that the antidepressant would lighten his spirits and help him adjust to his new life. Every morning for six months he was given the drug hidden inside his favorite food, raisin nut bread. Slowly, over many months, his pacing stopped entirely. It took a few weeks to wean him off the drug, and then the sanctuary staff released Abdi into the large enclosure with twenty-eight other bears, something that would have scared him to death less than a year earlier.

Today, more than a decade later, Abdi is doing very well. As cliché as it sounds, he's a curious bear. A recent photo shows him at the edge of a pond intently investigating a fallen log. The sanctuary staff wrote me to say, "Getting over such great trauma was not easy for Abdi, of course. He couldn't look at others for a long time. Maybe he didn't want to, maybe he was afraid. He chose loneliness. Only after a long and successful socialization process could he understand that he was one of the others. He is now a part of the group, although we can never totally erase the memories of the past."

Pet Pharm

The biggest animal consumers of psychotropic drugs are not zoo or sanctuary animals but the ones that live with us most intimately: our pets. Just like the shared rice-gruel remedies of the early twentieth century, we are now giving our cats, dogs, and canaries the same medications we take ourselves. A survey of prescription drug trends among 2.5 million insured Americans from 2001 to 2010, found that one in five adults is currently taking at least one psychiatric drug. Americans spent more than $16 billion on antipsychotics, $11 billion on antidepressants and $7 billion on drugs to treat attention-deficit hyperactivity disorder (ADHD) in 2010. And according to a recent study by the Centers for Disease Control, 87 percent of people who visit a psychiatrist's office leave with a prescription.

The Prozac prescriptions for Anna Nicole Smith's Sugarpie, Jacques Chirac's Sumo, and most recently, Lena Dunham's rescue dog, Lamby, are indicative of a thriving marketplace for animal pharmaceuticals—psychopharm and otherwise. The U.S. market for pet pharmaceuticals is large and growing, from $6.68 billion in 2011 to a projected $9.25 billion by 2015. Zoetis Inc. is the world's largest maker of animal medicines. Once a subsidiary of Pfizer, it went public in January of 2013 and raised $2.2 billion in its initial public offering, the largest IPO deal for an American company since Facebook. Elanco, a pet pharma company owned by Eli Lilly, has $1.4 billion in annual sales and is the fourth-largest animal health business in the world. Growth in Lilly's animal division recently outpaced its general pharmaceutical division for humans. Yearly sales of Pfizer's animal pharmaceuticals are worth roughly $3.9 billion, with companion animal meds representing 40 percent of the total.

Total sales of pet behavioral medications like fluoxetine are difficult to quantify because many pet owners buy their animals generic human versions from pharmacies like CVS or Walgreens. The Prozac,

Valium, Xanax, and other drugstore sales for dogs, cats, and parrots are therefore lumped in with the sale of the same drugs for humans.

The pet pharmaceutical industry is also touted as being recession-proof. Americans may even spend *more* on their pets during tough economic times. One market research firm recently claimed that people's love for their pets was an "excellent insulator against recessionary cutbacks." The same firm reported that many pet owners, both affluent and middle class, were less likely to limit spending on their animals than on their human family members during crises. This has proven to be true not only during the most recent economic downturns but also during the Great Depression, when, as the historian Susan Jones has suggested, families made great sacrifices in order to secure food for their pet dogs and cats.

Psychopharmaceuticals are particularly profitable. The most lucrative human drugs in 2012, after cancer treatments, were antidepressants, mood stabilizers, and other mental health drugs. People spent more on psychopharmaceuticals than on drugs to treat pain, and the market has steadily increased, 10 to 20 percent per year globally, even during the most recent financial crisis. The markup on these drugs is more than several thousand percent, and as David Healy has argued, they're worth more than their weight in gold.

The scale of investment in the development and marketing of these psychotropic blockbuster drugs, for both humans and other animals, is tied to popular ideas about the illnesses they're used to treat. The industry that produces these drugs works hard to guarantee their financial success and this means encouraging more people to use them, for themselves and their pets. Two key historical decisions in particular helped ensure the current popularity of pharmaceutical use in the United States. The first took place in 1951, when the FDA (via the Humphrey-Durham amendments to the Food and Drug Act) declared that new medications would be available by prescription only. Before this, people largely medicated themselves, buying what they needed over the counter. Critics of the FDA's decision argued that it harmed regular citizens by making them completely beholden to a small group of people with the power to prescribe, who themselves were now dependent upon the pharmaceutical industry. A second FDA decision, in 1997, relaxed regulations limiting direct-to-consumer advertising and opened the floodgates to the pharmaceutical marketing machine that would quickly begin to publicize signs and symptoms of disorders that were easily treatable with compounds like the new Prozac.

One of the most vocal proponents of the use of psychopharmaceutical drugs for other animals is the veterinary behaviorist Nicholas Dodman, of the Tufts Animal Behavior Clinic. "One of the things that people called me was the Timothy Leary of the veterinary profession," Dodman said. Like Leary, Dodman acted as a sort of pied piper, drawing other veterinarians and pet owners to his methods via textbooks and peer-reviewed articles and workshops like "The Well-Adjusted Cat."

Dodman argues that studies he has overseen, some of which are supported by drug companies like Eli Lilly, prove that Prozac eases separation anxiety and compulsive disorders in animals and also reduces aggression and other "problem" behaviors. He has published research on the use of antidepressants and psychotropic meds to treat everything from compulsive Doberman Pinschers and tail-chasing Terriers to corral-biting (or cribbing) horses and cats who pull out their fur.

In his book *The Well-Adjusted Dog,* Dodman argues that psychopharmaceuticals treat a dog's problems "from the inside out," though he believes that drugs are more successful if they are used in conjunction with behavior-modification training. The goal, he has said, should be to taper off the medication as soon as possible. In some cases, those in which withdrawing the meds causes the animal's anxiety, depression, fear, or aggression to return, he suggests an indefinite drug regime.

Dodman prescribes a wide variety of psychopharmaceuticals,

as do the many vets who have incorporated his ideas into their own practices. He uses tricyclic antidepressants (Elavil and Tofranil) for depression, phobias, and in some cases of aggression in dogs. But he calls SSRIs like Prozac, Zoloft, Paxil, Celexa, Lexapro, and Luvox the closest thing there is to a silver bullet for treating behavioral problems in animals. He used to believe Valium was helpful for treating anxiety, but he's now convinced that, like alcohol, the drug can reduce inhibitions, and in dogs prone to aggression, this can make them vicious. It is also addictive. But Dodman still finds it useful for treating acute fear—in cases like Oliver's thunderstorm panic.

Dodman didn't always work as a Merry Prankster to the dog legions. An Englishman, he spent the 1970s working as a roving country vet in the United Kingdom, a sort of contemporary James Herriot, the fictional alter ego of James Wight, the British veterinarian. Dodman moved to the United States in 1981 to be a professor of anesthesia at Tufts' veterinary school. There he began to wonder if psychotropic medicines could change veterinary practice in the same ways they were transforming human psychiatry. Dodman shared his ideas for the first time at a veterinary conference in the late 1980s and later told a journalist for the *New York Times* that "he saw jaws drop around the room. It was like, 'Who is this strange masked man?'" Thirty years later, largely through his many publicized success stories, the animal psychopharmaceutical industry is thriving in the United States.

Dodman remembers hearing the former dean of the Tufts vet school call Prozac the behavioral equivalent of the popular wide-spectrum deworming medication ivermectin. "Before ivermectin," he said, "vets had to carefully choose which dewormer to use to treat intestinal and other worm infestations in dogs, cats, and farm animals. After ivermectin, vets could reach for this one medication to deal with practically all of these problems. All I can say is thank heavens for Prozac and other SSRIs."

Dodman's colleague Nicole Cottam says that 50 to 60 percent of the people who come to the Tufts clinic want drugs for their dog, cat, or bird. "Most of our clients don't call or come back after the initial appointment. Unless it's to get refills. When people leave with a prescription and behavioral exercises, they tend to only use the pills."

The idea of a pill for pet problems is simply too seductive and, frankly, it's often useful. I know this from experience.

Oliver received his first Valium prescription after his jump, at the vet hospital. The second one came from his behaviorist. As I mentioned, we were supposed to give Oliver the Valium thirty minutes before a storm hit so that by the time the thunder and lightning descended he would be too blissed out to notice. We were also supposed to play him recorded sounds of thunder and rain. While we did this we were to pet him, but only when he reacted calmly. The duration of the faux-storm training was supposed to increase in single-minute increments until Oliver could listen to the CD peacefully for hours. The behaviorist also prescribed Prozac for his separation anxiety, telling us that it would take a few weeks to take effect and to let her know if there was any change in his behavior. We watched him, anxious now ourselves, but Oliver didn't seem to be happier or calmer.

The Valium, however, helped. It dulled his thunderstorm anxiety. The only problem was that Jude and I both worked outside of the house, and in D.C. summer thunderstorms tend to happen in the afternoon. Neither of us could return home thirty minutes before a storm hit in order to drug the dog. Five days a week Oliver was on his own when it came to afternoon weather anxiety. We tried playing the recorded sounds of storms to desensitize him, but the CD was less helpful than the drugs. Oliver simply wasn't bothered by the fake thunder. He endured the listening sessions with benign disinterest.

The behaviorist also suggested we use Valium to treat Oliver's separation anxiety, telling us to dose him thirty minutes before Jude and I left the house. And she urged us to retrain him around our leave-taking behavior.

The behavioral therapy or training process the vet outlined was supposed to start by Jude and I approaching the front door but not leaving, not even touching the knob. We were to do this repeatedly, until Oliver stopped acting anxious. The next stage was going up to the front door and touching the door knob. When this bored Oliver into not reacting, we were supposed to turn the handle and open the door but not walk through it. This was meant to happen in stages so that in the end, she promised us, we would be able to leave the house without Oliver caring at all. The problem was that this sort of training took weeks, if not months—and we still had to go through the door in the meantime.

We tried to do the exercises. We gave it our best shot. Or to be honest, we gave it our best shot for a while. But it was exhausting, for us and for Oliver. He was so finely attuned to the various stages Jude and I had for getting ready to leave that as soon as we tried to decouple one cue from his "they are leaving me" anxiety, picking up our keys, for example, Oliver would figure out another, such as making our lunches or putting on our work clothes. He may have been dysfunctional and disturbed, but he wasn't stupid.

Sometimes I stored my computer bag in our building's shared hallway because even the sight of it would make Oliver start vigilantly watching for our departure, panting heavily and pacing. He also reacted to the sight of suitcases. And the putting on of shoes. And the opening of the coat closet. Possibly, if Jude and I had left for work naked, through a window, with no lunches, no keys, no bags, no shoes, and at odd hours, we could have avoided triggering Oliver's anxiety.

Like me, a lot of people simply can't spend the kind of time it takes to retrain themselves and their pet. Or it doesn't work. Sometimes, as with Oliver's Prozac, the drugs don't help either, or the effects aren't dramatic enough. This unfortunately tends to be the end of the line for most animals. They are given up or forced into what the veterinary behaviorist E'Lise Christensen calls "the big sleep" and David Sedaris has referred to as visiting the "Youth-in-Asia." Behavioral drugs, if they work, can help stave off such mortal consequences.

Dodman argues that most dogs and cats are taken to shelters or put down for "being difficult." Indeed, 6 to 8 million dogs and cats are given up every year. According to the ASPCA, 3.7 million of them were euthanized in 2008. Aggressive and insecure dogs who menace visitors, or cats that won't stop spraying on the bedspread are the kinds of creatures most often left at shelters. Dodman argues that the great salvation of these ill-behaving dogs and cats is medication. Though I'm doubtful that medication-only therapeutic regimes are as effective as behavior therapy or at least behavior therapy *and* medication, psychopharm for pets can be a useful way station on the road to recovery, or a stopgap measure on the way to the gas chamber.

E'Lise Christensen is a proponent of psychopharmaceuticals because, she says, "Unlike in human medicine, we don't have inpatient treatment facilities for our patients. If you have a dog that's jumping out windows into traffic, for example, you have to heavily dose them so they don't hurt themselves."

Heavy doses for dogs are not quite the same as heavy doses for humans; dogs' livers can handle a lot more medication. This is why many dogs end up on dosages of antianxiety drugs that could kill a person. "I have a golden retriever patient right now," Christensen said, "that's taking 80 milligrams of Valium every four hours." This would make a human limp, verging on catatonia, but it's keeping the retriever from full-blown panic attacks.

Despite the fact that many of her clients are on the same drugs as their pets, Christensen hasn't seen too many people borrow from their pets' stash. She believes this is because she's very clear with her clients about how much higher dog dosages are and the fact that they've come to her because they want to help their animal. "What's much more common," she said, "is people sharing their own psych medications with their pets." Thankfully it doesn't usually harm or help because the human doses are so much lower. Her physician clients use their own prescription pads to prescribe for their pets. "Not my psychiatrist clients, interestingly enough. They tend to wait for me to prescribe."

And yet for long-term improvements in her patients' well-being, Christensen doesn't think the drugs are enough. The gold standard of care, she's convinced, is medication combined with behavioral training. This means keeping the animal from whatever triggers their fear and anxiety, whether it's being left alone or hearing a vacuum cleaner, while therapy is taking place. "If the animal isn't being exposed to their triggers," she said, "you can work with them to make whatever scares them less scary." I told her about my experience trying to retrain Oliver about the front door and she admitted that all of this is easier said than done.

Recently she treated a nervous dog who lived in Brooklyn. He was prone to biting strangers out of fear and anxiety. The dog became stressed simply walking down the sidewalk and his owner, a young woman, had a hard time keeping people at a safe distance. "She would tell passersby that the dog would bite and to keep walking but people actually got mad at her for not letting them pet the dog." As a solution, the woman moved to a house with a yard in White Plains and now commutes to work in the city. Her dog is much more relaxed. Christensen realizes this is too much to ask of most dog owners. "If I could change one thing about New York City," she said, "it would be that people would treat a dog on the sidewalk just like anyone else they didn't know. If you were walking down the street with a human family member, for example, only the creepiest strangers would think it's okay to come up and stroke them."

This made me think of self-identified "dog people." Many men and women who describe themselves this way will bend down into the personal space of dogs they don't know and extend an overconfident hand within muzzle reach, or aggressively ruffle fur on heads or hindquarters. These people are a bit like self-described ladies' men. That is, if you have to say you're a dog person, it probably isn't true.

Christensen counsels many of her clients to act as buffers for their nervous animals on city sidewalks, stepping between the dogs and would-be strokers. The humans become, in essence, therapy animals for their own pets.

When she was a veterinary student and resident at Cornell University in rural upstate New York, it was easier for Christensen to ask her clients to keep their pets away from their triggers during behavioral training. A dog who couldn't be left alone without panicking, for example, could accompany their human more places. In New York City, her clients can't keep their dogs with them all the time. Like many urban veterinarians, she is then more reliant on behavioral drugs to blunt the animals' fear and anxiety while they're still in therapy, a process that can take a very long time. "By the time many dog owners get to my office," she said, "they are verging on emotional bankruptcy. These people are desperate and exhausted. They're trying hard."

She has learned that most are capable of no more than four or five minutes of behavioral training per day. Ideally she would have people do fifteen minutes twice a day, but most of them can't do that. I told her about my disappointing results trying to retrain Oliver and what my failsafe method was when I reached my wit's end. I used the car. Oliver always calmed when I left him in our Subaru. I even daydreamed about opening a boarding kennel for anxious dogs that consisted of a parking lot staffed with nice people to deliver food and water to the cars and take the dogs on walks.

To my surprise, she didn't think I was crazy. "Many behaviorists actually encourage their clients to use cars to keep their dogs calm when they have to leave them alone," she said. "If you live in a temperate climate, are really careful, and it's not illegal in your city or state, this can be a good solution." The reason, she told me, that Oliver and so many other nervous dogs feel better in cars is because we trained them to be comfortable there without knowing it. Few people leave their dogs alone in the car for long, at least at first. So the dog learns, incrementally, to stay in the car for longer periods of time and no matter what, his or her humans always come back.

Doggy's Little Helpers

Dogs exist as they do, emotionally and physiologically, because they tend to like being around us. The dogs we have fed, treasured, and bred over the past fifteen thousand years are precisely the kinds of creatures that are more likely to suffer when separated from their human companions for most of the day. Contemporary canine anxiety disorders like Oliver's are the result of a trait we prize highly and have selected for in dogs: they enjoy being around people, particularly their own people, and are glad to spend time with us.

Today's pet dogs are a bit like Ham the chimpanzee, sent into space in 1961 to find out if humans could do it too. That is, many dogs living in contemporary urban and suburban households are occupying alien lands. They simply haven't had enough time to evolve into creatures who do well left alone all day with little exercise, social time, and ability to express their doggishness—something the Germans refer to as *Funktionslust*, taking pleasure in what one does best: a cheetah sprinting at full speed or a bat pinging his sonar through the night. Dogs are built to run and sniff and chase and hump indiscriminately. Most of them are happiest rolling in dead fish, pulling tampons out of the trash, and licking their genitals or someone else's.

Many dog owners are content to meet a dog on human terms but are unwilling to do so on the dog's terms. That is, we are thrilled when dogs are excited to see us at the end of the workday, but we don't want them to be running and jumping in circles, tails wagging explosively, paws everywhere, when we're at work. Instead we hope they're sleeping soundly, calmly grooming themselves, or perhaps taking a gander through the living-room window, not in longing but just to see what's there. This expectation isn't fair. It reminds me of times

I've fallen for men whose idiosyncrasies intrigued and captivated me at first, but as time wore on, those same traits began to drive me nuts. The fault was mine. You can't blame a man for being the kind of man you fell in love with. And you can't blame a dog for being a dog.

Most urban and suburban dogs are only encouraged to be themselves for a small fraction of the day. In my neighborhood just outside of San Francisco, the early evening, right before sunset, is that fraction. You can feel the collective wags of thousands of tails, the expectant panting at the door, the anticipation of the click of the leash on the collar, and then the overwhelming joy of going out. Out! They flood the sidewalks around my house with their pent-up frustrations, pissing and smelling and dragging their people along behind them like water-skiers. At the park the humans stand around tossing balls or chatting idly or calling their dog off another's rump. A half hour or an hour later, it's back to the house for dinner, some petting, maybe some television with the humans, and then bed. But this is not enough time for dogs to do dog things, even if they get to do it in the morning too.

The alternative for many people is simply not having a dog. For most of us, if you live and work in a city, you can't move to a farm just because your dog would like it better. If your dog hates being alone, you can't simply quit your job to stay home with her. There are other options, of course, but none is simple. You could hire a dog walker to visit one or more times a day, but they're expensive. You could move to an apartment near a park where the dog could be off leash every day, but perhaps the market's bad and you can't sell your condo. You could get another dog to keep your first dog company, but maybe your landlord allows only one pet. You could spend a lot of money on dog toys to stuff with peanut butter or marrow bones to freeze and then leave hidden around your house like a macabre Easter egg hunt, but you forget. This is life. We love our dogs, we try our best, but we often fail them. The truth is that all the squeaky plush toys in the world can't compare to a leashless life of daily stimulation and plenty of time with dogs, humans, and other animals. That is the kind of life dogs had before most people went to work in offices, factories, shops, and other places canines aren't welcome. And it is precisely the kind of life that keeps most of them from licking their paws into oozy messes or tearing up the sofa.

When dogs spend long hours doing nothing, they have too much energy to curl up into happy, satisfied rounds at the end of the bed. Their energy has to go somewhere, and for the less sturdy, or simply the more prone to anxiety or compulsion, that somewhere is Crazytown. This town has myriad diagnoses and one of the more common is separation anxiety. Pharmaceutical companies have taken note and even helped to shape ideas of the disorder.

The humans who own the more than 78 million pet dogs in the United States are a large market for companies like Pfizer and Eli Lilly. In 2007 Lilly launched Reconcile, chemically identical to Prozac (which went off patent in 2001), but unlike Prozac, it is beef-flavored, chewable, and FDA-approved for treating separation anxiety in dogs. Simultaneously the company released results from a Lilly-funded study arguing that 17 percent of U.S. dogs suffer from separation anxiety. A 2008 study estimated that 14 percent of American dogs have some degree of the disorder.

Lilly's Reconcile website is the virtual equivalent of getting licked by a puppy, one that is good at making her owners feel guilty. Phrases like "I wonder if he tears things up to get even with me. I wonder if it's my fault" appear in flash animation at the top of the screen. A video clip of a veterinarian with a southern drawl runs through a list of separation anxiety symptoms: drooling, destructive chewing, pacing, depression, anorexia, excessive barking, and "licking of coat." As the vet speaks, a young Golden Retriever mauls what looks like an expensive high-heeled shoe.

An older version of the site had a large banner that read "Separation is inevitable. Now anxiety isn't" and offered dog-bark ringtones and a glum-looking Beagle screensaver. It also linked to a study on the effects of Reconcile, used along with behavioral training, on separation anxiety. The study, on 242 dogs, was published in *Veterinary Therapeutics* in 2007 and paid for by Lilly. The dogs were split into two groups: one received a beef-flavored placebo pill and the other received the drug. Both groups also went through behavioral training. By the end of the study, anxiety symptoms in all of the dogs had diminished—by 72 percent in the Reconcile group and 50 percent in the placebo group. While the study does suggest some drug efficacy, it demonstrates the importance of behavioral training far more effectively.

There is also Clomicalm, introduced by Novartis in 1998. The active ingredient in the drug, clomipramine, is identical to the main ingredient in the company's antidepressant/OCD drug for humans, Anafranil, but this version was FDA-approved just for animals. Novartis describes Clomicalm as a medication for treating separation anxiety in dogs, but, as is the case with the human version, clomipramine is also frequently used to treat other signs of distress. In one strange experiment twenty-four Beagles were sent traveling in a truck for an hour, on three different occasions, to see if the drug helped ease travel anxiety. The results were inconclusive, but the Beagles drooled less than they normally did. The drug has been more successful in lessening tail-chasing in dogs and feather picking in cockatoos.

Depending on the dosage, the Clomicalm box features a Yellow Labrador, a Golden Retriever, or a Jack Russell Terrier. The dogs seem happy and alert, tongues lolling. Medicating a small dog costs roughly $600 per year. Larger dogs need larger doses, which cost more. The website allays consumer fears, and perhaps owner guilt, by assuring visitors that Clomicalm tablets are not tranquilizers or sedatives and won't affect your dog's personality or memory. Instead, the drugs helps animals "return to a normal life."

* * *

One of the most outspoken critics of psychopharm for pets is the veterinarian and behaviorist Ian Dunbar, who runs a canine-training empire called Sirius Dog Training and has written a number of books, among them *How to Teach a New Dog Old Tricks.* He leads training classes and workshops around the world and hosted a series on British television called *Dogs with Dunbar.* He says that he has never had to resort to drugs to treat a behavioral problem: "Drugs are simply unnecessary. They're touted as a quick fix, a panacea for all problems, but it's not true."

He believes that medicating dogs with psychotropic drugs mirrors an irresponsible approach to human health care. Instead, Dunbar argues that pet owners should use behavioral modification training and alter their own behavior so as not to reward their animal's disturbed or disturbing activities. "When people have problems with their dog," Dunbar told an interviewer, "I usually tell them 'The problem is your friend. You are going to learn so much from this.'"

Some problems, however, are no one's friend. When it came time to fill Oliver's Prozac and Valium prescriptions, I didn't give him any of the FDA-approved drugs for dogs, even if Reconcile was beef-flavored and chewable. My local drugstore carried the generic versions and could make up the appropriate dosage in fifteen minutes. I knew Oliver didn't care what flavor they were and would eat the pills if I stuck them into chunks of cheese. I took the veterinary prescription to the drop-off window at Walgreens. When the pharmacist called "Oliver Braitman" to the pickup window to pay for his prescription, I laughed. She handed me an unmarked bag (for patient privacy) and wanted to know if I had any questions about the medication.

"It's for my dog," I told her.

"Oh," she said. "That happens a lot."

Prozac Ocean, Prozac Nugget

Debating whether dosing other animals with psychopharmaceuticals is a good or bad idea may increasingly be beside the point. In a way, we may not have much of a choice. These drugs now suffuse our environment and parts of our food supply. More than 200 million prescriptions for antidepressants were written in the United States in 2010. Many of the active ingredients in these drugs are excreted in people's urine or flushed down the toilet in the form of extra pills. Wastewater treatment plants are not equipped to filter out pharmaceuticals, so the meds end up where our treated water does: in oceans, rivers, lakes, and our water supply. A recent study in the journal *Environmental Toxicology and Chemistry* demonstrated the presence of a range of antidepressants and their metabolites in drinking water, river water, and in the bodies of minnows. A few researchers are attempting to understand what this means for aquatic life.

In one experiment, bass exposed to Prozac stopped eating and eventually began to float vertically inside their tank. Another study looked at the effects of Prozac on shrimp. Wastewater concentrates in river estuaries and coastal areas where shrimp like to live, meaning that they and other creatures who live there are floating in the excreted drugs of whole towns and cities. Shrimp exposed to antidepressants were five times likelier to swim toward light than away from it, making them far more susceptible to predation by fish or birds.

Another recent study, published in *Environmental Science and Technology*, found an array of psychopharmaceuticals in the feathers of farmed chickens. Feather meal is a dietary supplement made of ground chicken feathers that is fed to pigs, cattle, fish, and even chickens themselves. In 2012, meal samples tested positive for antibiotics like Cipro, banned from animal feed in 2005. Just as disturbingly, one-third of the feather meal samples also contained fluoxetine (Prozac), acetaminophen (the active ingredient in Tylenol), and antihistamines (the active ingredient in Benadryl). Many poultry farmers feed their birds Benadryl, Tylenol, and/or Prozac to calm them down and reduce anxiety. Harried, stressed chickens don't grow as fast or produce meat as tender as content ones. Caffeine, in the form of green tea powder or coffee pulp, is also fed to chickens so that they'll feel more energized and stay awake longer to eat and lay. It's possible that these birds need anxiety-reducing drugs to counteract the stimulants.

According to the journalist Nicholas Kristof, poultry farmers don't always know what they feed their birds. The large agribusinesses require the poultry farmers who supply them with chickens to use proprietary food mixes, and the farmers may not know what the mixes contain. As with the shrimp, the effect that all of this has on the birds, and ultimately those of us who eat them, is unknown and unsettling.

Chapter Six

If Juliet Were a Parrot

> I'm beginning to understand that when we want to kill ourselves, it is not because we are lonely, but because we are trying to break up with the world before it breaks up with us.
>
> Pam Houston, *Contents May Have Shifted*

> It is the lark that sings so out of tune.
>
> William Shakespeare, *Romeo and Juliet*, Act 3, Scene 5

Charlie was a blue and gold macaw raised in Florida. As a baby she went everywhere with her human family and was treated as one of their own. Until she wasn't. When Charlie was five or six, her primary owner died, and she was given to a parrot breeding facility. Parrots, especially macaws, can live fifty or more years and are often orphaned in this way. Sometimes the loss of their people is devastating to the birds. Charlie, however, seemed to have survived this first loss fairly well; she bonded quickly with another parrot at the breeding center.

Not long after she embarked on this new friendship, the two birds were stolen. Charlie was eventually recovered and returned to the breeding center, but her companion was not. Clearly shaken, she began to pluck out her feathers. Her plucking was so relentless that, a few months later, Charlie was completely bald, save for a few feathers on her tail and head. The breeding center donated her to the Tampa Zoo.

Ann Southcombe was a keeper at the zoo. The same slim woman with a childlike voice and small hands who cared for the young Gigi, Ann has a kind, focused manner around animals. I once watched her calm a skittish squirrel in seconds. Ann had rescued the fluffy-tailed rodent after she fell from her nest as an infant and named her Mary. Mary now lives in a squirrel palace in Ann's spare bedroom. Over the past thirty-five years Ann has worked with an arklike range of animals, not just as a zookeeper but also as a research assistant and a wildlife rehabilitator. She helped raise Chantek the orangutan inside a trailer at the University of Tennessee as part of an anthropology and ape language study in the 1970s and 1980s. She taught him to use sign language, watched as he poured his own glasses of milk and swept out their trailer, and occasionally tracked him down at the campus bookstore, where he would go to eat candy bars. According to the study's lead researcher, Lyn Miles, three decades later Chantek still refers to himself, in sign, as "Orangutan person."

Ann also worked for a time with the signing gorilla Michael, the less-famous companion to Koko. She once taught an injured otter how to hunt in her front-yard fountain, using store-bought goldfish, and has taken care of a number of orphaned baby black bears, a lynx, a young chimpanzee, a golden eagle with a prosthetic beak, a series of owls and rabbits, and too many squirrels to count. She has dealt with her share of parrots too. None of them were as pathetic as Charlie.

"Charlie was plucked so bald," Ann remembers, "that she looked like a chicken ready to go in the roaster. I decided to take her home from the zoo and see if I could make her feel better."

First Ann bought a collar for Charlie, thinking it might get in her way and slow down the plucking. It didn't work. She also tried acupuncture and giving Charlie herbs. Charlie kept plucking.

"At night," Ann said, "I would let her sleep in my room. She had nightmares. I am sure of it. It was late at night and she was on her perch asleep, with her eyes closed. But she would make these anguished little squawks."

During the day, Ann took her outside. "I would bring her to this big old tree in my backyard and put her up on the lower branches while I did other things around the house—just so she could enjoy being outdoors. She loved it. But every once in a while she would fall off the tree. Since Charlie didn't have any feathers, she couldn't fly. But then she would walk along the ground back to the tree trunk and climb all the way back up again."

Except once. Ann left Charlie alone on the tree while she went shopping. She was gone for less than an hour.

"When I came back, she was dead."

Ann was puzzled as she recounted the story. "It was just so sad—she was impaled on a thin metal pole sticking out of the ground. The great irony was that metal pole. I used to have a pink plastic lawn flamingo, and you know how they are perched on those little metal legs that stick into the ground? Well, at some point, the plastic flamingo body disappeared and only those little metal stakes were left. While I was at the store, Charlie had fallen directly onto one of those stakes and it had gone completely through her chest. And, you know, I know this might sound crazy, but I think Charlie might have just had enough. She was a very intelligent bird. And that tree she was on was gigantic, so was the yard, and she played in both all the time. If she had wanted to really hurt or kill herself, she might have done it on purpose. Who knows? I know that's really anthropomorphizing, but it just seems like such a weird coincidence. Of all the places she could have fallen off, she fell directly on the tiny stake."

Parrots, like people, do not usually die from overplucking. Charlie's story is intriguing not just because she plucked herself bald but because the circumstances surrounding her death are so odd.

Whether other animals have the capacity to commit suicide may be the most vexing aspect of animal madness. The subject has interested philosophers since antiquity. Aristotle told the story of the Scythian stallion, who was said to have thrown himself into an abyss when he realized that he had been tricked into mating with his mother. And while Christianity frowned on suicide in general, pelicans—who reportedly tore out their own flesh to feed their young—served as animal avatars for Christ, avian symbols of self-sacrifice. Even John Donne wrote about the pelican as an emblem of the "natural desire of dying" in the seventeenth century.

Since its first documentation in the English language in 1732, the word *suicide* has meant the intentional harming of oneself with the intent to die, but defining what the act itself entails has been harder to pin down. The *DSM-V* does not include suicide, but it does define suicidal behavior disorder. To be diagnosed, someone must have tried to kill him- or herself in the previous two years and either abandoned the attempt or been stopped before going through with it. According to the *DSM*, the suicide cannot have been undertaken for political or religious reasons. A diagnosis of *nonsuicidal* self-injury is applied to people who have intentionally cut, burned, stabbed, hit, or "excessively rubbed" themselves to injury without seeming to want to die.

Self-destructive behaviors like these among nonhuman animals—from acts of drastic self-biting and rubbing to repeated head banging—have been widely documented by behaviorists, veterinarians, physiologists, psychologists, and other researchers. One paper, published by the *Psychiatric Clinics of North America* in 1985, "Animal Models of Self-Destructive Behavior and Suicide," suggested that suicidal humans and self-injuring animals were similar enough to make researching self-destructive impulses in both possible. The lead author, Jacqueline Crawley, the chief of behavioral neuropharmacology at the National Institutes of Health (NIH), and her coauthor, the chief of clinical studies at the NIH, argued that suicide is a distinctly human behavior that requires complex cognition but that other animals can mortally injure themselves too, in the wild and the lab.

"Although self-destructive and suicidal behaviors are not synonymous," they wrote, "the boundary between the two is often blurred."

In the twenty-five years since this study was published, other researchers have attempted to use animal models to understand self-harming behaviors and possible treatments in humans. One of these studies, completed in 2009, again by researchers at the NIH, tackled the issue of suicidal tendencies: "Suicide is a complex behavior that is, at best, complicated to study in humans and impossible to fully reproduce in an animal model," they wrote. "However, by investigating traits that show strong cross-species parallels as well as associations with suicide in humans, animal models may elucidate the mechanisms by which SSRIs are associated with suicidal thinking and behavior in the young." That is, research on lab animals could help us understand the possible connection between certain antidepressants and suicidal thoughts in young people. The traits and behaviors in lab animals that the authors considered indicators included aggression, impulsivity, irritability, hopelessness, and helplessness.

These studies are, in some sense, a twenty-first-century nod to the work of William Lauder Lindsay, Charles Darwin, George Romanes, and other Victorian naturalists who saw an easy continuity between human and other animal emotional lives. Contemporary investigations hinging on animal irritability or rodent hopelessness suggest that at least some researchers at lauded institutions like the NIH see self-harm as unfolding along a continuum, with purposeful self-killing located at one point and less lethal behaviors like self-biting and cutting located elsewhere, without actually granting the capacity for suicide to other animals. Humans, it seems, do not have a monopoly on self-destructiveness, even if the ways that we can reflect on or plan to harm ourselves are unique.

As far as Western psychiatry, psychology, and the science of mental health has been concerned, killing oneself on purpose implies a particular form of self-consciousness that we know exists in humans but can't be proven to exist in other animals. Yet Charlie may have been self-aware in her way. To what *degree* Charlie was self-aware, however, is a mystery, not to mention whether she understood that her fall from the tree would end her life. Despite these unknowns, it is still quite possible that Charlie felt her situation was so insufferable that she made the cognitive leap to do something because she no longer cared to preserve herself.

Beatriz Reyes-Foster is an anthropology professor at the University of Central Florida with a doctorate in sociocultural anthropology. As a graduate student at UC Berkeley, she studied suicide prevention efforts in the Mayan communities of Mexico's Yucatán peninsula. Later, she spent time in an acute ward of a public psychiatric hospital to observe the ways that Maya patients interacted with their psychiatrists.

I met Beatriz at what I suspected might be the most depressing conference I'd ever attend. The Suicide and Agency Workshop at the Max Planck Institute for Social Anthropology took place in the frigid darkness of late November in Halle, a gritty German city famous for its chocolate factory. To my surprise, Beatriz and most of the suicide-focused anthropologists in attendance were bright-eyed, funny, and only mildly chagrined about their chosen research topic. One afternoon Beatriz pulled me aside to say that my interest in animal suicide reminded her of something she'd observed in the Yucatán.

Most Maya families keep chickens, turkeys, and dogs. Wealthier families might also have pigs, cattle, and certain birds that are considered more or less useless, like ducks, geese, or pigeons ("For some reason, Yucatecans don't like to eat these kinds of fowl"). Because the majority of Maya live in conditions of poverty, it's virtually unheard of that anyone would spend money on veterinary care for their chickens, turkeys, or dogs. When one of these animals sickens and stops eating or becomes listless or apathetic, people will often say that the animal *se puso triste*, or became sad. When a creature's death seemed imminent, Beatriz often overheard family members announce that the animal *no tiene ganas*, or had no desire [to live]. "It makes me think of life posited as a form of struggle from which one eventually tires out," said Beatriz. "Death is understood as commonplace and not the result of neglect or mistreatment. When an animal becomes *triste* and loses its *ganas*, people simply think that its hour has come."

Occasionally, something similar happens among the Maya themselves. Beatriz mentioned the case of an elderly man in a small village in central Yucatán where she used to work. Known as Uncle Tomás to his family, he suffered a stroke in his late seventies that left him paralyzed and unable to work or attend to his corn fields. He began to tell everyone that he was useless. One day he attempted to choke himself to death with own hands but he wasn't strong enough. "Uncle Tomás's family reacted with sadness," said Beatriz, "but no particular help was sought to address his behavior." They did try to prevent future suicide attempts, but one day he began to refuse food. At first, the family tried to coax him to eat. When it became clear that Uncle Tomás was never going to give in and accept the food they offered, they stopped trying to convince him.

"Someone more judgmental than me might see this as an appalling example of how the elderly are treated once they are thought to be useless," Beatriz said. "But I think that view would be colored by a very modernist understanding of the need to preserve life at all costs. Uncle Tomás's life was not going to improve, in much the same way that the domesticated animals' lives are not going to improve. I think people can realize the inherent hardness of life, and this feeds into the belief that at some point we all become *triste* and lose our *ganas de vivir* because *ya nos llegó la hora*. We all have a time to die, this kind of death is only natural."

Not everyone is as clear about their motivations as Uncle Tomás. Abrupt deaths can be even more confusing. According to the American Association of Suicidology, only one in five or six people leave suicide notes. Friends, family, and mental health professionals are left to wonder why someone did what he or she did and even if it was intentional: "When he drove into the tree was it an accident?" "Was she admiring the view when she slipped, or did she jump?" "Did he mistakenly take too many pills or did he want to overdose?"

When a cogent adult like Uncle Tomás refuses to take his medication or eat, he is still committing suicide, although very slowly. Other people who act perhaps more impulsively, throwing themselves from high places or before oncoming traffic, may not have clearly calculated whether the impact will be fatal. Do these men and women intend to die or were they, a little like Oliver, surging with panic and the overwhelming urge to do something, anything, to end their pain? Because of the mystery that swirls around so many human deaths, my point isn't to prove that other animals kill themselves. Instead it is to suggest that certain creatures should be given the benefit of the doubt.

The Horse in Court and the Self-Stinging Scorpion

A compelling look at the phenomenon of animal suicide in history was published in 2010 by two British historians of science and medicine, Edmund Ramsden and Duncan Wilson. Their paper, "The Nature of Suicide: Science and the Self-Destructive Animal," generated a torrent of publicity despite the fact that the authors never specified whether animals were capable of suicidal acts. Instead they argued that tales of animal suicide told throughout history reflect prevailing human attitudes toward self-destruction.

"Scientists and social groups," they wrote, "have used animal suicide to understand and define self-destructive behaviour . . . to redeem or condemn and address the relation between humans and the natural world." That is, animal suicide accounts gave scientists, natural historians, and the general public a means of reflecting on the concept of human self-destruction as well as ideas about humanity's relationship to nature without always having to talk about people. Writing and thinking about animal suicides, just as we saw with the cases of animal heartbreak and homesickness, gave people a way to ponder their own afflictions, even if they were doing it unconciously.

The Victorian period was an extremely interesting time for this sort of inquiry. Fascination with romantic self-annihilation and scandal was on the rise. Suicides were a source of anxiety and disgrace. In England, the families of suicidal people made great efforts to hide evidence of these deaths because suicide was not only illegal and considered immoral but the property of those "convicted" reverted to the Crown and their bodies were banned from church cemeteries.

The fascination extended to suicide in other species. William Lauder Lindsay devoted an entire chapter of *Mind in the Lower Animals* to the topic in 1879. He believed that there were nine good reasons for an animal to commit suicide: old age, hurt feelings, physical pain, combined mental and bodily suffering, desperation, the frustration of captivity, melancholia, human cruelty, and self-sacrifice (usually by parents for the benefit of their young). He collected more than two dozen stories of purported suicide in sixteen animal species, including dogs, horses, mules, donkeys, camels, llamas, monkeys, seals, deer, scorpions, spiders, storks, roosters, and a canvas-back duck. Lindsay's work reflects a larger shift in Victorian attitudes toward suicide, one in which the act was becoming not just a moral issue but also a medical one.

Not every Victorian naturalist was as taken with the idea of self-destructive animals, however. Animal suicide was discredited most famously, at least within British scientific circles, by Conwy Lloyd Morgan in 1881. Morgan decided to test whether scorpions actually killed themselves when surrounded by fire. He designed a series of experiments that were "sufficiently barbarous . . . to induce any scorpion who had the slightest suicidal tendency to find relief in self-destruction." He heated the scorpions in bottles, burned them with acid, shocked them with electricity, and subjected them to other "general and exasperating courses of worry." Morgan watched the scorpions strike their own backs with their poisonous stingers, but he explained this behavior as an instinctive way of removing the irritation and, rather haughtily, accused anyone who thought otherwise of being "not accustomed to observation."

Morgan's efforts to disprove animal suicide were largely a reaction against the work of natural historians like Lindsay and George Romanes, who thought animals were capable of intelligence and reason. A friend of Darwin's, a devoted proponent of evolutionary theory and the first person to use the term *comparative psychology*, Romanes wrote about both Darwin and Lindsay in his studies of animal minds. Two years after Morgan's scorpion experiments, he published *Mental Evolution in Animals* and included an essay on instinct that Darwin wrote before his death. Another chapter, on imagination, covered a variety of examples of animal intelligence and creativity, such as "the well known cunning of the fox and wolf in eluding the hounds," and cites Lindsay on the topic of animal dreams and delusions. These unconscious escapades observed in sleeping horses, birds, hunting dogs, and elephants, who twitched, squawked, whinnied, or ran while asleep, were proof to Romanes of imagination. That is, a hunting dog whose paws flicked or nose twitched while sleeping seemed to be *imagining* hunting. He also discussed what he called "imperfect instinct" or "derangement of instinct," a sort of mental malfunction that could result in strange behaviors in other animals, such as a pigeon who developed a strong attraction to a glass bottle and showered it with courting behavior. Romanes, like Lindsay and Darwin, believed that insanity was "not an uncommon thing among animals."

Meanwhile, the idea of suicide as a crime was giving way to the concept of suicide as a disease and one that could be brought about by people's living conditions. By the late 1890s the Italian psychiatrist Enrico Morselli was also arguing that suicide could be motivated by the unconscious. He wrote, in *Suicide: An Essay on Comparative*

Moral Statistics, published in 1879, that people's motivations to do themselves in could have "secret causes" that even the suicide victim was not aware of.

The issue of nonhuman suicide was pushed further from the centers of academic inquiry in 1897 when the sociologist Emile Durkheim published his landmark book, *Suicide: A Study in Sociology*. Based on suicide statistics of the time, Durkheim suggested that suicide was more a result of social problems than people's inner turmoil. Even though he'd done no original research on the topic, Durkheim argued that all animal suicide cases previously reported, from the self-stinging scorpions to forlorn, food-refusing dogs, did not include evidence of sufficient purposefulness or forethought: The scorpion was not using his tail like a rifle, the dog was not using starvation like a noose.

The very same year, in an article for the journal *Mind*, the psychiatrist Henry Maudsley criticized Lindsay for falling prey to anthropomorphic reasoning in his animal suicide work, seizing on Lindsay's example of a cat who supposedly strangled herself in the fork of a tree after her kittens were drowned.

In 1903 Morgan reiterated his earlier stance on animal mental capacities in what would come to be known as "Morgan's canon," perhaps the most famous caution against anthropomorphism in modern history and a powerful influence on the radical behaviorists, who, by the 1930s, would see animal behavior as a function of largely unconscious processes. In a slight twist on Occam's razor, Morgan wrote, "In no case is an animal activity to be interpreted in terms of higher psychological processes, if it can be fairly interpreted in terms of processes which stand lower in the scale of psychological evolution and development." He was not referring to animal suicide specifically, but he refused nonhuman animals the capacity for any action that could otherwise be explained by instinct.

Despite this growing skepticism of animal suicide among behavioral and psychological researchers, the British and American public at the turn of the century was as hungry as ever for self-destructive animal stories. Accounts of animal suicides appeared in national newspapers and popular books (fiction and nonfiction) before, during, and after the time that Morgan, Durkheim, and Maudsley were denying animal suicide. In fact, they persisted well into the twentieth century.

One early article, published in the *New York Sun* in 1881, was typical of many such stories. It was a lengthy investigation into the suicidal impulses of ants, scorpions, spiders, snakes, hogs, dogs, and starfish; the last were said to immediately commit suicide on capture by dropping their limbs. The author argued that most animal suicides could be traced to wanting to avoid capture and pain. These reported tales were reflections of human opinions about socially appropriate ways to die, moral lessons about self-sacrifice, appropriate gender roles, and the ethics of capture and captivity. They were also, almost always, efforts to explain confusing animal behavior in terms that made sense to the people watching.

Rex the lion was one of these puzzling creatures. A member of the Ringling Brothers Circus, Rex was discovered hanging over the side of his cage in 1901, asphyxiated by his neck chain. According to his keeper, the lion's suicide was inspired by an embarrassing brawl with a younger male: "You know, lions are as vain as a society woman. To be thrown down and mauled around the way he was before everybody, after he had been boss for so long, just about broke his heart. I gave him the best of everything and petted him a whole lot but it was no use. He didn't hold his head up in the proud way of old times any more. He just moped around in the back part of the cage, and finally one day he committed suicide."

The most commonly reported suicidal animals at the turn of the century in both Britain and the United States were dogs and horses. Among the general population, dogs competed with nonhuman primates for the coveted spot of being most closely related to humans. Dogs and horses were also the easiest to observe, and there was increasing interest in protecting their well-being. In the late nineteenth and early twentieth centuries, organizations like the Royal Society for the Prevention of Cruelty to Animals popularized stories of self-sacrificing dogs because it lent credence to the fight for more humane treatment of animals. As for horses, popular natural history writers often characterized them as noble, sometimes even nobler than humans, making it perhaps less odd that they might suffer from the same sorts of emotional problems as their riders.

In February 1905 a superior county court in southern Washington State ruled that a horse had committed suicide. The horse, who had been rented from a local livery stable, was pulling a carriage along a muddy road when it became mired in the muck. According to the driver, he first attempted to free the horse himself, but the horse reportedly made little effort to fight its way out. The driver left to find help, but while he was gone the road flooded and the horse drowned. The horse's owner sued the driver, charging him with negligence, but admitted in front of the court that "for some time past the horse had seemed to take no interest in life." The court then ruled that it was a "plain case of suicide on behalf of the animal," and the horse's owner lost his suit. Other horses supposedly jumped in front of buses on purpose or tried to launch themselves off train trestles.

These tales certainly reflect moral lessons about self-sacrifice, masculinity (at least in the case of Rex), the ethics of capture and captivity, and opinions about acceptable ways to die. But they are recordings of actual animal behavior as well, however distorted and convoluted by observers. Since an animal's death was usually attributed to a cause, the stories also extended a kind of rationality to the suicidal dogs, horses, or lions. Despite the best efforts of animal suicide naysayers, people continued to identify with self-destructive animals.

Suicidal Flipper

Ric O'Barry is an outspoken former dolphin trainer turned animal advocate and has been arguing since the 1970s that dolphins commit suicide. A polarizing character even among certain activists, he was the subject of the Academy Award–winning documentary *The Cove*, about the dolphin slaughters of Taiji, Japan. Once, protesting at a meeting of the International Whaling Commission, he wore a TV monitor playing footage of dolphin killing. He has gone to jail for attempting to free captive dolphins used for performances, research, and by the U.S. Navy.

O'Barry is also at least partially responsible for contemporary dolphin and whale shows at amusement parks like SeaWorld, in which the cetaceans dance on their tails, flip and twist, squeak on command, and tow their trainers around to thumping pop music, before Jumbotron screens playing cheesy animations of pixilated splashes. O'Barry was one of the first celebrated dolphin trainers in the United States. In the 1960s he trained a series of dolphins to play Flipper on the popular TV series, which ran for three years. Reruns played in dozens of countries for another twenty. O'Barry even lived in the house featured in the show, at the edge of the man-made lagoon where the dolphins were kept. Flipper soon became the world's most famous dolphin, and, a little like Kleenex, his name came to stand for the creature itself.

There was, though, no single Flipper. In his book *Behind the Dolphin Smile*, O'Barry writes that the part of Flipper was played by five different trained dolphins. Flipper was both an illusion and an embodiment of 1950s and 1960s family entertainment, comforting plots in which the bad guys were caught by the end of every episode. The dolphin actor that played Flipper most often was a female named Kathy. She and O'Barry were particularly close.

During the first years of the show, O'Barry didn't entertain any doubts about Kathy's or the other dolphins' welfare. He enjoyed his new fame and burgeoning bank account, and mounted capture expeditions to collect free-living dolphins that he sold to the new performing-dolphin facilities popping up all over the globe as interest in Flipper blossomed. "I was probably the highest-paid animal trainer in the world during that period. . . . It's very easy to lull yourself into complacency when you're getting a new Porsche every year. . . . To be perfectly blunt, I was as ignorant as I could be for as long as I could be."

Everything changed for O'Barry when, shortly after the show ended, he received a call from Miami Seaquarium, where Kathy was being kept. She was not doing well. Her daily schedule had changed, and so had the staff caring for her. She was now isolated in a steel tank, away from the other dolphins that she knew. When O'Barry arrived at Seaquarium, he found Kathy covered in black blisters from sun exposure (she'd been floating listlessly at the top of the tank), barely breathing, and extremely weak. O'Barry jumped into the water with his clothes still on. According to him, Kathy swam into his arms, stopped breathing, and died. "Kathy died of suicide. . . . Dolphins and whales are not automatic air breathers," O'Barry said. "Every breath they take is a conscious effort. So they can end their life whenever they want to and that's what Kathy did. She chose to not take that next breath and you have to call that suicide, or self-induced asphyxiation in a steel tank. That's the thing that turned me around."

Kathy died on April 22, 1970. It was the very first Earth Day. Twenty million people gathered for demonstrations across the United States, many of them wearing gas masks in protest and carrying homemade posters of Planet Earth, the oceans hastily drawn in with blue pens and paint. A week later, inspired by this new environmental groundswell and his wrenching experience with Kathy, O'Barry was in the Bahamas attempting to cut the wires of a pen so that a dolphin that he captured off the coast of Miami and later sold to Bimini's Lerner Marine Lab could go free. That dolphin refused to swim through the hole O'Barry slashed in the side of her pen. He was caught and arrested, but that didn't deter him.

Whether Kathy killed herself on purpose is unknowable. But the trauma of watching her die changed O'Barry's life. Today there are no more Porsches, and he told me that he's spent the last forty years dedicated to eradicating the very business he helped start, hoping to prevent deaths like Kathy's at other marine parks and aquariums. I was moved by O'Barry's story, but I wanted to know if anyone else who worked with cetaceans believed that dolphins could drown themselves on purpose. I called Naomi Rose, then chief marine mammal scientist for the Humane Society of the United States. She earned her doctorate studying the social dynamics of male orcas in British Columbia, is a member of the International Whaling Commission's Scientific Committee, and has served on various national and international boards to evaluate marine mammal health and the impacts of whaling, environmental change, and more.

She believes that suicide in captive whales and dolphins is possible, and she locates it on a continuum with the other self-injurious things they do, like swimming in compulsive patterns or bashing their heads against the sides of their tanks. When I asked her if she thought that Kathy could have committed suicide, she said that it was certainly possible. She also told me something that I had never considered: the dolphins and whales we see in captivity are probably the most psychologically sturdy. They are the only ones who survive.

"The whole spectrum of open-ocean species like Stenella [*Stenella coeruleoalba*, or the striped dolphin] are good examples. They live in groups of a thousand individuals. You can capture them and put them in a tank but you will come in one morning and find them all floating. Pilot whales are similar. They will live a year or two at most and then die. Bubbles, a pilot whale who lived at SeaWorld for twenty years, was a rare exception. The animals you commonly see at the aquarium or marine park, the bottlenose dolphins, the orcas and the belugas, those are the resilient ones. The ones who put up with captivity." Rose is convinced, though, that from time to time even these sturdy creatures can simply give up on life; this may take the form of a depression so deep they refuse to eat or engage with others, killing themselves slowly.

The death certificate of a depressed man who does not seek treatment for quickly multiplying tumors may attribute his death to cancer, but in fact the underlying cause may be debilitating depression. As I mentioned earlier, human suicide includes a range of behaviors, from the passivity of a person who stops eating or doesn't take his medication, or refuses to see a doctor for a disturbing lump, to the actively volitional deaths of people who shoot themselves or plunge from skyscrapers and bridges. Just as the methods and time frames vary, so too do the motivations, justifications, and explanations. Nonhuman animals also have their own continuum of self-destruction. They may have fewer tools available to them to inflict mortal wounds and also lack humanity's sophisticated cognitive abilities with which to plan their own ends, but they can and do harm themselves. Sometimes they die.

People have tended to interpret animal behavior as suicidal when it seems to be related to horrible and inescapable living conditions, when it looks like self-sacrifice, or, as in the turn of the twentieth century examples, when it's a convenient way to shore up dominant societal values. But suicide has also been invoked when other animals' behavior seems puzzlingly illogical and there is no scientific consensus to explain it. These suicide stories reached their zenith not in isolated newspaper accounts of individual animal suicides or in tales like that of Kathy the dolphin but in the repeated human observations of dolphins and whales dragging their heaving bodies onto beaches around the world.

Mass Suicide

The fact that Kathy died on the first Earth Day may not be the coincidence that it initially seems. Other twentieth- and twenty-first-century accounts of suicidal animals aren't just mirrors of social attitudes toward self-destruction, they are also reflections of a different set of anxieties, stemming from reports of links between environmental toxins and mental illness, the unforeseen consequences of military sonar use on marine mammals, and the unknown effects of a warming climate.

Stranding events—two or more marine animals coming ashore while still alive—have been noted since antiquity, recorded in drawings, photographs, newspaper articles, and more recently, videos on YouTube. One Flemish engraving from 1577 shows three immense sperm whales in various stages of dying, beached on a sandy coastline. Their mouths are open and they seem to writhe in agony. More whales are in the waves, heading toward shore, spouting as they come. Small knots of people watch from cliffs above the beach and a few tall ships bob in the distance. Twenty years later a painting of a stranding event in Holland shows a beached whale the size of a house, surrounded by women in Elizabethan collars, men in tights, curious dogs, and at least one nobleman looking on with concern from the back of a horse. Scientific statistics on stranding are much more recent, dating only to the late nineteenth century.

From the 1930s on, there were relatively frequent comparisons made between human and cetacean suicide in the American press. In one representative account from 1937, in an article titled "Enigma of Suicidal Whales," a reporter struggled to understand why fifty false killer whales in South Africa would strand themselves. Ten years later forty-four whales swam out of a rough ocean and "deliberately beached themselves . . . in an apparent mass suicide." Again, the *New York Times* invoked suicide: "Several times in the past whales have beached themselves along the Florida coast in unexplained 'hari kari' fashion." These reportedly suicidal whales were not just self-destructive; they also reminded reporters and their readers of Japanese soldiers.

A particularly large mass stranding in Scotland in 1950 inspired eyewitnesses to invoke suicide because the deaths seemed so deliberate. According to one observer, 274 pilot whales thrust themselves onto the beach, where they "piled up like gigantic rocks in shallow water. . . . The big black-skinned mammals gasped out their lives, thrashing wildly with their tails and trumpeting eerie cries." More than a dozen six-foot-long calves kept returning to shore to be with the adults, even though fishermen towed them back out to deeper water. "The babies kept churning back 'like torpedos,'" the observer said. Their shrill cries seemed like a call and response with the deeper roars of the adults onshore. A representative from the Natural History Museum in London attempted to dismiss notions of cetacean suicide, saying that pilot whales are a "gregarious lot," and perhaps the pod leaders ran aground by accident and everyone else blindly followed.

As time went on, cetologists increasingly reacted to reports of dolphin and whale suicide with skepticism, arguing instead that the deaths should be attributed to unknown causes. When twenty-four pilot whales stranded near Charleston, South Carolina, in 1973, for example, scientists from the Smithsonian arrived to autopsy the bodies. The institution's curator of mammals told a reporter, "The suicide theory is conceivable, [but] at this point there is no evidence one way or another." The only consensus among marine mammal scientists was that there was no consensus.

Nonscientists were less skeptical. Reports of dolphin and whale mass suicides continued to appear in American newspapers throughout the 1960s and 1970s, part of a trend in increasing reports of strandings more generally. One reason for the general public's openness to stories of suicidal cetaceans might have been, at least partly, related to growing convictions about nonhuman animals being worthy of protection. Beginning in the mid-1960s, whales and dolphins became the perma-smiling visages of the new environmental movement through "Save the Whale" campaigns, popular recordings of whale songs, and critical coverage of modern whaling. The historian Etienne Benson has argued that widespread compassion for these animals influenced debates about the capture of dolphins and whales from the wild for display, the creation of the Marine Mammal Protection Act in 1972, and the deployment of tracking methods and devices still being used in cetacean science. The role these animals have played as empathetic figureheads—perhaps similar to the dogs and horses of the late nineteenth century, whose close affinity with humankind made them more deserving of empathy and also seemingly capable of killing themselves—may have made reports of cetacean suicide more likely, whether or not the animals were actually doing it.

Almost forty years later, doubt and confusion surrounding the reasons for strandings have not quite gone away. Credible theories of potential causes include harmful ocean noise from military sonar, oil exploration, or heavy ship traffic; pollutants that effect the animals' health and behavior; large-scale changes in climate and associated shifts in winds, currents, and ocean temperatures; illness and disease; or topography that confuses the animals into coming onshore or becoming stuck in shallow water. These stressors, combined with recent research on cetacean sociality, culture, personhood, and communication, may be the most powerful explanation thus far. One theory is that strandings are predicated on really tight social bonds, which might encourage otherwise healthy animals to strand alongside members of their pod or social group who have been harmed by ocean noise, pollutants, disease, or something else.

If there was such a thing as an animal suicide conference it would be the annual Hawaiian Monk Seal and Cetacean Responders Meeting. Attendees are largely unpaid volunteers who live near common stranding sites on America's Pacific or Atlantic coasts, or in Alaska or the Hawaiian Islands. They run phone trees of volunteers and go to the beach late at night or leave work on their lunch breaks to investigate oddly behaving seals, dolphins, and whales. Responders who work for marine sanctuaries, the National Oceanic and Atmospheric

Administration, the National Marine Fisheries Service, the Coast Guard, or the University of Hawaii, are paid, but most are not. These men and women will spend hours trying to keep a stranded dolphin cool and hydrated with wet towels and buckets of water or by floating the animals in kiddie pools. They set up protective boundaries with stakes and tape, and stand alongside grouchy and exhausted monk seals, asking beachgoer after beachgoer to move a few feet farther back and speak in quieter voices.

I attended the 2010 meeting in Hilo in order to find out if marine mammals can kill themselves. But as soon as I arrived, I felt as if I were at a contemporary conference of family therapists to inquire about hysteria. None of the scientists and stranding responders I spoke to wanted to hear the word *suicide*. To them it reeked of an anthropomorphism more rancid than a warming seal carcass. I'd ask an attendee—most of them with tan faces and arms, dressed in whale and dolphin T-shirts and adventure sandals—if he or she believed a stranded marine mammal was committing suicide. They cocked their heads, looked skeptically at the nametag that identified me as a doctoral student at MIT, and at least one responder simply walked away.

I stayed anyway and in listening to the days of talks, including a lengthy report on a sperm whale carcass that was lodged on the rocks in front of Neil Young's house, I learned a few surprising things about strandings. Primarily, that everything I thought I knew was wrong. A stranded whale or dolphin might actually be choosing the better of two awful fates. As one responder told me, "Imagine that you were trying to cross a freeway and you got hit by a bus but were able to drag yourself to the side of the road to rest. Would you want someone to come by and drag you back onto the freeway?" Contrary to what I used to think, a stranded dolphin or whale should never be dragged back into the water. The beach may be acting like a terrestrial life jacket, holding them up to breathe. Dolphins and whales don't naturally float, and if they are weak they are liable to sink since it takes effort to hold themselves up at the surface to breathe. If they can't manage this, they can drown. Even though beaching themselves may eventually prove fatal, sinking and running out of air inevitably is, and more quickly. Exhausted, sick, and/or injured cetaceans often strand on rocks or beaches because it is an alternative to drowning. Some will eventually recover and return to the water. And yet, as I mentioned earlier, occasionally a pod of dolphins or whales strands and stays stranded, even if only a few of them appear to be ill.

The cetologist Richard Connor has argued that stranding may have something to do with the way these social animals have adapted to their vast marine environment. No other group of mammals has evolved in a place so devoid of spots to hide from predators. Dolphins and whales don't retreat to dens or burrows; they don't climb trees or hide in caves. In the face of danger, they are able to hide only behind one another. This may have affected the evolution of their social worlds, making their ability to trust, communicate, and cooperate with one another even more important. This may also explain why some strandings include otherwise healthy individuals. These healthy dolphins and whales may strand simply because their social bonds with their ailing fellows are too powerful to allow them to swim away. Reflected in these explanations is an interesting tension: acknowledging the animals as sentient, intelligent, and purposeful, but perhaps not sufficiently so for marine mammal scientists to be comfortable calling their stranding behavior suicide.

Only one of a pod of nineteen white-sided dolphins that stranded in Ireland in 1997 was ill. This sick dolphin, who succumbed to congestive heart failure and other illnesses, was also the oldest and largest in the group. It may have been that all of these stranded animals were ill, but if so, their symptoms were invisible. Or it could be, as the cetologist Hal Whitehead suggests, that the healthy animals stranded out of empathy for or solidarity with their elder companion.

The case may also be that whales and dolphins don't think of themselves as humans do—that is, as individuals with an individual consciousness, self, and body. Perhaps to be a false killer whale, for example, is not to be an "I" but to be a "we." Perhaps stranding alongside a sick companion is not a conscious choice in the same way that it would be for a human.

As I sat in the audience in Hilo, listening to the researchers and volunteers bat around ideas about how to educate the general public that stranded whales or dolphins might be choosing to strand instead of drowning at sea and shouldn't be pushed or dragged back into the water, I thought about what Beatriz told me about *ganas de vivir*. The presenters may not have said the word *suicide*, but they were recognizing that dolphins and whales were *choosing* to strand, which is why they should be left alone. This certainly seemed like an admission of dolphin and whale intentionality if not explicit intentionality to die.

Since the Hilo conference in 2010, a number of studies have been released linking the use of military sonar to stranding behaviors. The U.S. National Marine Fisheries Service and the U.S. Navy released a report later that year announcing that between 2010 and 2015, naval activities in the Northwest Training Range, a marine region the size of California, would result in an estimated 650,000 cases of harm to marine mammals. This prompted a coalition of environmental and Native American groups, including Earthjustice and the National Resource Defense Council, to sue the National Marine Fisheries Service for failing to protect marine wildlife. Other, similar legal battles are now under way to protect marine mammals from naval testing off the coast of California. The latest study on these animals and anthropogenic sound, published in the *Proceedings of the Royal Society* in July 2013, argued that blue whales flee from military sonar, which interrupts their feeding and other activities, and are subsequently more likely to strand.

These studies, court filings, and media articles don't mention suicide, but they don't really need to. They are another iteration of the conversation of concern that has been unfolding about and around cetaceans since they became the focus of conservation, as opposed to whaling, campaigns. It is perhaps many of the same motivations and empathies that undergirded earlier reports of suicidal cetaceans that are now inspiring efforts to document the impact of noise and climatic changes on their behavior. "Save the Whales" may have become "Save the Whales from Sonar," but the underlying message continues to be "Save the Whales from Humans," whether they are suicidal or not.

Chris Parsons, a marine biologist at George Mason University, has been studying cetacean behavior for over a decade. During a recent stranding of pilot whales in Florida, casting about for a way to explain the whales' actions, he said, "It would be like someone walking in front of a car: Why did they do that? Were they sick? Were they distracted? Or could they not hear or see the car coming?" Perhaps they wanted to die.

Mad Hatters of the Sea

Mercury, a potent neurotoxin, was introduced into the hat-making trade in the seventeenth century. Rabbit pelts were dipped in hot mercuric nitrate to soften the stiff outer hairs and make the layers of fur pack together more easily during felting. The hatmakers worked in poorly ventilated rooms and were exposed to large amounts of the neurotoxin. By the latter part of the nineteenth century, symptoms of mercury poisoning were so common among these men that terms like "the hatter's shakes" and "mad as a hatter" slipped into everyday speech. But the hatters didn't just shake; they were excessively shy, lacked self-confidence, and felt very anxious. They were also known to be pathologically fearful and exploded with anger when criticized.

Today the most widespread source of mercury exposure in humans isn't hat making, it is the fish we eat. Inorganic mercury occurs naturally in the environment; emitted by volcanoes, for example. But the majority is produced by human activities, like coal burning. Bacteria, fungi, and phytoplankton ingest inorganic mercury once it settles in the ocean and then turn it into methyl mercury. In time, these microorganisms are consumed by fish and other marine creatures, who are then consumed themselves. The mercury becomes biologically magnified as it moves up the food chain in progressively larger animals until it reaches its most toxic concentrations in the biggest and longest-living marine predators, such as dolphins, sharks, and some whales.

Almost all of this ingested mercury is absorbed by the gastrointestinal tract, where it enters the bloodstream and is then distributed throughout the body. In humans and other animals, mercury crosses the blood-brain barrier easily and builds up in the brain. It also crosses the placenta and accumulates in the blood, brain, and body of fetuses. In adult humans, mercury damage is focused and somewhat limited, causing a loss of neurons in the visual cortex and cerebellum. In the developing brain the damage is more diffuse and destructive. High exposure levels in fetuses, as well as in babies and young children, can cause deafness, blindness, cerebral palsy, mental retardation, and paralysis. Even limited exposure can cause subtle but disturbing problems, like difficulties learning, speaking, and paying attention. It also leads to psychiatric problems. Chronic mercury poisoning can result in anxiety, excessive timidity, and, according to a recent article in the *Journal of Neuropsychiatry and Clinical Neurosciences* that backed up centuries-old lore dating back to the hatters, "pathological fear of ridicule."

The effect of mercury on marine mammals hasn't been researched in great detail, but there are a few studies on tissue samples taken from dolphins and whales during stranding incidents, from healthy whales in the open ocean, and from harbor seals and arctic seals. It has also been possible to work backward from studies of human populations, like those in the Faroe Islands, who eat whale and dolphin meat, to find out about mercury levels in the animals themselves. Over the past few years toxicologists have shown that the bodies of toothed whales, seals, sea lions, and polar bears—many of whom subsist on diets of mercury-laden fish—contain high levels of the nerve toxin. In harbor seals this contamination is also linked to compromised immunoresponses. Since mercury exposure in humans is thought to cause nervous-system problems, some with psychiatric effects, it is possible, as Ric O'Barry has argued, that marine mammals suffer from their own forms of neurological upset too, some of which may encourage stranding behaviors.

Mercury isn't the only environmental toxin that has been linked to mental illness in humans or other animals. These studies are part of a larger body of toxicology research plumbing possible links between environmental toxins and mental health in humans and other animals, if not suicide directly.

Lead, manganese, arsenic, and organophosphate insecticides have all been associated with increased incidence of mental illness in people and abnormal behaviors in laboratory animals. In one human study, factory workers exposed to lead suffered from higher rates of depression, confusion, fatigue, and anger. Studies in lead-exposed children showed increases in antisocial behavior and attention problems. Manganese poisoning is thought to result in anorexia, insomnia, and weakness. There have also been reports of people exposed to the toxin who laugh or cry endlessly and feel strong impulses to run, dance, sing, or talk impulsively. Chronic arsenic intoxication has been associated with everything from dizziness and diarrhea to depression and paranoid delusional thinking.

Experiments on lab rats and mice have shown that exposure to lead, arsenic, mercury, and other toxic substances made the rodents behave in strange ways well before they sickened and died. Trying to understand the potential relationship between these toxins and odd behaviors in wildlife is more difficult. There are, however, a few well-documented cases of self-destructive behaviors in other creatures caused not by toxins but by parasites.

Self-Destruction Infection

Beginning in the early 1990s the Czech scientist Jaroslav Flegr began to wonder if his occasional risky behavior—wandering into the path of honking cars on busy streets, brazenly sharing his disdain for his country's communist leadership when it was still dangerous to do so, or remaining calm as gunfire exploded around him on a research stint in Turkey—might not be a reflection of his personality but instead an infection. Flegr had just read a book by the evolutionary biologist Richard Dawkins that included the natural history of a flatworm that infects ant nervous systems, turning them into self-destructive drones, which ensures the flatworms' own reproductive cycle. Ants infected with the worm become extremely reckless, climbing to the top of blades of grass and locking on with their mandibles, as opposed to their normal behavior of burying their heads in the ground to avoid predators. Passing sheep and other grazing animals are much more likely to eat the ants when they're locked onto the tips of the grass. Once the flatworm finds its way inside the grazers, it is able to reproduce.

Flegr puzzled over whether he might be the human version of a reckless ant. When he joined a biology department specializing in the study of a similarly manipulative parasite, *Toxoplasma gondii*, and tested positive for the protozoa himself, he decided to focus on the study of the parasite's life cycle and its possible behavioral affects. *T. gondii*, or toxo, as it's often called, is shed in the feces of infected cats. Rodents, pigs, cattle, and other creatures then pick up the parasite from the soil as they graze or scavenge for food. Toxo will spread throughout the body, dispersing into the brain and other tissues. Humans can harbor the parasite too and are most likely to be exposed via cat litter boxes, by drinking water or eating produce contaminated by animal feces, or by consuming the uncooked meat of infected animals. Flegr discovered that the French, who eat more rare meat than Americans, have toxo infection rates as high as 55 percent in some areas. Americans, according to a Michigan State University study, have infection rates between 10 and 20 percent. The parasite can't complete its life cycle inside humans, rats, pigs, or other creatures, however. It needs to find its way back into a cat. This is where the toxo story becomes truly bizarre.

Flegr discovered that the parasite transforms its host animals into cat-delivery systems by changing their behavior, making rodents more active and likely to attract the attention of cats, for example. Building on Flegr's research, a parasitologist at London's Imperial College, Joanne Webster, discovered that infected rats and mice not only become more reckless in the presence of predators but are actually *attracted* to the smell of cat urine. Together with another parasitologist, Webster demonstrated that toxo increases the production of dopamine (a neurotransmitter associated with pleasure, but at high levels also with brain damage and schizophrenia) in the rodents' brains. When the researchers gave the infected rats antipsychotic medication, which blocks dopamine reception in the brain, the rodents' attraction to cats plummeted. Meanwhile, at Stanford University, the neuroscientist Robert Sapolsky and a postdoctoral student in his lab demonstrated that the parasite also dismantles fear responses in rat brains and encourages new connections, explaining their self-destructive behavior but also extending the theory to show that toxo actually makes the rodents' attraction to cat urine sexual in nature. Oddly, toxo also makes the infected males more attractive to female rats, a nifty biological trick since the parasite can also spread like an STD, traveling from the male rodent's sperm into the female's womb, where it infects her and her pups.

For decades we have known that pregnant women infected with the parasite are more likely to have miscarriages or give birth to stillborn infants or infants with abnormally large or small heads. But Flegr wondered if this parasite could also cause behavioral effects in humans, making them somehow more attracted to cats or perhaps more reckless with their own lives, like the rodents. He discovered that people who had been exposed to the parasite were more than twice as likely as others to have had a major car accident, showing, perhaps, that the infected people were riskier drivers. Subsequent studies in Turkey and Mexico had similar findings. More recently, Flegr published the results of a study that suggested infected human men like the smell of cat pee.

The psychiatric effects in people infected with toxo are even more startling. A number of studies, going back to the 1950s, have shown a correlation between toxo infection and schizophrenia, though not a causal link. One 2011 study of suicidal women in twenty European countries pointed to a possible correlation with each population's infection rates. This research echoes other investigations linking the presence of toxo to suicidal thinking and the incidence of suicidal acts in humans and possibly homicide. Yet toxo itself may not be a protozoan puppeteer, actively urging people to kill themselves. Instead the culprit may be the neurochemical response of the human body to the parasite, the damaged brain tissue it leaves behind, some aspect of the parasite's dopamine production, or a combination of these factors. That is, toxo, or anything else that causes neurological damage, may also make people more likely to kill themselves or suffer from schizophrenia.

A 2012 Michigan State University study published in the *Journal of Clinical Psychiatry* found that brain inflammation resulting from infection might be the most likely explanation. One of the study's lead investigators, Lena Brundin, wrote, "Previous research has found signs of inflammation in the brains of suicide victims and people battling depression, and there also are previous reports linking *Toxoplasma gondii* to suicide attempts." The study indicated that people who tested positive for the parasite were seven times more likely to attempt suicide.

In 2010 the toxo story took another interspecies twist. Otters started dying off in large numbers off the California coast. Even stranger, however, was that the number of otters attacked by sharks doubled over a twenty-five-year period. While recovering shark populations in the otters' range could partly account for the spike in attacks, researchers weren't convinced that this was the sole explanation. A few years earlier, a professor and veterinary parasitologist at the UC Davis School of Veterinary Medicine, Patricia Conrad, and her coauthors discovered that 42 percent of live otters and 62 percent of dead otters were infected with *Toxoplasma gondii*.

Conrad also found that otters who lived closer to cities were more likely to be infected. It seemed that the otters were contracting the parasite when runoff containing cat feces washed into coastal waters. Otters with moderate to severe brain inflammation, a symptom of toxo infection, were roughly four times more likely to die from shark attacks. A 2011 study of marine mammals in the Pacific Northwest also found high rates of toxo infection in dolphins, seals, and sea lions. Not only did these animals have *T. gondii*, but many were also infected with another parasite, *Sarcocystis neurona*, shed in the feces of opossums. The marsupials have been steadily expanding their range into the Pacific Northwest, where the possum poop, like cat feces, is washed into the marine environment during the area's frequent rainstorms. Researchers at the National Institute of Allergy and Infectious Diseases theorize that *S. neurona* exacerbates the symptoms of toxo in infected animals by weakening their immune system. There haven't been any studies, not yet anyway, on whether these seals and dolphins are also engaging in risky behavior—taunting sharks, playing chicken with oncoming boats, or putting themselves in danger in some other mysterious fashion.

Crazy Like a (Sea) Lion

Another strange phenomenon involving seemingly self-destructive wildlife is taking place in many of the same waters where sea otters may be acting like tiny, furry drunken sailors. Along the Pacific Coast,

California sea lions, *Zalophus californianus*, are suffering from what seems to be a strange form of environmentally induced madness.

One wet winter and spring I spent my Mondays volunteering at the Marine Mammal Center in the sage-covered hills of the Marin Headlands, overlooking San Francisco Bay. The only marine mammal hospital of its kind, it treats hundreds of sea lions, harbor seals, and elephant seals every year. The center's veterinary staff are some of the most experienced marine mammal specialists in the world.

On one of my first volunteer days at the hospital I walked past a young California sea lion. Her eyes darted frantically past mine as she tried to climb the twelve-foot chain-link fence of her pen, scrambling against the base of it repeatedly, failing to get a flipperhold. Someone had written "Frenzy" on the nametag hung on the outside of her enclosure. The name was fitting. While there have been some escapes at the center—once, before the facility was rebuilt, a sea lion escaped from his pen and the next morning's crew found him relaxing on the couch in the break room—sea lions do not normally attempt to climb chain-link fences.

Frenzy had come ashore roughly twenty-five miles south of Santa Cruz. Observers saw her seizing and acting sluggish and called for help. When I met Frenzy she had been at the Marine Mammal Center for just under a week. Her seizures had been brought under control with diazepam (Valium) but she was wriggly and in constant motion. Then she would suddenly stop, look around, and, as if triggered by something I couldn't see, rush to the door of her pen, barking and looking from side to side—behaving as the other sea lions do when it is feeding time and they see a volunteer or staff member with a bucket of fish. Only there was no one with a bucket of fish and she couldn't see me from where I stood.

Frenzy seemed to be responding to a sort of ghost stimulus. Her desires to bark, climb up the wall of the pen, and dive the length of the pool passed over her like clouds in a brisk wind. These behaviors weren't abnormal, but the way that she performed them was strange. It was a bit like watching my grandmother, who has dementia, try to cook dinner. While she is still able to turn on the stove and fill a pot with water, her actions unfold jerkily and not necessarily in order. You cannot quite pinpoint what is wrong, but clearly something is.

With a quick glance at Frenzy's chart I saw that the vet staff had decided that her symptoms were caused by algae. Surfers, swimmers, and other beachgoers along the Pacific Coast often see notices for red tides, algal blooms that tint the water red or an orangey brown, accompanied by warnings not to go in the water. Most of these blooms are harmless to animals, but a few dozen species of phytoplankton and cyanobacteria produce toxins. One of these is the diatom *Pseudo-nitzschia australis*, a tiny, narrow creature that looks like a nail file and produces domoic acid, a neurotoxin that affects humans and certain other animals. The acid accumulates in shellfish, sardines, and anchovies that eat the diatom during blooms. They are, in turn, eaten by sea lions, otters, cetaceans, and some humans.

Domoic acid toxicosis was first diagnosed by Frances Gulland, then chief scientist at the Marine Mammal Center, in 1998, when hundreds of oddly overconfident, seizing, and dehydrated California sea lions stranded on state beaches. In humans, exposure to the neurotoxin causes a condition known as amnesic shellfish poisoning. Sufferers can vomit and have diarrhea after eating contaminated mussels or other shellfish. In some cases, people may also become confused, lose their memory, or feel disoriented for a while. Rarely sufferers fall into a coma; for a minority of people, usually the elderly, the very young, or people with diabetes or chronic kidney disease, domoic acid can cause permanent cognitive problems.

Depending on where their primary hunting grounds are, sea lions like Frenzy can have long-term exposure to the toxin. If the domoic acid bathes their brains for a sustained period, it can literally drive them crazy, producing what looks like self-destructive behavior. For many sea lions, this is a short-term problem. Animals picked up by the Marine Mammal Center are rehydrated and fed uncontaminated fish to flush the toxin from their body and are treated with antiseizure medications. As long as they regain their strength and don't overly bond with their human caregivers (volunteers are asked not to look the extremely social and curious creatures in the eye, touch them, or speak to them directly in case they become too friendly), they can quickly be released. Sea lions who have had prolonged exposure to the acid, though, can't be returned to the ocean. Many of these animals have lost their sense of direction, are unable to dive as deeply or for as long as they once could, and may suffer from other problems, like pathological bravery. Researchers have located the source of these problems in the animals' hippocampi—a brain region associated with learning, memory, spatial navigation, and other functions. These sea lions' hippocampi are damaged, a possible explanation for their bizarre behaviors and psychiatric effects.

Recently researchers at the Marine Mammal Center and Moss Landing Marine Laboratories released a number of sea lions who had had chronic exposure to the toxin in order to better understand its long-term effects on behavior. They fit the animals with radio transmitters and tracked their movements. The results were depressing. Few of them could dive normally, and they swam in strange directions. One sea lion headed straight out into the open ocean without pausing to eat or rest and got halfway to Hawaii before he stopped transmitting; the researchers assumed that he was too exhausted to go on and died of starvation. Another, a female, swam two and a half miles up the Salinas River toward the vast inland stretches of artichoke fields and lettuce farms. She spent ten days just swimming in circles.

There is also the sea lions' weirdly overconfident behavior. One sea lion released from the Marine Mammal Center, known as CSL 7096, was extremely aggressive and hassled surfers at a surfing competition as they entered the water. Another sea lion, nicknamed Wilder, came ashore in the San Francisco Marina. He climbed on top of a police car and stayed there for forty-five minutes.

In humans, the hippocampus plays an important role in regulating our responses to anxiety in addition to helping us navigate our surroundings. It is highly sensitive to repeated exposure to stress hormones, and a number of brain-imaging studies have shown that people with certain psychiatric disorders, like PTSD, major depressive disorder, and borderline personality disorder, also have hippocampal abnormalities. What this means for sea lions is hard to know, but unfortunately for them, with every passing year there seem to be more opportunities to plumb the relationship between red tides and their mental health. According to the World Health Organization and the National Oceanographic Atmospheric Administration (NOAA), rising ocean temperatures may be increasing the number of *Pseudo-nitzschia* blooms.

Sea lions like Frenzy may be the new, flippered, barking version of coal miners' canaries, at least when it comes to possible effects of climate change on mental and physical health. Until the mid-1980s miners carried canaries in small cages down into the dark of the pit as alarms for deadly carbon monoxide levels. If the birds died, the men (if they didn't surface immediately) would too. Even a canary falling off her perch was considered a danger sign. Perhaps a sea lion resting unabashedly on top of a patrol car should be too.

Epilogue: When the Devil Fish Forgive

> There cannot be one companion species; there have to be at least two to make one.
>
> Donna Haraway, *The Companion Species Manifesto*

Occasionally what helps another animal the most is a form of common sense. Sometimes, however, we need to travel to the unlikeliest of places in order to be reminded.

I met Pi Sarote while drenched in sweat, heaving buckets of water across a clotted field alongside a small group of foreign volunteers, in the northeastern Thai province of Surin. We were planting bamboo seedlings that, we were told, would eventually feed the hundreds of hungry elephants who lived in the village of Baan Ta Klang. I was doubtful, though, that anyone was going to irrigate these pathetic little plants after the foreigners left. The irrigation project had the feeling of make-work. It was hot, and I was beginning to wonder why I was there. I wanted to learn about elephants, not bamboo plants. Specifically I wanted to learn about problem elephants, creatures who were difficult or killed people, and how the men and women of Baan Ta Klang, who were famous for their skills with the animals, made them more peaceable and content. When Pi Sarote appeared as one of the mahouts assigned to look after our group, I noticed that he carried no elephant hook or stick. I had not seen this before. I stopped complaining to myself and stared at him, as he made his way toward us across the field.

Two elephants walked calmly alongside him, pausing every once in a while to tear a tuft of grass from the hard ground or snatch a mouthful of leaves from an overhanging branch. One of the elephants, a seventy-eight-year-old female named Mae Bua, had soft eyes and powerful ears that snapped like flags. Later, I would learn that she was born outside of Pi Sarote's grandfather's house in 1932. When he died, Sarote's father inherited Mae Bua, and when he eventually died, too, Mae Bua passed to Sarote. If she lives long enough, the elephant will go to his sons, whom she already knows well. Sarote refers to her as "Grandmother" and hasn't lived a day of his life without her nearby, just like his father. Mae Bua has never been hit or trained to do tricks, and she has given birth to six calves, all of whom stayed in Sarote's family and were raised by her, together with five or six elephant nannies in their village.

The other elephant with Sarote that day, a six-year-old female named Noon Nying, was as rambunctious and loud as the older elephant was calm, observant, and measured. She repeatedly tried to unscrew the top of my water bottle until I showed her it was empty, and clambered upstairs into a makeshift vet clinic to steal unattended bananas. Noon Nying was also wildly protective of Sarote. If you wanted to walk with him, you had to be careful to walk on the side opposite from Noon Nying; otherwise she would push her large head, then her entire body, between you and him until she had reasserted herself at his side. Sarote's wife is the only person he can embrace in front of the elephant without making her upset.

At forty-three, Sarote has deep lines around his eyes from squinting into the sun after his elephants. He cracks jokes constantly, often about the animals, and when he talks, he rests his weight idly against one of his elephants, as another man might lean on a bar. The elephants respond by curling their trunks into his hands or lightly leaning their weight into him. The day that I met Pi Sarote, Noon

Nying had been with him only four months. They communicated in a mostly wordless manner that made it seem as if they had always known each other. The elephant followed him everywhere, squeaking at him constantly, a sort of elephant sonar, pinging back his attention and affection.

When Noon Nying and Sarote made their way to a muddy pond for her daily swim, he would turn to face her and in a barely audible whisper say, "Noon Nying, go ahead, swim." Then she would calmly head off into the water. When it came time to leave the pond, the other mahouts called, cajoled, and sometimes waded into the water to retrieve their elephants, a few of them waving metal hooks in the air. Sarote stayed onshore and cocked his head at Noon Nying, or whispered her name, and she came trotting out of the water to his side.

As we all worked together, planting, or filling water buckets, through those bright afternoons, I would look over from time to time and catch sight of Sarote. He'd be relaxing in the only shade that existed in the open field, the rectangle of earth directly below Noon Nying's belly, between her four legs. She grazed in place while he sat cross-legged beneath her, the top of his hat brushing against her stomach.

Four months earlier no one would have dared to get this close to Noon Nying, let alone sit directly beneath her. Noon Nying was so aggressive that her last mahout was terrified she would kill or badly injure him. Born in the district of Cha'am in southeastern Thailand to an elephant owned by Sarote's cousin, Noon Nying spent her first year with her mother. As is customary, she chose her own name by picking one of three different pieces of sugarcane that each signified a different name proffered by the elephant monk. Sarote lived nearby, and though he wasn't her mahout, he visited the calf and her mother often. When she was a year old, her training began. Noon Nying was taught to paint, holding the paintbrush in the tip of her trunk, and to hula-hoop with a giant plastic circle. She was being groomed to be a circus elephant, and learning the tricks was a grueling process. For a while she performed in a local show, painting pictures or kicking a soccer ball in a dusty, dilapidated circus ring in the center of the village; then, when she was four years old, Sarote's cousin rented her to an elephant camp in the North, just outside the city of Chiang Mai, not far from Elephant Nature Park, where Jokia and Mae Perm live.

Noon Nying worked at the camp for two years with a mahout hired by the camp staff. It did not go well. By the time she was six, the family back in Surin was receiving reports that Noon Nying was dangerous and becoming a liability to both her mahout and the tourists who came through the camp. She didn't like doing the tricks she was trained for and often flat-out refused to do the show. In frustration, the camp called Sarote's cousin, who was ill, and demanded that she be picked up. Knowing Sarote's skill with elephants, he asked him to go north to retrieve Noon Nying and see if anything could be done to make her more tractable and less upset. Sarote remembered her as a playful baby and had a feeling that she wasn't a killer, just miserable and lonely. When he arrived at the camp, he found an elephant who was thin and small for her age but happy to see him. He made the long trip back to Surin with the young elephant, hoping he would be able to afford to feed her without making her work in the circus.

When they arrived back home, the first thing Sarote did was introduce Noon Nying to Mae Bua, trusting that the companionship of the older elephant would be good for her. Not long before, Mae Bua had helped another anxious elephant. A female in the village gave birth to her first calf and not only refused to nurse but tried to kill the baby. The night she was born, it took forty mahouts carrying sharp sticks to keep the mother from hurting the calf. As soon as they had chained her, she would break the chains and rush toward her baby, trying to crush the small, terrified creature. After hours of this, the men tried putting the baby back with the mother, but she swung at the calf again, knocking her unconscious, and the men had to resuscitate the tiny elephant. The owner of the elephants, an experienced mahout named Pi Pong, asked Pi Sarote if they could use Mae Bua, who was known for her mothering skills and calm presence.

The men chained the mother elephant to a tree and gave the baby to Mae Bua, who stood just beyond the mother's reach. The men stepped back and watched Mae Bua stroke the calf and then slowly lead the baby in a circle around her mother. Guiding the calf with her trunk, Mae Bua then brought the baby to her mother's breast, heavy with milk, and urged her to drink. She did. Mae Bua stayed nearby, guarding the baby and comforting both elephants until the baby could drink on her own. Eventually Pi Pong unchained the mother. For two weeks Mae Bua stayed with the pair day and night, and from then on, she was never far from the mother and calf. Two years later, when Pi Pong sold the two elephants to the government in another province, Mae Bua broke her own chains trying to go to them. For days after they left, she would dash to where she last saw them, calling and calling.

"When a mother and calf are separated they get sad," Pi Pong said. "Aunties too. You need to find them new elephant friends or you won't be able to sleep through all the crying."

In the days after Noon Nying arrived back in the village, she, Pi Sarote, and Mae Bua walked together to the water and grazed in the dry forest. In the afternoon, Sarote bathed Noon Nying and led her around the village. At sunset he chained her next to Mae Bua and gave them both piles of grass or pineapple tops to eat. On cold nights he lit a fire to warm them; on buggy nights he lit a fire to smoke away the insects. Every day Pi Sarote stroked Noon Nying, encouraged her, fed her treats, shared his own snacks, laughed when she did something funny, and teased her gently. His children visited her and did the same. So did his friends and cousins. He spoke to her warmly and sometimes sternly, but always with affection. Soon Noon Nying began to put on weight and the squeaking started, her way of expressing her affection. She never quite bonded with Mae Bua, but she settled happily into her new life and routine.

"Now she knows she's important," said Sarote. "She loves me and I love her."

She also trusts him. Sarote believes that after affection, developing trust is the most important aspect of helping a disturbed and unhappy elephant become a happy one. "She doesn't like being chained, for example, but she knows this is what I have to do at night so I can sleep and see the rest of my family. She understands that every morning I will always be there to unchain her."

One night, as we sat around a fire drinking the local rice wine, a milky, fiery liquid called *satoh*, out of tin cups, Sarote's daughter lay sleeping at the edge of the fire. "When I was a child my daughter's age," Sarote said, "this was an entirely different place." As he spoke, I could hear the soft crunching of Mae Bua and Noon Nying behind him in the darkness.

Sarote and the other inhabitants of Baan Ta Klang are Guey, an ethnic group that has lived in this part of Southeast Asia for centuries. Elephants are the center of Guey culture—as family members, sources of income, and sacred beings whose births and deaths are important community events. For thousands of years Guey elephant shamans rode captive elephants into the forest and, using lassoes made from boar skin, captured between thirty and forty wild elephants per year. These multiday expeditions were tightly controlled and the men chosen carefully; all members of the party were forbidden to speak Guey during the expedition, communicating instead with a secret language known only to them that took years to learn and could only be spoken in the forest. The captured elephants were used by the Guey as transportation or were sold as war elephants to the kings of Siam. Later the Thai government and British companies bought the animals for use in logging.

The region's vast forests were home to not only large herds of wild elephants but also two species of rhinoceros (Javan and Sumatran), water buffalo, wild cattle (banteng, guar, and kouprey), tigers, leopards, Asian wild dogs, and many small herbivores. There were so many animals that the Guey built elevated houses to protect their rice harvests from hungry wildlife.

Until the last wild elephant was captured in 1961, Surin had supported healthy herds of the animals. By the end of the twentieth century, however, the forest, along with the animals it supported, was gone. Gigantic rice paddies came to dominate the landscape, dotted only with the occasional surviving tree. When Pi Sarote was a teenager, for the first time in Guey history the community had to buy food to feed their elephants, who had always grazed in the forest. Many people sold their elephants; those who remained had to be chained in place because there was no longer any forest to let them loose in.

The Guey's cultural shifts were as dramatic as the changes to the landscape that surrounded them. Generations of men had captured and trained wild elephants; now they farmed small-scale rice paddies and hunted game, found work on the large rice farms, or applied their immense skill with elephants to training the animals for trekking or circus shows. Some of these men took to the streets with their animals, begging in the nation's cities, where urban Thais pay to touch the elephants or to feed them sugarcane. That was the route that Sarote took with his large-tusked male Jan Jou, an elephant he'd been working with since he was eleven years old. The only other option available to Sarote would have been to sell or rent Jan Jou to a tourist camp, but he refused to do this. So for ten years he traversed the entirety of Thailand on foot with Jan Jou, sleeping under a tarp with other men from his village at night, the elephants asleep on their feet nearby. As he walked, Sarote learned to speak four languages fluently. He doesn't say so, but elephant is a fifth. He also traveled with his father and Mae Bua, offering rides on her back to paying customers. Sarote talks of these years sorrowfully. He stopped street-begging entirely after Jan Jou fell ill and died when they were far from home.

Few of these Guey men, descended from long lines of elephant capturers and shamans, want to take their elephants to the city to beg. The work is demeaning, dangerous, and uncomfortable for both the men and the elephants. They're far from their families and it's often difficult to find food and clean water for the elephants. Sarote returned to Baan Ta Klang as soon as he could, but he laments the lack of jobs and the kind of life this entails for the elephants. "Men stop begging in the cities and then they come home, but here the elephants must be chained all the time so the man can get a job in the rice paddies. This is not good for the elephant."

It occurred to me that the different lives of Pi Sarote's elephants mirrored the drastic changes that had unfolded around them. Mae Bua came of age at a time when she worked only occasionally. When needed, she helped the people who treated her and her babies like family. When she wasn't working she was let loose in the nearby forest, raising her calves as she saw fit and socializing with other elephants.

Despite being born to the same human family, Noon Nying entered a drastically different world. And yet, regardless of the many challenges that Sarote and his elephants faced, he was able to help her. He brought the young elephant back from the emotional and physical brink and restored her to health, saving her from becoming aggressive and isolated, encouraging her instead to be affectionate and sociable.

Almost two years after I first visited the village, I returned to see Sarote and his elephants. I arrived in a pickup truck full of pineapple tops, bananas, and cucumbers, and the elephants chained on either side of the road lifted their trunks, sniffing the air as we passed. Sarote met me with a big grin, teasing that I had finally come back to take over as Noon Nying's mahout. He explained that he'd sent Mae Bua south to stay with friends in Pattaya, an area thick with pineapple fields. She was ready to retire and he felt she would have a good retirement in a place where the elephants are given the run of the fields after the harvest, free to eat as many discarded pineapples as they chose.

Noon Nying had grown at least a foot since I'd last seen her and gained so much weight that she looked like she'd been inflated. She was taut, round, and strong, just like a seven-year-old elephant should be. She rumbled happily, standing inches from Sarote, still refusing to let anyone between them. We'd been chatting a few minutes when she let out a sudden sharp squeak and another elephant began running from across the road. He was the same size as Noon Nying, healthy and fat with a wide head and a thunderous gait. He headed straight for us and fast. I could feel the ground shaking.

Pi Sarote saw me flinch and laughed. "That," he said, "is Teng Mo. Thai for 'watermelon.' Don't tell me that you're scared of a watermelon."

As he spoke, I realized that the elephant wasn't barreling toward us at all but toward Noon Nying. They met with an explosion of trumpets.

The reticent, skinny elephant who, less than two years earlier, was largely indifferent to other elephants had made an elephant friend. For the last six months, Sarote told me, Teng Mo and Noon Nying had been inseparable. Teng Mo was five years old, sweet, and slightly impetuous but extremely affectionate with Noon Nying. He had been trained to sit and raise his leg on command, but his owners didn't want him to work in a circus, so they hadn't taught him to paint, play soccer, or hula-hoop.

Instead of leaving Sarote for her new elephant companion, Noon Nying now goes everywhere with her two males. She squeaks for both of them and doesn't like being separated from either one. When Sarote turned and walked toward a nearby pond, the two elephants fell into step on either side of him, extending their trunks to each other over his head and rumbling with contentment. The pachyderm-man sandwich moved slowly into the eucalyptus forest, Sarote invisible except for his feet, flashing between the eight others as they walked.

I remembered a conversation I'd had with Sarote on my last visit. We sat cross-legged in the shade next to a three-week-old elephant calf, her mother standing over her watchfully as she slept. The mother's stance reminded me of the way Noon Nying stood over Sarote as he rested in the field. Occasionally the calf squeaked in a dream or moved her feet as if she were running, and the mother reached her trunk down to stroke her sleeping baby.

"The elephant is very important here," Sarote said. "Like family. If we didn't have elephants this town wouldn't exist. I would not exist. It is how it's always been. And we help each other. Like when the river flooded last year. The water was so deep that people couldn't cross the rice paddies to bring the rice back. The elephants helped us. It was like it was before we had trucks. We worked together."

I wondered if there was something beneficial in this odd partnership, and if that something had helped Noon Nying. Was being in a relationship with *both* humans and her own kind therapeutic? Once, I asked Sarote what enabled him to heal a troubled elephant like Noon Nying. He said, "*Jai dee,*" and placed a hand on his chest. Literally translated, this means "good heart," but it means more than that, too. It signifies good intentions, wholeheartedness, and something else more mysterious that I was never quite able to pin down.

"If you have *jai dee,*" he said, "the animals will know it, and they will have *jai dee* too."

The reverse is also true. Many people in Surin believe that if you do not have *jai dee,* then your elephant may not be happy or loving. And if an elephant is mean and unkind for no good reason, if he or she is not goodhearted, this makes the mahout mean, unkind, and unhappy too. Sarote and the rest of the Guey elephant men, along with many other mahouts, veterinarians, and trainers I spent time with in other parts of Thailand believe that the border between humans and elephants is a porous one, at least in the context of mental health. They assume that feelings, intentions, and empathy can be transferred in ways that heal or harm. This belief in shared emotional experiences suffuses daily life for Sarote, his colleagues, and family members, affecting practical decisions such as whether a mahout gets a job with a particular elephant or whether he keeps that job if he and the elephant seem emotionally incompatible.

The elephant monk had told me, "In order to understand other animals, first you have to understand yourself." What I didn't know when I sat on the monastery steps hoping for some sort of grand pronouncement about animal minds, was that this also works the other way around. For centuries, humans have observed, worked with, befriended, poked, caged, trapped, fed, celebrated, prodded, denigrated, feared, identified with, suspected, hassled, petted, studied, dosed, and healed other animals—often with the aim of better understanding ourselves, our brain chemistry, behavior, thought processes, emotions, and struggles for sanity.

Perhaps, as the Guey elephant men convinced me, the divisions between humans and other animals are far more permeable than we think, at least when it comes to mental health and how we make sense of it. This is, in a way, not so unlike the dog trainer Ian Dunbar's assertion that the unhealthy lives we often lead as contemporary Americans are reflected back to us in the unhealthy lives of our cats and dogs. It's also reminiscent of the nineteenth-century idea of insanity as something that could be passed freely between humans and other animals. *Jai dee* is the other, brighter side of this coin. As Pi Sarote told me, "Anyone can have *jai dee*." Wholehearted happiness might be just as contagious as its opposite. Thank dog.

One of the most encouraging aspects of animal mental illness is that, against all odds, many creatures thrive, or at the very least, exhibit the kind of behavior that looks a lot like resilience. Brian the bonobo improved with the help of Lody, Kitty, and the keeper staff, a lift from pharmaceuticals, and a tightly controlled environment. Gigi became more resilient with the support of other female gorillas and the hard work of the humans who care for her. Mosha continues to hop around after Ladee, squeaking her delight throughout the quiet afternoons, while farther south Teng Mo and Noon Nying wrestle together in the mud, tangling with their affection for each other and for Pi Sarote. Recently, Mac the donkey has improved a bit as well. He's not biting himself or the metal bars of his corral as much, and he seems more relaxed. This is because my mom and her partner enlarged his enclosure, giving him a wider area in which to graze that keeps him busy eating thistles. They've also taken to drinking happy-hour beers at the top of the hill at the edge of Mac's range. He stands there, at dusk, just out of reach of the picnic table but a part of the human activity he's so curious about.

Some of these creatures have given their human caretakers a second chance, or a fourth one. I know Oliver did this for Jude and me. His first family disappointed him but he gave us his affection anyway. If dogs can hope, then perhaps that was what he was doing, or maybe his personality was simply one of desperate friendliness. Whatever anxieties, compulsions, or fears plagued him, they didn't keep him from expressing his dogged version of love.

A few years after Oliver died I went to Baja, Mexico, with the behavioral and wildlife biologist Toni Frohoff to dangle my hands over the side of a small fiberglass boat called a *panga*. I was hoping to meet a whale. Frohoff, the same researcher who consulted on the disturbed dolphin in a shopping mall, focuses on marine mammal behavior and communication, particularly among "solitary sociable cetaceans." These dolphins and whales choose not live with a pod of their own species but to be on their own, often socializing with humans more than their own kind. Frohoff has been all over the world studying these oddly behaving cetaceans, such as the orphaned baby beluga named Q in eastern Canada who preferred to sing and play with people. I went to Baja with Frohoff not to see solitary sociable cetaceans like Q but to see a whole group of them, the *ballenas amistosas*, or friendly gray whales. She promised me that if I splashed my hands in the water over the side of our little blue and white boat in the calving lagoons, the whales would come to check us out, and if we were lucky, perhaps even offer themselves up to be touched.

California gray whales summer in the Arctic, feeding on small invertebrates. In the late fall and early winter female whales of calving age and males old enough to breed travel five thousand miles to the warm, shallow lagoons on Baja's Pacific Coast to give birth, nurse their calves, and mate. Throughout the mid-nineteenth century and again in the first half of the twentieth, this seasonal congregation of whales was a target. Whaling captains like Charles Melville Scammon, who used the latest in whaling technology, studied the animals exhaustively, and wrote detailed accounts of their natural history. He used this to his advantage, going to the lagoons to harpoon the calves, a ploy to attract the real oil prizes, the mothers, who rushed the boats in an attempt to rescue their babies from the whalers.

The grays fought back so intensely that whaling captains and crews called them "devil fish." The animals killed men, splintered their boats into shards, and prompted one San Francisco reporter to write in 1863, "As many men are lost catching them as in all the other whaling grounds put together."

The whales' ferocity couldn't save them. By the early 1900s there were fewer than two thousand California gray whales left. Protections were put in place in the 1930s and 1940s, and slowly the population began to recover. Then one winter morning in 1972, something puzzling happened.

The Mexican fisherman Francisco Mayoral, known to his friends as Pachico, was in the middle of Laguna San Ignacio fishing from a *panga*. Suddenly, he and the friend he was with felt their boat stop moving. It was as if they had beached themselves on land, only the boat was still in the middle of the lagoon. The fishermen realized that they had actually beached themselves on the back of an immense female whale. She slid farther beneath their boat and lifted them a terrifying few inches in the air and gently set them back down again. She raised her head to the surface right alongside Pachico, eyeing him. He waited a moment, then reached out to touch her, first with his finger and then his whole hand. After a few seconds, she slowly sank back down.

Soon fishermen all over the lagoon began sharing similar accounts. The whales seemed curious, as if they wanted to communicate. And they were playful. Reports of whales acting the same way in the other Baja calving lagoon, Ojo de Liebre, filtered in.

Pachico's son Ranulfo, now a seasonal whale-watching guide like most fishermen in the area, told the journalist Charles Siebert that before his father was approached, "everyone went out of [their] way to avoid the whales." After that first meeting with an *amistosa,* though, everything changed. The fishermen started putting their hands in the water all over the lagoon, and the whales swam up to them, placing their giant heads underneath the men's hands. The whales swiveled to make eye contact, they blew misty breath out of their blowholes onto the boats, they lifted the *pangas* up gently and set them back down again. Forty years later the fishermen aren't surprised when the whales do this. They're used to it. Today roughly 10 to 15 percent of the whales in the lagoon, predominantly mothers with calves, are sociable. During the winter months, when the whales are in the lagoons, the fishermen stop fishing and work as whale-watching guides and boat drivers instead, earning a good living by taking small groups of ecotourists and researchers out to see the "friendlies." They tell each of the tourists who come to see the whales to take off their sunglasses because the greys like to make eye contact.

One early March afternoon in Laguna San Ignacio, I spent more than an hour with an adult whale and her calf. The forty-foot-long mother swam slowly toward our *panga*, her month-old calf bobbing alongside, and then in a single, swift movement, she pushed her baby onto her head and then thrust him toward the boat and our waiting, splashing hands. Again and again she did this and as the calf rolled off the wide expanse of his mother's head, he shot bubbles, breathed loudly, and turned to look us in the eye. Sometimes he opened his mouth—exposing his baleen, still bright, new, unstained—and then shifted so that we could stroke the side of his head, his wide gums, his long jaw. Stroking the calf felt a little like touching one of those plastic-covered foam stadium seats, or boat keychains that are meant to float. His skin was smooth but squishy and even though the water was cold, he was warm. Twice his mother gently lifted the boat and set it back down again. At least half a dozen times the young whale came up under our hands just to be patted and rubbed, making long, unbroken eye contact with each of us. When he exhaled he covered me in whale snot and seawater, his eyes shining. Was it a joke? I couldn't tell. But his interest, his playfulness, was unmistakable. He seemed like the world's largest toddler.

Searching for a friendly whale in the vast expanse of Laguna San Ignacio is impossible; the whales can be anywhere in the lagoon, including an area off limits to whale-watching boats. The guides know that the only way to meet the whales is to motor to the middle or drive around slowly, waiting for an interested whale to approach *them*.

What's particularly startling about all of this is that gray whales live a long time, up to eighty years and perhaps even longer. Those first friendly whales who approached Pachico and the other fishermen may have been old enough to remember whaling. They, their mothers, or their fathers could have fought for their lives in the lagoon when it was churning with hunters and their boats, the water red with the blood of whales and men.

"The first time I came down here," Frohoff told me, "the first whale to approach the boat came up to my side of the *panga*. I put my hand in the water and she slid under it, and I saw a harpoon scar on her side. I was blown away that this whale knew what it meant to be attacked by a human and she was approaching me anyway."

The reasons the whales do this remain a mystery. A few theories float and bob tentatively. One is that the whales may be using the boats and human hands like loofahs, to rub off barnacles. But the animals don't rub hard enough to scrape anything off, and a whale rubbing with any force on a boat would capsize it, something that hasn't happened since whaling ended. Another theory is that the fishermen are secretly feeding the whales. But the females don't eat while they're nursing their calves, and the calves drink only their mother's milk. The whale-watching guides and fishermen, who spend all day every day in late winter and early spring maneuvering around the animals and getting to know the behavior of individuals, have their own theories.

Jonas Leonardo Meza Otero has been taking people to see the creatures for his entire adult life, and he believes he has at least part of the answer. One evening, as we sat on folding chairs on the beach, drinking Dos Equis and watching the whales spout a few hundred yards offshore, he said, "I think they're curious. And also they know that they are safe here. I believe the mothers are showing their calves what humans are. She is teaching them a lesson. Also, maybe they're a little bored since all they are doing here is nursing, and we offer them something else to do."

It's true that these whales' friendly interactions take place only in the Baja lagoons. The same whales are spotted off the coast of the United States and Canada as they make their way north, but they don't interact with people anywhere else as they do here.

There is another theory: that the mothers are teaching their calves about boats. Besides orcas, who prey on gray whale calves, collisions with boats along the whales' migration routes are the biggest threat to their survival once they leave the lagoon.

"Some naysayers," Frohoff said, "might claim that these whales don't have the intelligence to know the difference between the peaceful climate in the lagoon today and what transpired in the past, that they're not smart enough to remember that humans can inflict pain and cause death. However, historical evidence, as well as the limited data we do have on these whales, compel us to think otherwise."

There are many accounts of whales learning to avoid certain areas—particularly dangerous spots where they might run into hungry orcas looking to pick off a calf, human hunters, or be more likely to be hit by boats. Frohoff has credited the whales' memories for this self-protective behavior and argues that in order to survive their lengthy migrations they have to be intelligent and make quick assessments and decisions.

It may be a stretch to say that the extreme violence of whaling created a kind of species-level psychological trauma in the animals. But perhaps not. As research on whale sociality, communication, and cognition has shown, many cetacean species have culture and language and belong to complex societies. They live a long time, and the ability to remember where they've been harmed and where they've felt safe is key to the whales' survival. Mass killings at the hands of humans were fundamental events in their natural history. Their choice to approach us in what was once a watery killing field is a fundamental event in ours.

We can call the whales' behavior resilience or recovery, or we can anthropomorphize it as a kind of human-directed forgiveness. At the very least, the whales are doing something that seems a lot like the expression of affectionate and playful curiosity. Watching a free-living calf swim out of the depths with his mother and, on her urging, look into my eyes while I looked into his is one of the most powerful and mystifying encounters of my life. I believe this is because it was born of choice. Unlike an aquarium beluga, a zoo-dwelling panda, or my neighbor's Chihuahua, who may make eye contact because there is nowhere else to look, because they hope to be fed or because they fear me, the Baja whales looked at me with, I'm convinced, something like the same wonder and curiosity I had for them.

Throughout the spring and summer after I left Baja, I thought about the young whales and their mothers as they headed north to the Arctic, maneuvering around container ships and navy vessels and pods of orcas. I wondered whether they passed dolphins heading in to strand, otters drunk with confidence, or frenzied sea lions heading for open water. More than anything, though, I thought about our encounters with other animals and wondered what we might do to make these interactions more like those between the humans and whales of Baja. Could we affect the mental health of both captive and wild animals for the better, not simply by striving to do no harm but by seeking to rectify our mistakes?

It's a bit of a generalization, but over the past century most of our thinking about wildlife has fallen into two opposing philosophical camps: leave wild animals totally alone, or hunt, hassle, extirpate, or domesticate them. Neither the hard-line conservation approach, such as passing laws to keep people far from wildlife or isolating their habitats from humans, or the opposite, allowing unfettered human access to wild animals and the places they depend upon, has worked. We cannot leave other animals completely alone because we have suffused the world with ourselves and our activities. We also like being around animals and some of them seem to like being around us.

Oliver taught me this. So did Mosha, Noon Nying, and even Gigi. The weight of all these accumulated stories convinced me that we should pay closer attention to the mental health of other creatures—because what is good for them is so often good for us. Many people have already taken on this responsibility, and the resulting observations—of monkey executives, nervous dogs, relaxed rats, demented sea lions, and more—have quietly influenced how we think about our own unraveling minds and what we might do to stitch them back together again.

Trying to understand Oliver also led me to be a bit kinder to myself and the humans and other animals around me. When we feel kinship with a pig or a pigeon, *really* feel it, we can't help but share a bit of that affection with our own animal selves. There are exceptions, of course. Adolf Hitler loved his German Shepherd Blondi so much that he risked his own life in the last few weeks of the war to leave his bunker to take her on walks. And Kim Jong-il reportedly spent hundreds of thousands of dollars on his Shih Tzus and Poodles, flying in a French veterinarian to treat them and feeding the dogs food scraps from his own plate (ensuring that they ate better than the majority of North Koreans). For most people, though, to selflessly love another creature is to be open to loving other humans, who are animals as much as pandas, cows, or Shih Tzus. This is why I never trust an animal rights activist who is misogynistic or thinks that *Homo sapiens* are, at heart, more rotten than any other species. Human rights activists are animal rights activists by default. The reverse should also be true. Oliver didn't teach me this as much as losing him made me teach it to myself.

I'm ashamed to admit it, but I'm not sure where Oliver's ashes ultimately wound up. Jude returned to Boston before I did, and he volunteered for the sad duty of picking up what was left of our dog, along with his smelly webbed collar printed with acorns and his round bed. I know Jude took the collar into a forest in western Massachusetts, set it on a rock, and walked away. I don't know which forest. I couldn't bear to ask.

This is how we love the people and other animals closest to us. When we lose them, the pain is crushing, sensory. I still have a tactile memory of Oliver's ears, of holding his paws—the pads rough and spreading, the fur between them soft and light. I remember the smell of his neck, feral but comforting, like the pine floors of our apartment in D.C.

For years after he died, thinking about Oliver was like visiting a raw, guilty country. I tried to avoid it. Instead I went to other, actual countries. I met elephants and parrots, cats and whales, horses and seals. Every time I reached for their hides, feathers, fur, or skin, I was reaching for him.

What I discovered is that the guilty country is crowded. So many of us are there looking for answers and blaming ourselves, wondering what would have happened if we'd taken the dog to the park more often, refused to adopt the second cat, who the first one despised, cleaned the iguana's tank more frequently, given the hamster more time in his plastic ball, or ridden the horse as much as we had first intended. Animal madness isn't our fault, though—not always, anyway. When it comes to caring for the creatures with whom we share our beds, couches, backyards, and deepest affections, most of us try our best to help them. We often try harder, in fact, than we ever could have imagined possible. Some of us break our own hearts trying, drain our savings accounts, put the vet visits on the credit card, hoping fervently for some sort of deliverance before the bill arrives. Our intentions are good. It's simply that falling short is the human condition, and some problems cannot be taken care of by hoping.

This should not let us off the hook. There are many structural elements of our lives with other creatures that cause needless suffering and could easily be done away with. We could stop teaching elephants to paint, dance, and play soccer, and casting chimps in commercials and giraffes in feature films. We could close our nation's zoos, or at the very least stop deluding ourselves that it's our right to see exotic wildlife like gorillas, dolphins, and elephants in every major American city. We could stop trying to convince ourselves that keeping animals in cages or tanks is the best way to educate and inform one another about them, especially since it often costs the animals their sanity. We could instead turn these zoos and other facilities into places where people might engage with animals, domestic and wild, who often thrive in our presence, creatures like horses, donkeys, llamas, cows, pigs, goats, rabbits, and even raccoons, rats, squirrels, pigeons, and possums. We could exchange the polar bear pools for petting zoos and build teaching farms, urban dairies, and wildlife rehabilitation centers where city-dwelling children and adults could volunteer or take classes on cheese making, beekeeping, gardening, veterinary science, wildlife ecology, and animal husbandry.

We could also stop leading the sorts of lives that cause large numbers of our pets to end up on psychopharmaceuticals. We could spend more time walking and playing with them and less time on our phones, checking email and watching television. We could stop bringing animals into our lives that deep down, we know we cannot care for, and we could recognize, in them and their crazy behavior, our own unhealthy habits reflected back to us.

We could also truly begin to acknowledge the other minds in the water. That is, we could accept the fact that dolphins, whales, and other marine life may literally be driven mad by our actions, and we could make more consistent efforts to protect their hearing, migration routes, water quality, and food sources, since in the end, that is also what is better for us.

We could stop eating mentally ill pigs, chickens, and cows, and do away with corporate farming practices so cruel they're often institutionalized torture. We could stop trimming our coats with the fur of compulsive mink, foxes, sable, and chinchillas and quit testing our drugs, cosmetics, and medical procedures on lab animals housed alone and in terribly uncomfortable conditions.

We could also, and most important, make a lasting peace with Darwin's belief that humans are just another kind of animal, different only by degree. This kind of change will not be easy or fast. It will take the self-transformative power of chameleons, the resolve of mules, the fortitude of migrating whales, and the ingenuity and compassion of humans. It will be worth it.

Afterword

In the year since this book was first published, I've been relentlessly surprised by everything that I didn't know. A lot of it can be boiled down to lessons from the dog trainer I hired to help me with Cedar, the stubborn, fuzzy Akita mix I adopted while on book tour in Portland, Oregon.

I picked Lisa Caper from a long list of trainers with business names like "Better Nature Dogs" or "Perfect Paws" because the first thing she said to me on the phone was "I don't teach dogs to do tricks," drawing out the word *tricks* disdainfully. "My goal is to make confident, calm dogs who know what's expected of them and therefore feel safe." When she arrived for our first appointment in a cherry-red convertible, her Catahoula mix, Loki, waited happily for her in the backseat. Even though the top was down. Even though our meeting took more than an hour.

Her fee was the better part of my monthly food budget but I would have paid her anything. There are few situations in life in which writing a check can guarantee peace and happiness. I was convinced, and still am, that this was one of them.

Lisa put Cedar and me into a private boot camp that involved us articulating clear expectations for each other—rules like no jumping up on people, staying out of the kitchen when I'm cooking, and refraining from licking the refrigerator door when hungry. I reinforce these rules with tiny bits of dehydrated liver dispensed from a hideous ripstop pouch affixed to my waist and the pressure of a prong collar that I affectionately refer to as Cedar's BDSM necklace.

It turned out that he did have a bit of separation anxiety, whining and yelping when he was left alone. With time (and training him to understand that a crate is a safe place to be), Cedar is much calmer. My own anxiety is a work in progress. For the first few months of our life together I was tempted to barricade the windows. It's humbling to admit this. It's even more humbling to admit that I had to hire someone to help me manage my anxiety about my dog's *potential* anxiety, all while travelling around the country talking to people about my inability to help my last dog with his anxiety. But this is how we learn.

By rights, Cedar has far more reason to be haunted, compulsive, and sad than Oliver. He was given up twice and I know he has loved other people, perhaps many of the same ones who gave him up. But he greets every new human as if they will never disappoint him. I can't decide if this makes him brave, naïve, or some mixture of the two, but I suppose it doesn't matter. I simply want to be more like him.

Once, at dusk, we were walking to the dog park and Cedar caught sight of an older man wearing a cowboy hat. He was out of his head with excitement, pulling on the leash, whining and barking, all in the direction of the cowboy. I let Cedar get close and, hearing someone panting heavily behind him, the man turned. Cedar froze, his tail mid-wag, and then his posture collapsed. He didn't look sad exactly—his attention turned back to me and the park—but the change in him was astounding. "It's you it's you it's you it's youuuuuuuu!" his whole body seemed to be saying, then nothing. He appeared quite zen about it.

Not me. I was filled with rage and gratitude: rage at the people who gave up this sweet creature and gratitude that they did. Then, Cedar tugged on the leash and urged me forward, as he always does.

Somehow, with me barely noticing it, I've become one of those people who says that her best friend is a dog. Like any friendship, ours is occasionally prone to miscommunication and frustration, but more often than not we take turns encouraging each other to leave something alone that isn't worth our time or attention, or pull each other toward something else that looks delicious or exciting. I may have been a service animal for Oliver, but Cedar and I are emotional supports for one another.

My friend Lon Hodge lives with a service dog named Gander, a golden-haired labradoodle with a wizard beard and dark eyes. From 1973 to 1981 Lonnie served in the military medical corps. But two years after he reentered civilian life he woke up one morning crippled with anxiety, petrified to leave his house. He couldn't work and he couldn't sleep; he suffered from panic attacks during the day and nightmares at night. He had a resting heart rate of 120 beats per minute and suicidal thoughts. When Lonnie went to his cardiologist, he was told he had PTSD. "Imagine yourself standing on the edge of a balcony on a high building," he said. "You know that jolt like maybe you might fall over? Now multiply that feeling by four and experience that all day, every day."

In 2012, after seeing a television special on service animals, Lonnie, who'd never had a dog of his own, wrote to Freedom Service Dogs, a nonprofit in Denver, Colorado. Gander was being trained by a female inmate in a Colorado prison and came to him seven months later. They got along from their very first day together and Lonnie's anxiety and suicidal thoughts quieted. He and Gander now travel the country advocating for mental health services for American veterans, ensuring that the Americans with Disabilities Act is being upheld in a way that supports service animals in public places, and visiting schools, hospitals, and community centers where Gander can be hugged, marveled at, and petted by people who need him. Two years into their relationship, Lonnie and Gander are very attuned to each other's emotions. Besides his PTSD, Lonnie has a fear of heights, possibly stemming from his time rappelling backward out of military helicopters. In the beginning, Gander kept Lonnie calm by standing between him and whatever precipice or building edge on which they found themselves. But slowly, Gander has started to exhibit a fear of heights too, something he learned by carefully watching his human. "The other day we were high up in a parking garage," Lonnie said, "and Gander got really uncomfortable. There was another time he got scared in a glass elevator."

What's interesting is that Gander's fear of heights isn't making him a bad service dog. In fact, it's the opposite. Lonnie is now so preoccupied with comforting Gander when he seems anxious that he doesn't have the energy or attention he once did to pay to his own fear.

"With PTSD, it's hard to lift your head above your anxiety," he said. "Gander breaks down your defenses. I've come back to myself and I've come back to the belief that the world is filled with a lot of good people. I want to be the person my dog thinks I am."

I can relate. I will try my hardest not to let Cedar down and if I occasionally do, despite my best efforts, I will remember what each of my dogs has taught me:

You can try to interview a dog but it won't tell you very much.

Healthy relationships between and among species are built on met expectations and consistency.

Resiliency may be learned, something innate, or a mysterious combination of the two, but all animals have it to varying degrees.

We are incredibly lucky to have each other.

Acknowledgments

There are not enough species of thanks to express my gratitude to the people and other animals that have made this book possible. I am especially indebted to the generosity of the zoo, shelter, veterinary clinic, and sanctuary staff, and veterinarians who answered my questions and occasionally introduced me to their animal coworkers. You are my heroes.

In particular, thank you, Dr. Mel Richardson, fearless champion and companion to nonhumans everywhere; Dr. Hayley Murphy at ZooAtlanta; Pat Derby and Ed Stewart at the Performing Animal Welfare Society Sanctuary; Daniel Quagliozzi at the SFSPCA; Nicole Cottam at the Tufts Animal Behavior Clinic; my crew leaders, fellow volunteers, and the entire staff of the Marine Mammal Center; the beluga caregivers at the Mystic Aquarium; the volunteer staff at the Congo Gorilla Forest at the Bronx Zoo; Ric O'Barry; Dr. Diana Reiss; Dr. Lori Marino; the elephant care staff at the Oakland Zoo; Katherine McCleod; the volunteer docents at the San Francisco Zoo; Dr. E'Lise Christensen and her patients; the International Marine Mammal Trainers Association; Dr. Joseph LeDoux; Ruth Samuels; Ann Southcombe; Dr. Donna Haraway; Phoebe Greene Linden and her flock; Barbara Bell and the bonobos at the Milwaukee County

Zoo; Mike Mease and the Buffalo Field Campaign; Pam Schaller and the penguins at the California Academy of Sciences; Dr. Nigel Rothfels; Dr. Beatriz Reyes Foster; Gail O'Malley; the entire staff of Franklin Park Zoo's tropical forest, especially Paul Luther; and dedicated friend and troop member of gorillas everywhere, Jeannine Jackle.

In Thailand, I am deeply grateful to Jodi Thomas, Pi Sarote, Lek Chailert, Dr. Preecha Phuangkum, Richard Lair, Gawn, Paladee, Pi Pong, Pi Som Sak, Silke Preussker, Mattie Illel, Pra Ahjan Harn Panyataro, Ann Tidarat Jitsarook, Jeff Smith, Dr. Pak, Jokia, Rara, Mae Perm, Mosha, Noon Nying, Mae Bua, and Teng Mo. Thank you to the staff at Elephant Nature Park, Friends of the Asian Elephant Hospital, the Thai Elephant Conservation Center, the Surin Project, the village of Baan Ta Klang, and the elephants who live there.

In Mexico, I'd like to thank Baja Discovery, the whale-watching guides, *pangeros*, and the many cooperatives dedicated to protecting the lagoon, the Mayoral family, Marcos Sedano, Lupita Murrillo, Molo, Dr. Toni Frohoff, Nina Katchadourian, and the whales who came out to meet me.

Human physicians, psychotherapists, and counselors to whom I am indebted include Dr. Cynthia Zarling, Catherine Keeling, Dr. Harry Prosen, Dr. Michael Mufson, Dr. Phil Weinstein, Dr. Ralph Nixon, Dr. Barbara Natterson-Horowitz, Maria Cimino, and Dr. David Jones.

I owe much to the archivists and librarians who granted me access to their materials and pointed me in fruitful directions, including the staff at the Wildlife Conservation Society archives; Barbara Mathe and the hardworking archivists at the American Museum of Natural History; Darrin Lunde at the Smithsonian; the librarians and research staff at the California Academy of Sciences; the archivists at Bethlem Mental Hospital; Hayden Library at MIT; Widener Library at Harvard; and the research assistance of Sharon Price, Matthew Christensen, Brooke LeVasseur, and Stella Smith-Werner.

At MIT, I am deeply grateful for the guidance of Dr. Harriet Ritvo, Dr. Stefan Helmreich, the faculty and staff of the History, Anthropology, and STS program, Karen Gardner, and my fellow graduate students. At Harvard University, I would like to thank Dr. David Jones, Dr. Janet Browne, and Dr. Sarah Jansen, as well as all of my students in Dogs and How We Know Them.

For reading various pieces of this book and offering invaluable insight, I thank the Neuwrite Group at Columbia University, Dr. Carl Schoonover, Jon Mooallem, the Headlands Writing Group, Eric Marcus, the Max Planck Institute for Social Anthropology, Dr. Etienne Benson, Doug McGray, and Carrie Donovan. For giving me the opportunity to test this material live and in person, thank you to Sina Najafi and *Cabinet Magazine*, *Pop-Up Magazine*, the TED Fellows Program, and the Headlands Center for the Arts.

For research support, I am grateful to the History, Anthropology and Science, Technology and Society Program at MIT, the National Science Foundation's IGERT Program, the Center for Advanced Visual Studies at MIT, the John S. Hennessey Fellowship for Environmental Studies, the MIT Presidential Fellowship, Colleen Keegan, and the History of Science Department at Harvard University.

For housing, metaphorical and physical, I thank Regine Basha and Gabriel Pérez-Barreiro; Ann Hamilton, Emmet, and Michael Mercil; Barbara Mathe; Andi Sutton and Colin Wilkins; Ann Hatch; Brittany Sanders and Robert Polidori; and Sharon Maidenberg, Holly Blake, and Brian Karl.

I would be a raving mad animal myself without Cal Peternell, Donna Karlin, Sharon Price, Ann Hamilton, Jill and Phil Weinstein, Rebecca Goodstein, Caitlin Swaim, Samin Nosrat, Nancy Moser, Maria Barrell, Auriga Martin, Quinn Kanaly, Brooke LeVasseur, Stefanie Warren, Catherine and Travis Keeling, Leyla Abou-Samra, Pamela Smith, Dario Robleto, Joanna Ebenstein, Kelly Dobson, Christina Seeley, Floor van de Velde, Emily Weinstein, Maria DeRyke, Aubree Bernier-Clarke, Travis Burnham, and Constance

Hockaday. Thank you also to Rigo 23, who taught me to ask forgiveness rather than permission. Amitav Ghosh invited me to his office hours and gave me my life. Kathleen Henderson's drawings make me a better animal. Thank you. I owe so much to the real Jude and his parents, Melanie and Terry, for loving Oliver so well. I am grateful, always.

Barney Karpfinger lights up the depths like an insightful anglerfish. There isn't a more generous human in the bestiary. Priscilla Painton, if editors were elephants, you'd be the smartest, strongest one. I'm so grateful to be in your herd. Thank you also to Jonathan Karp, Sydney Tanigawa, Anne Tate Pearce, Dana Trocker, Sophia Jimenez, Kellyn Patterson, and the rest of the book lovers at Simon & Schuster.

Finally, Lynn and Howard Braitman, Rob Moser, and Jake and Alice Braitman: letting me keep a donkey in the house may have had something to do with my career choice. Thank you for this and for everything else. There are no words without you.

Notes

Introduction

4 *"There is a perfect gradation between sound people and insane"*: Gruber, "Darwin on Man," quoted in Roy Porter, *Mind Forg'd Manacles: A History of Madness in England from the Restoration to the Regency* (London: Athlone Press, 1987), 37, 268.

Chapter One: The Tail Tip of the Iceberg

11 *This idea of animals as machines:* William Coleman, *Biology in the Nineteenth Century: Problems of Form, Function and Transformation,* Cambridge Studies in the History of Science (Cambridge, UK: Cambridge University Press, 1978), 121–22.

11 *identifying humanlike emotions or consciousness in other animals:* Lorraine Daston and Gregg Mitman, eds., *Thinking with Animals: New Perspectives on Anthropomorphism,* new ed. (New York: Columbia University Press, 2006).

11 *He believed that the similar emotional experiences of people and other creatures:* Mental illness is also a key part of *Expression* because Darwin thought that the insane (as he called them) were a purer source for the study of emotion. Like any good Victorian he was preoccupied with all manner of social mores and inhibitions, and he felt, perhaps rightly, that many people in insane asylums had been loosed from the shackles

of proper emotional control and expressed themselves more authentically.

Yet Darwin did not see these people as morally bankrupt, as many physicians of his time did. Instead he saw the insane as simply not self-conscious, as unaware of themselves and lacking an idea of self. Since they weren't self-conscious, they couldn't embarrass themselves and thus were unchecked in their expression of emotion. This, Darwin believed, made the insane into perfect study subjects for what despair, anger, fear, and more *really* felt and looked like. And so he devoted a lot of space in his book to covering the phenomenon of insanity in human beings, discussing such things as upset mentally ill humans raising the hair follicles on their heads just as dogs do their hackles (this last point not having withstood the test of observation) and poring over photos of people in insane asylums. Janet Browne, "Darwin and the Expression of the Emotions," in *The Darwinian Heritage*, ed. David Kohn (Princeton, NJ: Princeton University Press, 1985), 307–26.

11 *"with the manner in which she then tried"*: Charles Darwin, *The Expression of the Emotions in Man and Animals* (London: John Murray, 1872), 120.

11 *"Not far from my house"*: Ibid., 58, 60.

12 *"a peculiar short snuffle"*: Ibid., 129. Darwin may not have observed happy pumas or tigers himself. Nor had he seen crying elephants. Instead he relied on letters and published observations of people who had, coupled with his own careful observation of the animals he lived among, saw in his travels, and observed at the Regent's Park Zoo.

12 *"Man and the higher animals"*: Charles Darwin, *The Descent of Man, and Selection in Relation to Sex* (London: John Murray, 1874), 79.

13 *Darwin doesn't seem to have done any original research*: This is based on searches performed in 2010 and 2011 in the Darwin Correspondence Project (http://www.darwinproject.ac.uk/), as well as personal communication with Janet Browne and David Kohn in 2009–10 and careful readings of citations in Darwin's published works.

13 *"I hope to prove that"*: Richard Barnet and Michael Neve, "Dr Lauder Lindsay's Lemmings," *Strange Attractor Journal* 4 (2011): 153.

13 *Lindsay believed that the minds of insane people*: W. Lauder Lindsay, *Mind in the Lower Animals, in Health and Disease*, vol. 2 (New York: Appleton, 1880), 11–13.

14 *Lindsay also wrote about feral children:* Ibid., 14.

14 *Insane humans were also compared to:* Porter, *Mind-Forg'd Manacles,* 121–29.

14 *some of the incurables "are kept as wild beasts":* John Webster, *Observations on the Admission of Medical Pupils to the Wards of Bethlem Hospital for the Purpose of Studying Mental Diseases,* 3rd ed. (London: Churchill, 1842), 85–86.

15 *He was even convinced that some human lunatics:* Lindsay, *Mind in the Lower Animals,* 18–19.

15 *a mother stork who "let herself" be burned alive:* Ibid., 131–33.

19 *"Being in love is our most common version":* Mark Doty, *Dog Years: A Memoir* (New York: Harper, 2007), 2–3.

21 *the past forty to fifty years of research:* R. W. Burkhardt Jr., "Niko Tinbergen," 2010, http://www.eebweb.arizona.edu/Courses/Ecol487/readings/Niko%20Tinbergen%20Biography.pdf (accessed August 5, 2012); Richard W. Burkhardt Jr., *Patterns of Behavior: Konrad Lorenz, Niko Tinbergen, and the Founding of Ethology* (Chicago: University of Chicago Press, 2005).

21 *Lorenz even described one of his geese as depressed:* Heini Hediger, *Wild Animals in Captivity* (London: Butterworths Scientific, 1950), 50.

22 *The neuroscientist Jaak Panksepp:* Jaak Panksepp, *Affective Neuroscience: The Foundations of Human and Animal Emotions* (New York: Oxford University Press, 2004), 3.

22 *YouTube videos . . . of Dr. Panksepp stirring:* http://www.youtube.com/watch?v=j-admRGFVNM (accessed May 1, 2013).

23 *Panksepp believes the happy sound:* Barbara Natterson-Horowitz and Kathryn Bowers, *Zoobiquity: The Astonishing Connection Between Human and Animal Health* (New York: Vintage, 2013), 95.

23 *"increasingly wanted to understand how the human mind":* "Science of the Brain as a Gateway to Understanding Play: An Interview with Jaak Panskepp," *American Journal of Play* 2, no. 3 (Winter 2010): 245–77.

24 *believes that rabbits, for example:* Panksepp, *Affective Neuroscience,* 13, 15.

24 *An explosion of recent research on dogs:* One representative example is Isabella Merola, Emanuela Prato-Previde, and Sarah Marshall-Pescini, "Dogs' Social Referencing towards Owners and Strangers," *PLoS ONE* 7, no. 10 (2012): e47653.

24 *studies of hormonal fluctuations in baboons:* Jonathan Balcombe, *Second Nature: The Inner Lives of Animals* (New York: Palgrave Macmillan, 2010), 47.

24 A *number of recent studies:* Jason Castro, "Do Bees Have Feelings?," *Scientific American*, August 2, 2011, http://www.scientificamerican.com/article.cfm?id=do-bees-have-feelings; Sy Montgomery, "Deep Intellect," *Orion*, November–December 2011, http://www.orionmagazine.org/index.php/articles/article/6474/; "What Model Organisms Can Teach Us about Emotion," *Science Daily*, February 21, 2010, http://www.sciencedaily.com/releases/2010/02/100220184321.htm; Balcombe, *Second Nature.*

24 *The results of these studies are changing debates:* There have been some distinctions made within the neurosciences and affective sciences more generally about the differences between "emotions" and "feelings." Antonio Damasio and Joseph LeDoux, for example, have argued that emotions are not necessarily conscious states, while feelings may be, and are the result of our minds trying to make sense of emotions. See, for example, Antonio R. Damasio, *Descartes' Error: Emotion, Reason, and the Human Brain* (New York: G. P. Putnam, 1994), 131–32, 143.

24 *As the neurologist Antonio Damasio has argued:* Ibid.

25 *"I think that emotions—although they are subject to selection":* Lori Marino, personal communication, May 4, 2011.

25 *The ethologist Jonathan Balcombe believes that emotions:* Balcombe, *Second Nature,* 46.

25 *the only animals to have been proven self-aware:* Although this is slightly different, I have never known a dog to be confused by his or her reflection in a mirror, sliding glass door, or the glossy surface of an oven. Perhaps some dimwitted dogs do bark or sniff at the reflection, but most don't. While this doesn't prove that they see the reflected dog as *themselves,* it doesn't prove that they don't.

African Grey parrots will use mirrors as tools to gather information about where food, toys, or playmates are, but they don't necessarily groom while looking at their reflection. A parrot may recognize herself but might care to do something else with the information contained in the mirror (for instance, that a human it knows is busy making a fruit salad behind her). Even among the great apes, the degree to which chimpanzees identified themselves in mirrors depended on the individual chimpanzee. The same held true with gorillas. D. M. Broom, H. Sena, and K. L. Moynihan, "Pigs Learn What a Mirror Image Represents and Use It to Obtain Information," *Animal Behaviour* 78, no. 5 (2009): 1037; I. M. Pepperberg et al., "Mirror Use by African Gray Parrots *(Psittacus erithacus)*," *Journal of Comparative Psychology* 109 (1995): 189–95; G. G. Gallup Jr., "Chimpanzees: Self-Recognition," *Science* 167 (1970): 86–87; V. Walraven, Van L. Elsacker, and R. Verheyen, "Reactions of a Group of Pygmy Chimpanzees *(Pan paniscus)* to Their Mirror Images: Evidence of Self-Recognition," *Primates* 36 (1995): 145–50; D. H. Ledbetter and J. A. Basen, "Failure to Demonstrate Self-Recognition in Gorillas," *American Journal of Primatology* 2 (1982): 307–10; F. G. P. Patterson and R. H. Cohn, "Self-Recognition and Self-Awareness in Lowland Gorillas," in *Self-Awareness in Animals and Humans: Developmental Perspectives*, ed. S. T. Parker and R. W. Mitchell (New York: Cambridge University Press, 1994), 273–90.

26 *In 2012 a group of prominent neuroanatomists:* Philip Low, "The Cambridge Declaration on Consciousness," ed. Jaak Panksepp et al., Cambridge University, July 7, 2012.

27 *Despite centuries of investigation by everyone from:* Panksepp, *Affective Neuroscience*, 13.

27 *The psychologist Paul Ekman put forth the most famous list:* Paul Ekman, "Basic Emotions," in *Handbook of Cognition and Emotion*, ed. Tim Dalgleish and Mick J. Power (New York: Wiley, 2005), 45–60; John Sabini and Maury Silver, "Ekman's Basic Emotions: Why Not Love and Jealousy?," *Cognition and Emotion* 19, no. 5 (2005): 693–712.

28 *This sort of circular reasoning is:* Jaak Panksepp cautions against circular interpretations of animal behavior in *Affective Neuroscience*, 13.

29 *Behaviorally, the disease is similar:* T. Satou et al., "Neurobiology of the Aging Dog," *Brain Research* 774, nos. 1–2 (1997): 35–43; Carl W. Cotman and Elizabeth Head, "The Canine (Dog) Model of Human Aging and Disease: Dietary, Environmental and Immunotherapy Approaches," *Journal of Alzheimer's Disease* 15, no. 4 (2008): 685–707.

30 *Instead, canine Alzheimer's seems to be due:* Dr. Ralph Nixon, Director of Center of Excellence for Brain Aging and Executive Director of the Pearl Barlow Center for Memory Evaluation and Treatment, New York University, personal communication. December 5, 2013.

30 *Learning about fear and responding to it involve neural pathways:* Joseph LeDoux, "Emotion, Memory and the Brain: What We Do and How We Do It," LeDoux Laboratory Research Overview, http://www.cns.nyu.edu/home/ledoux/overview.htm (accessed June 7, 2012).

31 *John Fulton performed the first frontal lobotomies*: Jack D. Pressman, *Last Resort: Psychosurgery and the Limits of Medicine* (Cambridge, UK: Cambridge University Press, 1998), 13, 48–65.

31 *he called out in clear Italian*: Shorter and Healy, *Shock Therapy*, 35–41.

31 *By 1947, nine out of ten*: Ibid., 78–80.

32 *the OCD symptoms fade after the operation*: P. Hay, P. Sachdev, S. Cumming, J. S. Smith, T. Lee, P. Kitchener, and J. Matheson, "Treatment of Obsessive-Compulsive Disorder by Psychosurgery," *Acta Psychiatrica Scandinavica* 87, no. 3 (March 1993): 197–207; E. Irle, C. Exner, K. Thielen, G. Weniger, and E. Rüther, "Obsessive-Compulsive Disorder and Ventromedial Frontal Lesions: Clinical and Neuropsychological Findings," *The American Journal of Psychiatry* 155, no. 2 (February 1998): 255–263; M. Polosan, B. Millet, T. Bougerol, J.-P. Olié, and B. Devaux, "Psychosurgical Treatment of Malignant OCD: Three Case-Reports," *L'Encéphale* 29, no. 6 (December 2003): 545–552.

32 *Recent magnetic resonance imaging (MRI) of dogs*: Gregory Burns, "Dogs Are People, Too," *New York Times*, Oct. 5, 2013. http://www.nytimes.com/2013/10/06/opinion/sunday/dogs-are-people-too.html?_r=0.

33 *a recent estimate by the National Public Health Service*: LeDoux, "Emotion, Memory and the Brain."

34 *changing levels of regulatory transmitters*: Jaćek Dębiec, David E. A. Bush, and Joseph E. LeDoux, "Noradrenergic Enhancement of Reconsolidation in the Amygdala Impairs Extinction of Conditioned Fear in Rats: A Possible Mechanism for the Persistence of Traumatic Memories in PTSD," *Depression and Anxiety* 28, no. 3 (2011): 186–93.

34 *"It's not the rat part of the rat"*: He argues that testing phenomena that have to do with the neocortex (the deeply grooved and wrinkled gray matter that is much larger in humans and other apes—as it is in whale, dolphin, and elephant brains—may allow for extremely complex thinking) wouldn't be as useful since there is no homologous structure in the rat. Joseph E. LeDoux, personal communication, January 28, 2010.

34 *LeDoux believes that feelings, as humans think of them*: Joseph E. LeDoux, "Rethinking the Emotional Brain," *Neuron* 73, no. 4 (Feb. 23, 2012): 653–676.

35 *Rats who have been shocked enough times to lose interest in food*: Email correspondence with Joseph E. LeDoux, November 7, 2009.

35 *They simply gave up*: Martin E. Seligman and Steven Maier, "Failure to Escape Traumatic Shock." *Journal of Experimental Psychology* 74, no. 1 (1967):

1–9. ; Bruce J. Overmier and Martin E. Seligman, "Effects of Inescapable Shock Upon Subsequent Escape and Avoidance Responding," *Journal of Comparative and Physiological Psychology* 63, no. 1 (1967): 28–33.

35 *"uncontrollable events can significantly debilitate organisms"*: Seligman, Martin E. "Learned Helplessness." *Annual Review of Medicine* 23, no. 1 (1972): 407–412.

36 *For the psychologist and cognitive researcher Diana Reiss:* Diana Reiss, *The Dolphin in the Mirror: Exploring Dolphin Minds and Saving Dolphin Lives* (New York: Houghton Mifflin Harcourt, 2011), 242–43.

36 *Diana's own dog:* Personal communication, Diana Reiss, February 5, 2014.

37 *Skinner wrote about superstitious animal behavior in 1947:* B. F. Skinner, "Superstition in the Pigeon," *Journal of Experimental Psychology* 38, June 5, 1947, 168–172.

37 *Professional athletes may be:* Justin Gmoser, "The Strangest Good Luck Rituals in Sports," *Business Insider*, October 31, 2013.

37 *"It is possible . . . that your simple man"*: Quoted in Marc Bekoff, *The Emotional Lives of Animals: A Leading Scientist Explores Animal Joy, Sorrow, and Empathy—and Why They Matter* (Novato, CA: New World Library, 2008), 122.

39 *"Just because I say a dog is happy or jealous"*: Ibid., 123.

39 *these antagonized primates:* Robert M. Sapolsky, *A Primate's Memoir* (New York: Scribner, 2001); R. M. Sapolsky, "Why Stress Is Bad for Your Brain," *Science* 273, no. 5276 (1996): 749.

39 *His attention to their psychodramas:* Robert M. Sapolsky, "Glucocorticoids and Hippocampal Atrophy in Neuropsychiatric Disorders," *Archives of General Psychiatry* 57, no. 10 (2000): 925–35; Robert M. Sapolsky, L. M. Romero, and A. U. Munck, "How Do Glucocorticoids Influence Stress Responses? Integrating Permissive, Suppressive, Stimulatory, and Preparative Actions 1," *Endocrine Reviews* 21, no. 1 (2000): 55–89; Robert M. Sapolsky, "Why Stress Is Bad for Your Brain," 749; Sapolsky, *A Primate's Memoir.*

39 *"I'm not anthropomorphizing"*: Robert Sapolsky, quoted in Bekoff, *The Emotional Lives of Animals*, 124.

40 *Between 1955 and 1960 he and his team bred enough baby rhesus monkeys:* Donna Haraway, *Primate Visions: Gender, Race, and Nature in the World of Modern Science* (New York: Routledge, 1989), 231–32.

40 *In a now infamous series of experiments:* Harry Harlow and B. M. Foss, "Effects of Various Mother-Infant Relationships on Rhesus Monkey

Behaviors," *Readings in Child Behavior and Development* (1972): 202; Haraway, *Primate Visions*, 231–32, 238–39.

40 *One series involved offering baby monkeys a choice*: Haraway, *Primate Visions*, 238–39; Harlow and Foss. "Effects of Various Mother-Infant Relationships on Rhesus Monkey Behaviors," 202.

41 *Another of Harlow's experiments demonstrated*: Harry F. Harlow and Stephen J. Suomi, "Induced Depression in Monkeys," *Behavioral Biology* 12, no. 3 (1974): 273–96; B. Seay, E. Hansen, and H. F. Harlow, "Mother-Infant Separation in Monkeys," *Journal of Child Psychology and Psychiatry* 3, nos. 3–4 (1962): 123–32; H. A. Cross and H. F. Harlow, "Prolonged and Progressive Effects of Partial Isolation on the Behavior of Macaque Monkeys," *Journal of Experimental Research in Personality* 1, no. 1 (1965): 39–49.

41 *the now extremely abnormally behaving monkeys*: Stephen J. Suomi, Harry F. Harlow, and William T. McKinney, "Monkey Psychiatrists," *American Journal of Psychiatry* 128, no. 8 (1972): 927–32; Stephen J. Suomi and Harry F. Harlow, "Social Rehabilitation of Isolate-Reared Monkeys," *Developmental Psychology* 6, no. 3 (1972): 487–96.

41 *psychoanalyst and psychiatrist René Spitz observed*: Rachael Stryker, *The Road to Evergreen: Adoption, Attachment Therapy, and the Promise of Family* (Ithaca, NY: Cornell University Press, 2010), 14–15; R. A. Spitz, "Hospitalism: An Inquiry into the Genesis of Psychiatric Conditions in Early Childhood," *Psychoanalytic Study of the Child* 1 (1945): 53–74; R. A. Spitz, "Hospitalism: A Follow-Up Report on Investigation Described in Volume I, 1945," *Psychoanalytic Study of the Child* 2 (1946): 113–17, quoted in "Attachment," Advokids, http://www.advokids.org/attachment.html (accessed March 23, 2012); John Bowlby, "John Bowlby and Ethology: An Annotated Interview with Robert Hinde," *Attachment and Human Development* 9, no. 4 (2007): 321–35.

42 *Spitz believed that the lack of human touch and affection*: Deborah Blum, *Love at Goon Park: Harry Harlow and the Science of Affection* (New York: Perseus, 2002), 50–52.

42 *Bowlby's and Spitz's research, combined with Harlow's experimental results*: Frank C. P. van der Horst, Helen A. Leroy, and René van der Veer, " 'When Strangers Meet': John Bowlby and Harry Harlow on Attachment Behavior," *Integrative Psychological and Behavioral Science* 42, no. 4 (2008): 370–88.

42 *"lasting psychological connectedness between human beings"*: Van der Veer, " 'When Strangers' Meet.' "

42 *Harlow's monkeys also ended up helping*: Friends of Bonobos, "The Sanctuary" (accessed February 2, 2012) http://www.friendsofbonobos.org/sanctuary.htm.

43 *Scotland's Overtoun Bridge, otherwise known as the "dog suicide bridge"*: "Why Have So Many Dogs Leapt to Their Death from Overtoun Bridge?," *Daily Mail*, http://www.dailymail.co.uk/news/article-411038/Why-dogs-leapt-deaths-Overtoun-Bridge.html (accessed January 9, 2014).

44 *"Military Dogs of War"*: James Dao, "More Military Dogs Show Signs of Combat Stress," *New York Times*, December 1, 2011; Lee Charles Kelley, "Canine PTSD: Its Causes, Signs and Symptoms," *My Puppy, My Self, Psychology Today*, August 8, 2012; Monica Mendoza, "Man's Best Friend Not Immune to Stigmas of War; Overcomes PTSD," Official Website of the U.S. Air Force, July 27, 2010, http://www.peterson.af.mil/news/story.asp?id=123214946 (accessed Aug. 1, 2013); Marvin Hurst, " 'Something Snapped': Service Dogs Get Help in PTSD Battle," *KENS5.com*, February 10, 2012; Catherine Cheney, "For War Dogs, Life with PTSD Requires Patient Owners," *Atlantic*, December 20, 2011; Jessie Knadler, "My Dog Solha: From Afghanistan, with PTSD," *The Daily Beast*, March 14, 2013, http://www.thedailybeast.com/articles/2013/03/13/my-dog-solha-from-afghanistan-with-ptsd.html (accessed Mar. 4, 2013).

44 *Pavlov became interested in canine neurosis*: According to Breuer, Anna's hysteria symptoms included (among other things) partial paralysis of her limbs, weakness, loss of motion in her neck, a nervous cough, lack of appetite, hallucinations, agitation, mood swings, destructive behavior, amnesia, tunnel vision, odd speech patterns (in which she did not conjugate verbs), and difficulty speaking German (but was still able to translate German texts into English). John Launer, "Anna O and the 'Talking Cure,' " *QJM* 98, no. 6 (2005): 465–66; G. Windholz, "Pavlov, Psychoanalysis, and Neuroses," *Pavlovian Journal of Biological Science* 25, no. 2 (1990): 48–53.

45 *The lab performed endless variations*: Michael W. Fox, *Abnormal Behavior in Animals* (Philadelphia: Saunders, 1968), 81; Windholz, "Pavlov, Psychoanalysis, and Neuroses."

45 *Pavlov's views on the similarities*: Fox, *Abnormal Behavior in Animals*, 85, 119.

46 *Pavlov had his critics:* H. S. Liddell, "The Experimental Neurosis," *Annual Review of Physiology* 9, no. 1 (1947): 569–80.

46 *argued that Pavlov's work was inferior to analysis:* In 1929 one Viennese psychoanalyst argued that Pavlov's work was far inferior to analysis when it came to understanding neurosis in humans. Pavlov countered that neurosis in both people and other animals was rooted in the collision of excitation and inhibition (the basic processes he was testing in the lab) and that if his dogs could speak they would most likely say that they could not control themselves and so they did what was forbidden and were punished. But none of these verbal dog reports, Pavlov continued, would add anything new to the knowledge gained from the experiments themselves. Windholz, "Pavlov, Psychoanalysis, and Neuroses"; Liddell, "The Experimental Neurosis."

46 *his ability to return his dogs to their normal, nonneurotic state:* He may have changed his mind somewhat. A few years later, while working with humans at a clinic focused on nervous diseases, Pavlov began to put far more stock in the process of psychoanalysis, at least when it came to understanding humans, and he spent the last few years of his life researching the causes and manifestations of what were then the nervous diseases of hysteria, neurasthenia, and psychasthenia in people. Windholz, "Pavlov, Psychoanalysis, and Neuroses"; Liddell, "The Experimental Neurosis."

47 *These ideas were adopted by the military:* Liddell, "The Experimental Neurosis."

47 *Pavlovian ideas of conditioning and deconditioning:* Other animals are part of this story as well, as is the idea of "traumatic memory," which was sometimes seen as related to nervous disorders. Our modern concept of traumatic memory dates back to the years leading up to World War I, stemming from experiments like Pavlov's, but also, in part, to the earlier work of two American physicians, George Crile and Walter Cannon, and their experiments on cats and dogs. One of their research foci was on nervous shock. Crile and Cannon believed that extreme fear could cause physical problems similar to surgical shock in both humans and cats (a potentially lethal condition that left surgical patients pallid, faint, cold, anxious, and with a weak pulse, among other symptoms). Cannon's experiments on felines, in which he destroyed the connection between the cat's cortex and the rest of its nervous system, caused an extreme emotional reaction that looked a lot like fearful arousal: the cats' hair stood

on end, sweat seeped from their toes, their heart rate and blood pressure spiked, and they eventually collapsed and died. Cannon called this "sham rage" and used the cats' experiences to explain the deaths of men who had been exposed to severe emotional shocks. Allan Young, *The Harmony of Illusions: Inventing Post-Traumatic Stress Disorder* (Princeton, NJ: Princeton University Press, 1997), 24, 42; Frederick Heaton Millham, "A Brief History of Shock," *Surgery* 148, no. 5 (2010): 1026–37.

47 *PTSD sufferers also experience a variety of symptoms:* "DSM-5 Criteria for PTSD," U.S. Department of Veterans Affairs, http://www.ptsd.va.gov/professional/pages/dsm-iv-tr-ptsd.asp (accessed July 1, 2013); "Post Traumatic Stress Disorder," *A.D.A.M. Medical Encyclopedia*, National Library of Medicine, March 8, 2013, http://www.ncbi.nlm.nih.gov/pubmedhealth/PMH0001923/ (accessed December 5, 2013).

48 *Doctors treating soldiers in the wake of World War I:* More recently, evolutionary psychologists have argued that today's PTSD may be an extreme adaptive behavior; that is, the soldier's harrowing response to the horror he witnessed could be an unconscious attempt to keep himself away from war in the future. I believe this is an oversimplification of a complex reaction to trauma and anxiety. See, for example, Lance Workman and Will Reader, *Evolutionary Psychology: An Introduction* (Cambridge, UK: Cambridge University Press, 2004), 229; Young, *The Harmony of Illusions*, 64.

48 *African elephant calves:* Hope R. Ferdowsian et al., "Signs of Mood and Anxiety Disorders in Chimpanzees," *PLoS ONE* 6, no. 6 (2011). See also G. A. Bradshaw et al., "Building an Inner Sanctuary: Complex PTSD in Chimpanzees," *Journal of Trauma and Dissociation: The Official Journal of the International Society for the Study of Dissociation* 9, no. 1 (2008): 9–34; G. A. Bradshaw, *Elephants on the Edge: What Animals Teach Us about Humanity* (New Haven, CT: Yale University Press, 2010).

48 *Chimps who have spent time at testing facilities:* Ferdowsian et al., "Signs of Mood and Anxiety Disorders in Chimpanzees," e19855. See also Bradshaw et al., "Building an Inner Sanctuary."

48 *an account of this sort of suffering:* Balcombe. *Second Nature*, 59.

49 *Whether these animals were indeed experiencing the same sorts:* Young, *The Harmony of Illusions*, 284.

49 *disorders such as "shell shock" and "war neuroses":* Ibid.

49 *Traumatized infants and pre-school-age children:* Judith A. Cohen and Michael S. Scheeringa, "Post-Traumatic Stress Disorder Diagnosis in

Children: Challenges and Promises," *Dialogues in Clinical Neuroscience* 11, no. 1 (March 2009): 91–99; M. S. Scheeringa, C. H. Zeanah, M. J. Drell, and J. A. Larrieu, "Two Approaches to the Diagnosis of Posttraumatic Stress Disorder in Infancy and Early Childhood," *Journal of the American Academy of Child and Adolescent Psychiatry* 34, no. 2 (February 1995): 191–200; Richard Meiser-Stedman, Patrick Smith, Edward Glucksman, William Yule, and Tim Dalgleish, "The Posttraumatic Stress Disorder Diagnosis in Preschool- and Elementary School-Age Children Exposed to Motor Vehicle Accidents," *The American Journal of Psychiatry* 165, no. 10 (October 2008): 1326–1337.

50 A *few search and rescue dogs exposed to the loud, dangerous:* Personal communication, Nicole Cottam, June 19, 2009, August 2, 2009; personal communication, Jim Crosby, March 14, 2011.

50 *He offers a checklist for people interested in diagnosing their own pets:* Lee Charles Kelley, "Canine PTSD Symptom Scale," http://www.leecharleskelley.com/images/CPTSD_Symptom_Scale.pdf.

50 *Kelley found that asking him to bark:* Lee Charles Kelley, "Case History No. 1—My Dog Fred," originally published in "My Puppy My Self," at PsychologyToday.com on July 10, 2012; http://canineptsdblog.blogspot.com/2013/02/canine-ptsd-case-history-no-1my-dog-fred.html.

51 *Of the roughly 650 American military dogs deployed:* Dao, "More Military Dogs Show Signs of Combat Stress"; Lee Charles Kelley, "Canine PTSD: Its Causes, Signs and Symptoms," *My Puppy, My Self, Psychology Today,* August 8, 2012; Monica Mendoza, "Man's Best Friend Not Immune to Stigmas of War; Overcomes PTSD," U.S. Air Force, July 27, 2010, http://www.peterson.af.mil/news/story.asp?id=123214946 (accessed November 5, 2012); Marvin Hurst, " 'Something Snapped': Service Dogs Get Help in PTSD Battle," KENS5.com, February 10, 2012; Cheney, "For War Dogs, Life with PTSD Requires Patient Owners."

51 *Dr. Walter Burghardt . . . believes the disorder applies to many dogs:* See, for example, Kelly McEvers, " 'Sticky IED' Attacks Increase in Iraq," National Public Radio, December 3, 2010; Craig Whitlock, "IED Casualties in Afghanistan Spike," *Washington Post*, January 26, 2011; James Dao and Andrew Lehren, "The Reach of War: In Toll of 2,000, New Portrait of Afghan War," *New York Times,* August 22, 2012; Malia Wollan, "Duplicating Afghanistan from the Ground Up," *New York Times,* April 14, 2012; Mark Thompson, "The Pentagon's New IED Report," *Time,* February 5, 2012; Ahmad Saadawi, "A Decade of Despair in Iraq,"

New York Times, March 19, 2013; Michael Barbero, "Improvised Explosive Devices Are Here to Stay," *Washington Post*, May 17, 2013; Terri Gross with Brian Castnor, "The Life That Follows: Disarming IEDs in Iraq," *Fresh Air*, National Public Radio, June 7, 2013.

51 *dog noses are still the most effective tools*: Spencer Ackerman, "$19 Billion Later, Pentagon's Best Bomb-Detector Is a Dog," Wired, October 10, 2010; Allen St. John, "Let the Dog Do It: Training Black Labs to Sniff Out IEDs Better Than Military Gadgets" Forbes, April 9, 2012, http://www.forbes.com/sites/allenstjohn/2012/04/09/let-the-dog-do-it-training-black-labs-to-sniff-out-ieds-better-than-military-gadgets/ (accessed April 10, 2012).

Chapter Two: Diagnosing the Elephant

59 *It wasn't included in the DSM in 1980*: Edward Shorter, *A Historical Dictionary of Psychiatry* (New York: Oxford University Press, 2005), 226–27.

59 *Sunita was born in a residential house*: Chris Dixon, "Last 39 Tigers are Moved from Unsafe Rescue Center," *New York Times*, June 11, 2004; Lance Pugmire, Carla Hall, and Steve Hymon, "Clashing Views of Owner of Tiger Sanctuary Emerge," *Los Angeles Times*, April 25, 2003. "Meet the Tigers," Performing Animal Welfare Society Sanctuary, www.pawsweb.org/meet_tigers.html_ (accessed December 6, 2011).

60 *Mel brought me to see Sunita*: "Tic Disorders," American Academy of Child and Adolescent Psychiatry, May 2012, http://www.aacap.org/cs/root/facts_for_families/tic_disorders (accessed February 11, 2013); John T. Walkup et al., "Tic Disorders: Some Key Issues for DSM-V," *Depression and Anxiety* 27 (2010): 600–610, http://www.dsm5.org/Research/Documents/Walkup_Tic.pdf.

63 *like attention deficit disorder*: Andrew Lakoff, "Adaptive Will: The Evolution of Attention Deficit Disorder," *Journal of the History of the Behavioral Sciences* 36, no. 2 (2000): 149–69.

63 *Dogs can be diagnosed with the disorder today*: "Separation Anxiety," DSM-V Development, http://www.dsm5.org/Pages/RecentUpdates.aspx (accessed July 15, 2013).

63 *It became a viable diagnosis in 1978*: Shorter, *A Historical Dictionary of Psychiatry*, 32.

64 *Many people, at least those who*: See Grier, *Pets in America*, 13–14, 121–30, 136.

64 *The historian Katherine Grier:* Grier, *Pets in America,* 156.

64 *eventually, similar brain chemistry:* See, for example, Cotman and Head, "The Canine (Dog) Model of Human Aging and Disease"; B. J. Cummings et al., "The Canine as an Animal Model of Human Aging and Dementia," *Neurobiology of Aging* 17, no. 2 (1996): 259–68; Belén Rosado et al., "Blood Concentrations of Serotonin, Cortisol and Dehydroepiandrosterone in Aggressive Dogs," *Applied Animal Behaviour Science* 123, nos. 3–4 (2010): 124–30.

65 *The American College of Veterinary Behaviorists currently certifies:* American College of Veterinary Behaviorists, http://www.dacvb.org/resources/find/ (accessed August 1, 2013).

65 *The* actual *number of vets diagnosing emotional problems:* Center for Health Workforce Studies, "2013 U.S. Veterinary Workforce Study: Modeling Capacity Utilization Final Report," *American Veterinary Medical Association* (April 16, 2013): vii.

66 *equine self-mutilation syndrome, which is similar, he says, to Tourette's:* N. H. Dodman et al., "Equine Self-Mutilation Syndrome (57 Cases)," *Journal of the American Veterinary Medical Association* 204, no. 8 (1994): 1219–23.

70 *The AKC's breed standard for Bernese Mountain Dogs:* "Get to Know the Bernese Mountain Dog," American Kennel Club, http://www.akc.org/breeds/bernese_mountain_dog/index.cfm (accessed March 1, 2002).

70 *As for cats, Siamese, Burmese, Tonkinese, Singapura:* Nicholas Dodman, *If Only They Could Speak: Understanding the Powerful Bond between Dogs and Their Owners* (New York: Norton, 2008), 260–62.

73 *no animal disorder is a perfect mirror of a human condition:* K. L. Overall, "Natural Animal Models of Human Psychiatric Conditions: Assessment of Mechanism and Validity," *Progress in Neuro-Psychopharmacology and Biological Psychiatry* 24, no. 5 (2000): 729.

74 *diagnosed in people who feel excessively anxious:* According to the *DSM*, these worries must also be out of proportion to the likelihood of the event itself, such as the relentless anxiety that you may be late to work, even if you're never late, or the recurring fear that your daughter will be kidnapped on her way home from school. Yes, these things do happen, but for most people, worrying about them is only a passing thought, not a fixation. American Psychiatric Association, *DSM-IV: Diagnostic and Statistical Manual of Mental Disorders,* 4th ed. (Arlington, VA: American Psychiatric Association, 1994), 432–33.

83 *Before World War II, roughly two-thirds of Thailand:* Larry Lohmann, "Land, Power and Forest Colonization in Thailand," *Global Ecology and Biogeography Letters* 3, no. 4/6 (1993): 180.

83 *Since many of the prime logging regions:* Richard Lair, *Gone Astray: The Care and Management of the Asian Elephant in Domesticity* (Bangkok: FAO Regional Office for Asia and the Pacific, 1997), http://www.fao.org/DOCREP/005/AC774E/ac774e00.htm (accessed December 28, 2011).

87 *In his book . . .* The Boy Who Was Raised as a Dog: Bruce D. Perry and Maia Szalavitz. *The Boy Who Was Raised as a Dog and Other Stories from a Child Psychiatrist's Notebook: What Traumatized Children Can Teach Us about Loss, Love and Healing* (New York: Basic Books, 2007); Robert F. Anda, Vincent J. Felitti, J. Douglas Bremner, John D. Walker, Charles Whitfield, Bruce D. Perry, Shanta R. Dube, and Wayne H. Giles, "The Enduring Effects of Abuse and Related Adverse Experiences in Childhood: A Convergence of Evidence from Neurobiology and Epidemiology," *European Archives of Psychiatry and Clinical Neuroscience* 256, no. 3 (April 2006): 174–186; B. D. Perry and R. Pollard, "Homeostasis, Stress, Trauma, and Adaptation: A Neurodevelopmental View of Childhood Trauma," *Child and Adolescent Psychiatric Clinics of North America* 7, no. 1 (January 1998): 33–51, viii; Bruce D. Perry, "Neurobiological Sequelae of Childhood Trauma: PTSD in Children," In *Catecholamine Function in Posttraumatic Stress Disorder: Emerging Concepts*, 233–255 (*Progress in Psychiatry* 42, Arlington, VA: American Psychiatric Association, 1994); James E. McCarroll, "Healthy Families, Healthy Communities: An Interview with Bruce D. Perry," *Joining Forces Joining Families* 10, no. 3 (2008), http://www.cstsonline .org/wp-content/resources/Joining_Forces_2008_01.pdf.

88 *There may also be lasting effects:* Perry and Szalavitz, *The Boy Who Was Raised as a Dog*, 19.

88 *In 2009, the U.S. Department of Health and Human Services:* Child Welfare Information Gateway, *Understanding the Effects of Maltreatment on Brain Development*, Issue Brief, U.S. Department of Health and Human Services, November 2009; Perry and Szalavitz, *The Boy Who Was Raised as a Dog*, 247.

88 *Perry's very first patient:* Over the course of three years, Perry was able to help Tina regain control over her stress response and make decisions after thinking them through instead of blindly reacting. Unfortunately he was not able to help her change her behavior entirely. Perry felt that, in

the end, her newfound control over her stress response only helped her to better hide her trauma. Perry and Szalavitz, *The Boy Who Was Raised as a Dog*, 22–28.

92 *When I look at a gorilla:* Dale Jamieson, professor of environmental studies and philisophy at NYU, has suggested that zoos also replicate problematic species distinctions and that they are woven into the very architecture (the confienment itself marking a false distinction between the animals who are caged and the human animals who are not). Dale Jamieson, Against Zoos," in *Morality's Progress: Essays on Humans, Other Animals, and the Rest of Nature* (Oxford: Oxford University Press USA, 2003), 166–175; and Dale Jameison, "The Rights of Animals and the Demands of Nature," *Environmental Values* 17 (2008), 181–189.

93 *Human stereotypies include:* Deivasumathy Muthugovindan and Harvey Singer, "Motor Stereotypy Disorders," *Current Opinion in Neurology* 22, no. 2 (April 2009): 131–136.

94 *Horses may take small, rhythmic gulps of air:* See, for example, Jean S. Akers and Deborah S. Schildkraut, "Regurgitation/Reingestion and Coprophagy in Captive Gorillas," *Zoo Biology* 4, no. 2 (1985): 99–109; M. C. Appleby, A. B. Lawrence, and A. W. Illius, "Influence of Neighbours on Stereotypic Behaviour of Tethered Sows," *Applied Animal Behaviour Science* 24, no. 2 (1989): 137–46; M. J. Bashaw et al., "Environmental Effects on the Behavior of Zoo-Housed Lions and Tigers, with a Case Study of the Effects of a Visual Barrier on Pacing," *Journal of Applied Animal Welfare Science* 10, no. 2 (2007): 95–109; Yvonne Chen et al., "Diagnosis and Treatment of Abnormal Food Regurgitation in a California Sea Lion *(Zalophus californianus)*," in *IAAAM Conference Proceedings* 68 (International Association for Aquatic Animal Medicine, 2009); Jonathan J. Cooper and Melissa J. Albentosa, "Behavioural Adaptation in the Domestic Horse: Potential Role of Apparently Abnormal Responses Including Stereotypic Behaviour," *Livestock Production Science* 92, no. 2 (2005): 177–82; Leslie M. Dalton, Todd R. Robeck, and W. Glenn Youg, "Aberrant Behavior in a California Sea Lion *(Zalophus californianus)*," in *IAAAM Conference Proceedings* 145–46 (International Association for Aquatic Animal Medicine, 1997); J. E. L. Day et al., "The Separate and Interactive Effects of Handling and Environmental Enrichment on the Behaviour and Welfare of Growing Pigs," *Applied Animal Behaviour Science* 75, no. 3 (2002): 177–92; Andrzej Elzanowski and Agnieszka Sergiel, "Stereotypic Behavior of a Female Asiatic Elephant *(Elephas maximus)* in a Zoo," *Journal of*

Applied Animal Welfare Science 9, no. 3 (2006): 223–32; Loraine Tarou Fernandez et al., "Tongue Twisters: Feeding Enrichment to Reduce Oral Stereotypy in Giraffe," *Zoo Biology* 27, no. 3 (2008): 200–212; Georgia J. Mason, "Stereotypies: A Critical Review," *Animal Behaviour* 41, no. 6 (1991): 1015–37; Edwin Gould and Mimi Bres, "Regurgitation and Reingestion in Captive Gorillas: Description and Intervention," *Zoo Biology* 5, no. 3 (1986): 241–50; T. M. Gruber et al., "Variation in Stereotypic Behavior Related to Restraint in Circus Elephants," *Zoo Biology* 19, no. 3 (2000): 209–21; Steffen W. Hansen and Birthe M. Damgaard, "Running in a Running Wheel Substitutes for Stereotypies in Mink *(Mustela vison)* but Does It Improve Their Welfare?," *Applied Animal Behaviour Science* 118, nos. 1–2 (2009): 76–83; Lindsay A. Hogan and Andrew Tribe, "Prevalence and Cause of Stereotypic Behaviour in Common Wombats *(Vombatus ursinus)* Residing in Australian Zoos," *Applied Animal Behaviour Science* 105, nos. 1–3 (2007): 180–91; Kristen Lukas, "An Activity Budget for Gorillas in North American Zoos," Disney's Animal Kingdom and Brevard Zoo, 2008; Juan Liu et al., "Stereotypic Behavior and Fecal Cortisol Level in Captive Giant Pandas in Relation to Environmental Enrichment," *Zoo Biology* 25, no. 6 (2006): 445–59; Kristen E. Lukas, "A Review of Nutritional and Motivational Factors Contributing to the Performance of Regurgitation and Reingestion in Captive Lowland Gorillas *(Gorilla gorilla gorilla)*," *Applied Animal Behaviour Science* 63, no. 3 (1999): 237–49; Avanti Mallapur and Ravi Chellam, "Environmental Influences on Stereotypy and the Activity Budget of Indian Leopards *(Panthera pardus)* in Four Zoos in Southern India," *Zoo Biology* 21, no. 6 (2002): 585–95; L. M. Marriner and L. C. Drickamer, "Factors Influencing Stereotyped Behavior of Primates in a Zoo," *Zoo Biology* 13, no. 3 (1994): 267–75; G. Mason and J. Rushen, eds., *Stereotypic Animal Behavior: Fundamentals and Applications to Welfare*, 2nd ed. (CABI, 2006); Lynn M. McAfee, Daniel S. Mills, and Jonathan J. Cooper, "The Use of Mirrors for the Control of Stereotypic Weaving Behaviour in the Stabled Horse," *Applied Animal Behaviour Science* 78, nos. 2–4 (2002): 159–73; Jeffrey Rushen, Anne Marie B. De Passillé, and Willem Schouten, "Stereotypic Behavior, Endogenous Opioids, and Postfeeding Hypoalgesia in Pigs," *Physiology and Behavior* 48, no. 1 (1990): 91–96; U. Schwaibold and N. Pillay, "Stereotypic Behaviour Is Genetically Transmitted in the African Striped Mouse *Rhabdomys pumilio*," *Applied Animal Behaviour Science* 74, no. 4 (2001): 273–80; Loraine Rybiski Tarou, Meredith J. Bashaw, and Terry L. Maple, "Failure of a Chemical Spray to Significantly Reduce Stereotypic Licking in a Captive Giraffe," *Zoo Biology* 22, no. 6 (2003): 601–7; Sophie Vickery and Georgia Mason, "Stereotypic Behavior in Asiatic Black and Malayan Sun Bears," *Zoo Biology* 23, no. 5 (2004): 409–30; Beat Wechsler, "Stereotypies in Polar Bears," *Zoo Biology* 10, no. 2 (1991): 177–88; Carissa L. Wickens and Camie R. Heleski, "Crib-Biting Behavior in Horses: A Review," *Applied Animal Behaviour Science* 128, nos. 1–4 (2010): 1–9; Hanno Würbel and Markus Stauffacher, "Prevention of Stereotypy in Laboratory Mice: Effects on Stress Physiology and Behaviour," *Physiology and Behavior* 59, no. 6 (1996): 1163–70.

94 *more than 16 billion farm and lab animals:* Naomi R. Latham and G. J. Mason, "Maternal Deprivation and the Development of Stereotypic Behaviour," *Applied Animal Behaviour Science* 110, nos. 1–2 (2008): 99; Jeffrey Rushen and Georgia Mason, "A Decade-or-More's Progress in Understanding Stereotypic Behaviour," in *Stereotypic Animal Behaviour*, ed. Jeffrey Rushen and Georgia Mason (CABI, 2006), cited in Temple Grandin and Catherine Johnson, *Animals Make Us Human: Creating the Best Life for Animals* (New York: Houghton Mifflin Harcourt, 2009), 15.

94 *This includes 91.5 percent of pigs:* Rushen and Mason, "A Decade-or-More's Progress in Understanding Stereotypic Behavior," 15.

94 *A large percentage of the roughly 100 million:* Number of lab animals: Balcombe, *Second Nature*. For more on animal stereotypy, see chapter 3.

94 *a strong correlation between lab, zoo, and farm animals:* Latham and Mason, "Maternal Deprivation and the Development of Stereotypic Behaviour," 84–108.

95 *"really intense stereotypies":* Grandin and Johnson, *Animals Make Us Human*, 4.

96 *She has categorized autism "as a way station":* Temple Grandin, "Animals in Translation," http://www.grandin.com/inc/animals.in.translation.html (accessed December 1, 2013).

96 *observed a wild coyote pup he called Harry:* Marc Bekoff, "Do Wild Animals Suffer from PTSD and Other Psychological Disorders?," *Psychology Today*, November 29, 2011, http://www.psychologytoday.com/blog/animal-emotions/201111/do-wild-animals-suffer-ptsd-and-other-psychological-disorders (accessed December 11, 2013).

96 *dosed them with a gut microbe,* Bacteroides fragilis: Hsiao et al., "Microbiota Modulate Behavioral and Physiological Abnormalities Associated with Neurodevelopmental Disorders," *Cell,* (accessed December 11, 2013).

97 *linked autism spectrum disorders to intestinal problems:* Sara Reardon, "Bacterium Can Reverse Autism-like Behaviour in Mice," *Nature News,* December 5, 2013, http://www.nature.com/news/bacterium-can-reverse-autism-like-behaviour-in-mice-1.14308 (accessed December 6, 2013); Natalia V. Malkova et al., "Maternal Immune Activation Yields Offspring Displaying Mouse Versions of the Three Core Symptoms of Autism," *Brain, Behavior, and Immunity* 26, no. 4 (May 2012): 607–16; Isaac S. Kohane et al., "The Co-Morbidity Burden of Children and Young Adults with Autism Spectrum Disorders," *PloS One* 7, no. 4 (2012): e33224.

99 *According to the AZA, the average zoo and aquarium visitor:* John H. Falk et al., "Why Zoos and Aquariums Matter: Assessing the Impact of a Visit to a Zoo or Aquarium," Association of Zoos and Aquariums, 2007, http://www.aza.org/uploadedFiles/Education/why_zoos_matter.pdf; "Visitor Demographics, Association of Zoos and Aquariums," http://www.aza.org/visitor-demographics/ (accessed March 10, 2013).

100 *In 2007 the AZA published the results:* Falk et al., "Why Zoos and Aquariums Matter"; Lori Marino et al., "Do Zoos and Aquariums Promote Attitude Change in Visitors? A Critical Evaluation of the American Zoo and Aquarium Study," *Society and Animals* 18 (April 2010): 126–38, http://www.nbb.emory.edu/faculty/personal/documents/MarinoetalAZAStudy.pdf.

103 *Most men and women pluck hairs:* "Trichotillomania," in *Diagnostic and Statistical Manual of Mental Disorders-V* (Washington, DC: American Psychiatric Association, 2013), 312.39.

104 *The fifth edition of the DSM:* Ibid.

104 *The habit might be a symptom of anxiety:* "Hair Pulling: Frequently Asked Questions, Trichotillomania Learning Center FAQ," trich.org/about/hair-faqs.html (accessed November 28, 2010); "Pulling Hair: Trichotillomania and Its Treatment in Adults. A Guide for Clinicians," the Scientific Advisory Board of the Trichotillomania Learning Center, at www.trich.org/about/for-professionals.html (accessed November 28, 2010); Mark Lewis and Kim Soo-Jeong, "The Pathophysiology of Restricted Repetitive Behavior," *Journal of Neurodevelopmental Disorders* 1 (2009): 114–32.

104 *Hair pulling has been reported in six primate species:* Viktor Reinhardt, "Hair Pulling: A Review," *Laboratory Animals,* no. 39 (2005): 361–69.

104 *"Tache seems unable"*: "Re: The Attempt to Save Noir from Barbering," discussion post, January 26, 2010, www.fancymicebreeders.com/mousefancie forum (accessed November 28, 2010).

105 *In fact it seems that their clients may* enjoy *it:* F. A. Van den Broek, C. M. Omtzigt, and A. C. Beynen, "Whisker Trimming Behaviour in A2G Mice Is Not Prevented by Offering Means of Withdrawal from It," *Lab Animal Science*, no. 27 (1993): 270–72.

105 A *few researchers have suggested:* Biji T. Kurien, Tim Gross, and R. Hal Scofield, "Barbering in Mice: A Model for Trichotillomania," *British Medical Journal*, no. 331 (2005): 1503–5. See also Joseph D. Garner et al., "Barbering (Fur and Whisker Trimming) by Laboratory Mice as a Model of Human Trichotillomania and Obsessive-Compulsive Spectrum Disorders," *Comparative Medicine* 54, no. 2 (2004): 216–24.

105 *Studies using mice as stand-ins:* The fact that humans tend to pluck themselves and mice tend to pluck one another hasn't stopped the use of mouse experimental models. Since both the barber mouse and her client engage in the process by choice, even though it must be at least a little painful, researchers have tended to assume, for better or for worse, that the behavior in people is simply spread between two individuals in the mice. Alice Moon-Fanelli, N. Dodman, and R. O'Sullivan, "Veterinary of Models Compulsive Self-Grooming Parallels with Trichotillomania," in *Trichotillomania*, ed. Dan J. Stein, Gary A. Christenson, and Eric Hollander (Arlington, VA: American Psychiatric Press, 1999), 72–74.

105 *An experiment conducted in 2002:* "Compulsive Behavior in Mice Cured by Bone Marrow Transplant," *Science Daily*, May 27, 2010, www.sciencedaily.com/releases/2010/05/100527122150.htm (accessed November 28, 2010); Shau-Kwaun Chen et al., "Hematopoietic Origin of Pathological Grooming in Hoxb8 Mutant Mice," *Cell*, 2010; 141 (5): 775; see also "Mental Illness Tied to Immune Defect: Bone Marrow Transplants Cure Mice of Hair-Pulling Compulsion," News Center, University of Utah, www.unews.utah.edu/p/?r=022210–3 (accessed November 28, 2010).

106 *birds pluck when bored, frustrated, or stressed:* Lynne M. Seibert et al., "Placebo-Controlled Clomipramine Trial for the Treatment of Feather Picking Disorder in Cockatoos," *Journal of the American Animal Hospital Association* 40, no. 4 (2004): 261–69. See also Lynne M. Seibert,

"Feather-Picking Disorder in Pet Birds," in *Manual of Parrot Behavior*, ed. Andrew U. Luesche (Oxford: Blackwell, 2008).

106 *Phoebe Greene Linden has lived with parrots:* Phoebe Greene Linden, personal communication, November 5, 2010.

107 *Joe roamed around the neighborhood:* Brian MacQuarrie and Douglas Belkin, "Franklin Park Gorilla Escapes, Attacks 2," *Boston Globe*, September 29, 2003.

107 *Primatologists like Frans de Waal and Jane Goodall:* Frans De Waal, *The Ape and the Sushi Master: Cultural Reflections by a Primatologist* (New York: Basic Books, 2001), 214–16; Bijal P. Trivedi, " 'Hot Tub Monkeys' Offer Eye on Nonhuman 'Culture,' " *National Geographic News*, February 6, 2004, news.nationalgeographic.com/news/2004/02/0206_040206_tvmacaques.html (accessed November 28, 2010).

Chapter Three: Family Therapy

115 *Harriman tells the story of an eight-year-old rabbit:* Marinell Harriman, *House Rabbit Handbook: How to Live with an Urban Rabbit*, 3rd ed. (Alameda, CA: Drollery Press, 1995), 92.

116 *One member of the Rat Fan Club wrote:* Angela King, "The Case against Single Rats," *The Rat Report*, http://ratfanclub.org/single.html (accessed April 12, 2013); Angela Horn, "Why Rats Need Company," National Fancy Rat Society, http://www.nfrs.org/company.html (accessed April 12, 2013). Also see Kathy Lovings, "Caring for Your Fancy Rat," http://www.ratdippityrattery.com/CaringForYourFancyRat.htm (accessed April 1, 2013).

116 *"Rats definitely notice":* Monika Lange, *My Rat and Me* (Barron's Educational Series, 2002), 58.

121 *They tried one last antipsychotic:* H. W. Murphy and M. Mufson. "The Use of Psychopharmaceuticals to Control Aggressive Behaviors in Captive Gorillas," Proceedings of "The Apes: Challenges for the 21st Century," Brookfield Zoo, Chicago (2000): 157–60.

121 *Sadly, this isolation period: From Cages to Conservation*, WBUR documentary, http://insideout.wbur.org/documentaries/zoos/ (accessed December 1, 2013).

125 *The expression "getting your goat":* The writer H. L. Mencken floated this idea. See Christine Ammer, *The American Heritage Dictionary of Idioms* (Boston: Houghton Mifflin Harcourt, 1997), 242. The *Oxford English*

Dictionary lists the first mention of the phrase: http://www.oed.com/view/Entry/79564?rskey=cKiv56&result=2&isAdvanced=false#eid.

125 *Before Seabiscuit was a champion:* Laura Hillenbrand, *Seabiscuit: An American Legend* (Random House Digital, 2003), 98–100.

125 *a racehorse named Miss Edna Jackson:* "Goat and Race Horse Chums: Filly at Belmont Park Won't Eat If Her Friend Is Away," *New York Times*, May 13, 1907.

126 *a horse named Exterminator:* Amy Lennard Goehner, "Animal Magnetism: Skittish Racehorses Tend to Calm Down When Given Goats as Pets," *Sports Illustrated*, February 21, 1994, http://sportsillustrated.cnn.com/vault/article/magazine/MAG1004875/index.htm (accessed November 15, 2009).

126 *Giving racehorses animal companions:* Online forums for horse breeders, riders, racing enthusiasts, and others are rife with discussions about animal companions for horses. See, for example, "Companion Animals," Horseinfo, http://www.horseinfo.com/info/faqs/faqcompanionQ2.html (accessed January 20, 2012); "Companion Animals for Horses," Franklin Levinson's Horse Help Center, http://www.wayofthehorse.org/horse-help/companion-animals-for-horses.php; "Readers Respond: Your Tips for Providing Horses with Companions," About.com. http://horses.about.com/u/ua/basiccare/companionridertips.htm (accessed January 20, 2012); "What Animals with a Horse?," Permies.com, http://www.permies.com/t/9560/critter-care/animals-horse; "Companion Animals for a Horse," Horse Forum, http://www.horseforum.com/horse-training/companion-animals-horse-45342/ (accessed January 20, 2012).

126 *John Veitch, an American Hall of Fame trainer:* Goehner, "Animal Magnetism."

126 *Another Hall of Fame trainer, Jack Van Berg:* Ibid.

126 A *man in charge of one of the Jumbotron screens:* Graham Parry, personal communication, June 28, 2011.

127 *Potbellied pigs may indeed be useful:* "Stable Goats Help Calm Skittish Thoroughbreds."

127 *the life of the Giant Pacific octopus:* Devin Murphy, "Brains over Brawn," *Smithsonian Zoogoer*, March 2011, http://nationalzoo.si.edu/Publications/Zoogoer/2011/4/Cephalopods.cfm (accessed April 7, 2011); Ellen Byron, "Big Cats Obsess over Calvin Klein's 'Obsession for Men,' " *Wall Street Journal*, June 8, 2010; "Phoenix Zoo Tortoise Enrichment," http://www

.phoenixzoo.org/learn/animals/Giant_tortoise_article_22.pdf (accessed June 10, 2010).

128 *Their website lists their library:* The Shape of Enrichment, http://www.enrichment.org/miniwebfile.php?Region=Video_Library&File=collection.html&File2=collection_sb.html&NotFlag=1 (accessed June 10, 2010).

128 *That year's amendments to the Animal Welfare Act:* "Environmental Enrichment and Exercise," USDA, http://awic.nal.usda.gov/research-animals/environmental-enrichment-and-exercise (accessed June 10, 2010).

129 *Recently, the Wilhelma Zoo in Stuttgart, Germany:* Allan Hall and Wills Robinson, "How about 'the Ape Escape'? Bonobos in German Zoo Have New Flat-Screen TV Installed Which Lets Them Pick Their Favourite Movie," *Daily Mail*, November 26, 2013; http://www.dailymail.co.uk/sciencetech/article-2514113/Bonobos-apes-German-Zoo-flat-screen-TV-installed.html (accessed June 10, 2010); "Bonobo Apes in Hi-Tech German Zoo Go Bananas for Food, Not TV Porn," NBC News, November 26, 2013; http://worldnews.nbcnews.com/_news/2013/11/26/21626507-bonobo-apes-in-hi-tech-german-zoo-go-bananas-for-food-not-tv-porn (accessed November 26, 2010).

130 *James Breheny, the director of the Bronx Zoo:* "Zoo Director (O.K. Be That Way)," *New York Times*, July 21, 2009.

131 *fastest growing retail sector:* Carol Tice, "Why Recession-Proof Industry Just Keeps Growing," *Forbes*, October 30, 2012; "2013/2014 National Pet Owners Survey by American Pet Product Association, American Pet Product Association, December 2013.

132 *Still, Donna Haraway, the philosopher of science:* personal communication, Donna Haraway, February 17, 2014.

135 *The cover of her 1995 book,* Getting in TTouch: Not unlike the phrenologists of yesteryear who believed that the contours of the skull could shed light on a person's character, Tellington has said that she can look at the different parts of a horse's face and find clues to its personality.

135 *Her patented TTouches have names:* Tellington Touch Training, http://www.ttouch.com/whatisTTouch.shtml (accessed February 5, 2012).

136 *She now works with all sorts of animals:* Tellington-Jones isn't the only person doing this. There are certification programs for lots of different types of animal massage. See, for example, International Association of Animal Massage and Bodywork/Association of Canine Water Therapy, http://www.iaamb.org/mission-and-goals.php (accessed February 5,

2012), or Chandra Beal, *The Relaxed Rabbit: Massage for Your Pet Bunny* (iUniverse, 2004), among many others.

136 *A variety of studies on humans have demonstrated the power of massage*: The role of massage in helping humans deal with anxiety has been evaluated in a few different contexts. See Susanne M. Cutshall et al., "Effect of Massage Therapy on Pain, Anxiety, and Tension in Cardiac Surgical Patients: A Pilot Study," *Complementary Therapies in Clinical Practice* 16, no. 2 (2010): 92–95; Tiffany Field, "Massage Therapy," *Medical Clinics of North America* 86, no. 1 (2002): 163–71; Melodee Harris and Kathy C. Richards, "The Physiological and Psychological Effects of Slow-Stroke Back Massage and Hand Massage on Relaxation in Older People," *Journal of Clinical Nursing* 19, no. 7–8 (2010): 917–26; Christopher A. Moyer et al., "Does Massage Therapy Reduce Cortisol? A Comprehensive Quantitative Review," *Journal of Bodywork and Movement Therapies* 15, no. 1 (2011): 3–14; Wendy Moyle, Amy Nicole Burne Johnston, and Siobhan Therese O'Dwyer, "Exploring the Effect of Foot Massage on Agitated Behaviours in Older People with Dementia: A Pilot Study," *Australasian Journal on Ageing* 30, no. 3 (2011): 159–61.

136 *Massage has also been used on dressage horses:* Kevin K. Haussler, "The Role of Manual Therapies in Equine Pain Management," *Veterinary Clinics of North America: Equine Practice* 26, no. 3 (2010): 579–601; Mike Scott and Lee Ann Swenson, "Evaluating the Benefits of Equine Massage Therapy: A Review of the Evidence and Current Practices," *Journal of Equine Veterinary Science* 29, no. 9 (2009): 687–97; C. M. McGowan, N. C. Stubbs, and G. A. Jull, "Equine Physiotherapy: A Comparative View of the Science Underlying the Profession," *Equine Veterinary Journal* 39, no. 1 (2007): 90–94.

136 *photos of men and women in barn jackets:* "Benefits of Equine Sports Massage," Equine Sports Massage Association, http://www.equinemassageassociation.co.uk/benefits_of_equine_sports_massage.html (accessed December 24, 2012).

137 *Frediani believes that TTouch relaxes muscle tension:* Mardi Richmond, "The Tellington TTouch for Dogs," *Whole Dog Journal*, August 2010; Jodi Frediani, personal communication, January 18, 2011, and May 9, 2012.

140 *Mosha triggered a land mine:* A much smaller number of mines are planted by nonstate groups. See "Burma (Myanmar)," Landmine and Cluster Munition Monitor, available at http://www.the-monitor.org/.

144 *The most recognized bonobo researcher:* Books by Frans B. M. de Waal include *Good Natured: The Origins of Right and Wrong in Humans and Other Animals* (Cambridge, MA: Harvard University Press, 1996); *The Ape and the Sushi Master: Cultural Reflections by a Primatologist* (New York: Basic Books, 2001); *Bonobo: The Forgotten Ape* (Berkeley: University of California Press, 1997), with Frans Lanting; *The Age of Empathy: Nature's Lessons for a Kinder Society* (New York: Crown, 2009).

145 *De Waal believes that we owe them:* Frans de Waal, "The Bonobo in All of Us," PBS, January 1, 2007, http://www.pbs.org/wgbh/nova/nature/bonobo-all-us.html; Frans B. M. de Waal, "Bonobo Sex and Society," *Scientific American* 272, no. 3 (1995); de Waal and Lanting, *Bonobo: The Forgotten Ape.*

145 *Brian "would vomit thirty, forty, fifty times a day": Primate Week,* interview with Barbara Bell and Harry Prosen, "Lake Effect with Bonnie North," WUWM Public Radio, February 20, 2012; http://www.youtube.com/watch?v=_SW0re1LGOs.

146 *On Prosen's first visit to the zoo:* Steve Farrar, "A Party Animal with a Social Phobia," *Times for Higher Education,* July 28, 2000, http://www.timeshighereducation.co.uk/story.asp?storyCode=152816§ioncode=26 (accessed June 1, 2010); personal communication, Dr. Harry Prosen, October 1, 2010.

146 *"I have had some difficult interview situations":* Harry Prosen and Barbara Bell, "A Psychiatrist Consulting at the Zoo (the Therapy of Brian Bonobo)," in *The Apes: Challenges for the 21st Century. Conference Proceedings,* Brookfield Zoo, 2001, 161–64.

147 *Brian's fisting habit developed:* Prosen and Bell, "A Psychiatrist Consulting at the Zoo."

147 *a reputation for healing distressed bonobos:* Jo Sandin, *Bonobos: Encounters in Empathy* (Milwaukee: Zoological Society of Milwaukee, 2007), 25–27.

148 *Once, when a younger male stole a mailing tube:* Ibid., 49–50.

148 *He was also very attached to his OCD rituals: Primate Week,* interview with Barbara Bell and Harry Prosen; http://www.youtube.com/watch?v=_SW0re1LGOs.

149 *"But the beauty of the drug therapy":* Ibid.

151 *He was impressed by their ability:* Kelly Servick, "Psychiatry Tries to Aid Traumatized Chimps in Captivity," *Scientific American,* April 2, 2013, http://www.scientificamerican.com/article.cfm?id=psychiatry-comes-to-the-aid-of-captive-chimps-with-abnormal-behavior.

151 *"That's why we populated the globe, not chimps"*: Ibid.; personal communication, Dr. Harry Prosen, October 1, 2010.

151 *Bell and Prosen share a slightly different belief about bonobos*: Prosen and Bell, "A Psychiatrist Consulting at the Zoo."

152 *By 2001, four years after Brian arrived:* Sandin, *Bonobos*, 59–68.

152 *"They still get along fine," says Bell:* Jo Sandin, "Bonobos: Passage of Power," *Alive*, Milwaukee County Zoological Society (Winter 2006), http://www.zoosociety.org/pdf/conserveprojects/WinterAlive06_BonobosPassageofPower.pdf (accessed June 10, 2010).

153 *After Lody died:* Paula Brookmire, "Lody the Bonobo: A Big Heart," *Alive Magazine*, Milwaukee Zoological Society, April 2012, 25.

153 *Over the last fifteen years both Prosen and Bell:* Ibid.; personal communication, Dr. Harry Prosen, October 1, 2010.

Chapter Four: Proxies and Mirrors

155 *physicians who treated various forms of insanity:* Edward Shorter, *A History of Psychiatry: From the Era of the Asylum to the Age of Prozac* (New York: Wiley, 1997) 53, 90–91, 113–14; Roy Porter, *A Social History of Madness: The World through the Eyes of the Insane* (London: Weidenfeld and Nicolson, 1987); Andrew T. Scull, *Hysteria: The Biography* (New York: Oxford University Press, 2009), 8–13.

155 *the word* madness *has meant many different things:* See "mad, adv.," *Oxford English Dictionary* Online, June 2011, http://www.oed.com/view/Entry/6512?redirectedFrom=mad.

156 *The disease was also scary because:* Harriet Ritvo, *The Animal Estate: The English and Other Creatures in the Victorian Age* (Cambridge, MA: Harvard University Press, 1987), 168–69.

156 A *mad dog could be anywhere:* Ibid., 177.

156 *The public's anxieties about mad dogs:* Examples include "Mad Dogs Running Amuck: A Hydrophobia Panic Prevails in Connecticut," *New York Times*, June 29, 1890; "Mad Dog Owned the House: Senorita Isabel's Foundling Pet Takes Possession," *New York Times*, June 18, 1894; "Lynn in Terror," *Boston Daily Globe*, June 27, 1898; "Suburbs Demand Death to Canines: Englewood and Hyde Park, Aroused by Biting of Children, Ask Extermination. Hesitation by Police. Say 'Mad Dog Panic' Order of 'Shoot on Sight' Would Sacrifice Fine Animals. Forbids Reckless Shooting. Victims of Vicious Dogs. Dog Disperses Euchre Party,"

Chicago Daily Tribune, June 7, 1908; "Mad Dog Is a Public Enemy," *Virginia Law Register* 15, no. 5 (1909): 409.

156 *The historian Harriet Ritvo:* Ritvo, *Animal Estate*, 175, 177, 180–81, 193.

157 *Infection was also thought to jump from dogs to other animals:* "Mad Horse Attacks Men: Veterinary Who Shot the Animal from Haymow Says It Had Rabies," *New York Times*, April 6, 1909; "Mad Horses Despatched: Soldiers' Home Animals Are Killed for Rabies: Equines Bitten by Afflicted Dog Are Isolated and, after a Few Weeks Show Signs of Having Been Infected. Death Warrant Quickly Executed. Other Horses Affected," *Los Angeles Times*, May 9, 1906; "Burro with Hydrophobia: Bites Man, Kills Dog and Takes Chunk from Neck of Horse," *Los Angeles Times*, March 16, 1911; "Career of a Crazy Lynx: The Mad Beast Killed by a Woman after Running Amuck for Thirty Miles," *Chicago Daily Tribune*, May 10, 1890; "Stampeded by Mad Cow: Animal Charges Saloon and Restaurant, People Fleeing for Safety," *Los Angeles Times*, June 5, 1909; "Bitten by a Mad Monkey: Little Mabel Hogle Attacked by a Museum Animal. While Viewing Curios with Her Father, George Hogle, 912 North Clark Street, the Beast Rushes upon the Girl. Lively Battle Ensues. Father Kicks the Animal Away and the Daughter Faints. Brute, Said to Be Mad, Is Killed. Wound Cauterized," *Chicago Daily Tribune*, November 28, 1897.

157 *when Oliver Goldsmith published the poem:* Oliver Goldsmith, "An Elegy on the Death of a Mad Dog," in *The Oxford Encyclopedia of Children's Literature*, ed. Jack Zipes (New York: Oxford University Press, 2006).

157 *One small dog, for example, discovered with a pig:* "Wreck in Midocean; a Mad Dog and a Little Pig the Sole Occupants of a Brig," *New York Times*, March 9, 1890.

157 *Animals could also go mad from a lifetime of abuse:* " 'Smiles,' the Big Park Rhinoceros: Bought at Auction for $14,000, She Needs Constant Care, Although She Has an Ugly Temper," *New York Times*, March 29, 1903.

158 *Maddened horses, as they were known:* "Dragged by Mad Horses: A Lady's Dress Catches in the Wheels. She May Recover," *Los Angeles Times*, April 30, 1888; "Runaways in Central Park: Two Horses Wreck Three Carriages and a Bicycle. M. J. Sullivan's Team Had a Long and Disastrous Run before a Park Policeman Caught It. Mrs. Crystal and Her Children Have a Narrow Escape. Julius Kaufman Gives a Mounted Policeman a Chance to Distinguish Himself," *New York Times*, June 4, 1894; "Mad Horses' Wild Chase: Dashed through Streets with 1,000

People at His Heels. Hero Who Saved Three Tots Hurt, Fatally Maybe, While Trying to Stop Him. Thrown by Trolley Car," *New York Times,* September 14, 1903; "A Mad Horse," *New York Times,* June 20, 1881.

158 *the monkey mascot of a New Orleans baseball team:* "Mad Monkey Scares Fans: Queer Mascot of New Orleans Ball Team Makes Trouble and Game Stops," *Los Angeles Times,* July 19, 1909.

158 *As late as the 1920s and 1930s:* On the set of a 1930s Paramount film being shot in the Santa Monica Mountains, actress Dorothy Lamour was attacked by a "mad ape" (the film's chimp actor, named Jiggs) and saved by a young man from the props department. A few years later a second mad monkey was on the loose in the Los Angeles suburb of Tarzana and ended up caged after its exploits ended in one angry neighbor's garage. "Mad Cats in Madder Orgy," *Los Angeles Times,* May 2, 1924; "A Mad Cow Mutilates Two People," *Los Angeles Times,* May 19, 1889; "Color of Mad Parrot Saves It from Death," *San Francisco Chronicle,* October 3, 1913; "Film Aide Saves Actress from Mad Ape's Attack," *Los Angeles Times,* July 7, 1936; "Monkey Caged after Biting Second Person," *Los Angeles Times,* January 8, 1939.

158 *just a few months before forging an alliance with Hitler*: "Mussolini Attacked by Mad Ox: African Fete Throng in Panic as Horns Barely Miss Premier," *New York Times,* March 15, 1937.

158 *many of the most enduring stories concern elephants:* The 1887 *Los Angeles Times* article "Bad Elephants" was a rap sheet of elephants gone mad and bad. Mogul was killed in 1871 during an effort to subdue him, and an elephant named Albert with Barnum's circus was shot by soldiers in New Hampshire after he killed his keeper. In 1901 Big Charley killed his keeper in Indiana by throwing him into a stream twice and then standing on top of him until he drowned. A few years later Topsy was electrocuted at Coney Island after killing three men in as many years, one of whom had fed her a lit cigarette. There was also Mandarin, Mary, Tusko, Gunda, Roger, and countless others who were shot, electrocuted, hanged, and strangled for striking out at their keepers, riders, grooms, or trainers, or at bystanders, often for very good reason. "Bad Elephants," *Los Angeles Times,* January 16, 1887; "Mad Elephants: A Showman's Recollections of Keepers Killed and Destruction Done by Them. Peculiarities of the Beasts," *Boston Daily Globe,* November 20, 1881; "Mad Elephants: The Havoc the Great Beast Causes When He Rebels against Irksome Captivity," *New York Times,* August 26, 1880; "Mad Elephants:

Big Charley Killed His Keeper at Peru, Indiana. Twice Hurled Him into a Stream and Then Stood upon Him," *Boston Daily Globe*, April 26, 1901; "Death of Mandarin: Huge Mad Elephant Strangled with the Help of a Tug and a Big Chain," *Boston Daily Globe*, November 9, 1902; "Bullet Ends Gunda, Bronx Zoo Elephant: Dr. Hornaday Ordered Execution Because Gunda Reverted to Murderous Traits. Died without a Struggle. His Mounted Skin Will Adorn Museum of Natural History and His Flesh Goes to Feed the Lions," *New York Times*, June 23, 1915; "Death of Gunda," *Zoological Society Bulletin* 18, no. 4 (1915): 1248–49; "British Soldier's Miraculous Escape from Death on Tusks of a Mad Elephant," *Boston Daily Globe*, September 19, 1920; "Mad Elephant Rips Chains and Walls: Six-Ton Tusko Wrecks Portland (Ore.) Building before Recapture by Ruse. Sharpshooters Cover Him. Thousands Watch Small Army of Men Trap Pachyderm with Steel Nooses Hitched to Trucks," *New York Times*, December 26, 1931.

158 *the mad elephant genre, published in the* New York Times *in 1880:* "Mad Elephants: The Havoc the Great Beast Causes When He Rebels against Irksome Captivity."

159 *Captive elephants have been known to suddenly explode into violence:* While it would end up being applied predominantly to animals and human children, in the seventeenth century the phrase *running amok* concerned Malaysians and opium. Traveling Portuguese described frenzied Malaysians as "Amucos," and many reports associated the state with being high. In his 1833 *Naval History of England* R. Southey wrote about the "pitch of fury which the Malays excite in themselves by a deleterious drug, before they run amuck" (perhaps ignoring the more obvious role of colonization and oppression in any frenzied behavior among Malaysians). Twenty-five years later it was being applied to animals. See "amok, n. and adv.," *Oxford English Dictionary* Online, June 2011, http://www.oed.com/view/Entry/6512?redirectedFrom=amok. Other elephants targeted no individual but rampaged around until they were caught and hanged, electrocuted, or otherwise punished. In 1902 a "huge mad elephant" named Mandarin was strangled with a chain in New York after running amok with the Barnum and Bailey Circus. In 1920 a story of the dramatic escape by a British soldier from the tusks "of a mad elephant" made the American papers. "Bad Elephants," *Los Angeles Times*, January 16, 1887; "Mad Elephant: Big Charley Killed His Keeper at Peru, Indiana"; "Death of Mandarin"; "British Soldier's Miraculous Escape

from Death on Tusks of a Mad Elephant"; "Mad Elephant Rips Chains and Walls."

159 *These accounts were commonplace:* See, for example, "An Elephant Ran Amok During the Shooting of a Film," *Orlando (FL) Sentinel*, March 7, 1988; "Woman Trying to Ride an Elephant Is Killed," *New York Times*, July 7, 1985; "Elephant Storms Out of Circus in Queens," *New York Times*, July 11, 1995; Phil Maggitti, "Tyke the Elephant," *Animals' Agenda* 14, no. 5 (1994): 34; Karl E. Kristofferson, "Elephant on the Rampage!," *Reader's Digest* 142, no. 854 (1993): 42; "African Elephant Kills Circus Trainer," *New York Times*, August 22, 1994.

159 *Campbell flopped to the side, limp:* "African Elephant Kills Circus Trainer," *New York Times*, August 22, 1994.; video footage of Tyke's attack, *Banned from TV*, http://www.youtube.com/watch?v=ym7MS4I7znQ (accessed January 7, 2014).

160 *"And the next thing I knew, it was running by me, bloody":* Will Hoover, "Slain elephant left tenuous legacy in animal rights," *Honolulu Advertiser*, August 20, 2004, http://archives.starbulletin.com/2004/08/16/news/story2.html.

161 *He claimed he was punishing her:* Rosemarie Bernardo, "Shots Killing Elephant Echo across a Decade," *Star Bulletin*, August 16, 2004, http://archives.starbulletin.com/2004/08/16/news/story2.html.

161 *she was suffering from skin abscesses:* Christi Parsons, " '93 Incident by Cuneo Elephant Told," *Chicago Tribune*, August 24, 1994, http://articles.chicagotribune.com/1994–08–24/news/9408240223_1_circus-officials-shrine-circus-tyke; "Hawthorn Corporation Factsheet," PETA, http://www.mediapeta.com/peta/pdf/hawthorn-corporation-pdf.pdf.; "A Cruel Jungle Tale in Richmond," *Chicago Tribune*, January 13, 2005, http://articles.chicagotribune.com/2005–01–13/news/0501130241_1_elephants-animal-welfare-act-hawthorn-corp.

161 *A year later the USDA:* Maryann Mott, "Elephant Abuse Charges Add Fuel to Circus Debate," *National Geographic*, April 6, 2004, http://news.nationalgeographic.com/news/2004/04/0406_040406_circuselephants_2.html.

161 *Chunee, a once docile Asian elephant:* Ritvo, *Animal Estate*, 225–27.

162 *Gunda too was once an approachable star elephant:* "Death of Gunda."

162 *Debates over what to do with him captivated New Yorkers:* Ibid.; William Bridges, *Gathering of Animals: An Unconventional History of the New*

York Zoological Society (New York: Harper & Row, 1974), 234–42; "Bullet Ends Gunda, Bronx Zoo Elephant."

162 *Forepaugh's shows included:* Advertisement for Adam Forepaugh's Circus in Athletic Park, Washington, D.C., *National Republican*, April 11, 1885, http://chroniclingamerica.loc.gov/lccn/sn86053573/1885-04-11/ed-1/seq-6/ (accessed May 1, 2012).

163 *Tip was "docile as a lamb":* "An Elephant for New York: Adam Forepaugh Presents the City with His $8,000 Tip," *New York Times*, January 1, 1889; "Tip Is Royally Received: Forepaugh's Gift Elephant Arrived Yesterday. Met at the Ferry and Escorted through the Streets by Thousands of Admirers," *New York Times*, January 2, 1889.

164 *For his first few years inside the Central Park elephant house:* "Tip's Life in the Balance: The Murderous Elephant's Fate to Be Decided Tomorrow," *New York Times*, May 8, 1894; Wyndham Martyn, "Bill Snyder, Elephant Man," *Pearson's Magazine* 35 (1916): 180–85; John W. Smith, "Central Park Animals as Their Keeper Knows Them," *Outing: Sport, Adventure, Travel, Fiction* 42 (1903): 248–54.

164 *"must reform or die":* "Tip Must Reform or Die: Central Park's Big Elephant on Trial for His Life," *New York Times*, May 3, 1894.

164 *One morning, as the keeper went to feed Tip his breakfast:* Ibid.; Smith, "Central Park Animals as Their Keeper Knows Them," 252–54.

164 *The elephant waited three years to try to hurt Snyder again:* "Tip Must Reform or Die"; "Tip's Life in the Balance"; Martyn, "Bill Snyder," 180–85.

164 *Daily newspaper articles covered his plight:* Charles David, a man who had worked for Forepaugh and known Tip for years, put it this way: "I think Snyder doesn't exercise Tip enough. . . . In circuses, when an elephant gets unruly, they punish him . . . or they walk him from one town to another, twenty miles or so. That takes the ugliness out of him. If Snyder cannot subdue Tip, then they ought to get another keeper." Apparently David was ignored."Tip Must Reform or Die"; "Tip's Life in the Balance"; "Tip's Life May Be Sacred: Mr. Davis Says the City Agreed He Should Not Be Killed," *New York Times*, May 7, 1894.

165 *He also may have been going through musth:* An elegant account of another elephant in musth is the centerpiece of George Orwell's 1936 essay "Shooting an Elephant." As a colonial officer in Burma, he was pressured to shoot a "mad elephant": "I watched him beating his bunch of grass against his knees, with that preoccupied grandmotherly air

that elephants have. It seemed to me that it would be murder to shoot him." George Orwell, *Shooting an Elephant, and Other Essays* (New York: Harcourt, Brace, 1950); Preecha Phuangkum, Richard C. Lair, and Taweepoke Angkawanith, *Elephant Care Manual for Mahouts and Camp Managers* (Bangkok: FAO Regional Office for Asia and the Pacific, 2005), 52–54.

165 *a new wave of animal advocates:* See Anita Guerrini, *Experimenting with Humans and Animals: From Galen to Animal Rights* (Baltimore: Johns Hopkins University Press, 2003); Carol Lansbury, *The Old Brown Dog: Women, Workers, and Vivisection in Edwardian England* (Madison: University of Wisconsin Press, 1985); Susan J. Pearson, *The Rights of the Defenseless: Protecting Animals and Children in Gilded Age America* (Chicago: University of Chicago Press, 2011); Keith Thomas, *Man and the Natural World: A History of the Modern Sensibility* (New York: Pantheon Books, 1983).

166 *the Central Park commissioners unanimously decided:* I was not able to verify whether Tip actually killed anyone, and I doubt the Parks Commission was able to either. "Tip Tried and Convicted; Park Commissioners Sentence the Elephant to Death," *New York Times*, May 10, 1894; "Tip Swallowed the Dose; But Ate His Hay with Accustomed Regularity in the Afternoon. Tried to Poison an Elephant. It Was Unsuccessful at Barnum & Bailey's Winter Quarters Yesterday. If It Fails To-day the Animal Will Be Shot," *New York Times*, March 16, 1894; "Tip to Die by Poison To-Day; Hydrocyanic Acid Capsules in a Carrot at 6 A.M.," *New York Times*, May 11, 1894.

166 *The park flooded with visitors:* "Big Elephant Tip Dead: Killed with Poison after Long Hours of Suffering," *New York Times*, May 12, 1894.

168 *The term* nostalgia *could be used interchangeably with* homesickness*:* Nostalgia was first diagnosed in 1678 by a Swiss doctor, Johannes Hofer, and was considered an "affliction of the imagination" caused by the desire to go back to one's native land. Soldiers, students, prisoners, exiles, or anyone barred from returning to their home could contract the disease, according to another Swiss doctor in 1720. The disease was characterized by the single-minded obsession with return. Nostalgia wasn't "demedicalized" until the turn of the twentieth century, when it lost its "bodily connotations and became even more linked with time." Jennifer K. Ladino, *Reclaiming Nostalgia: Longing for Nature in American Literature* (Charlottesville: University of Virginia Press, 2012), 6–7.

168 *During the Civil War, for example*: Susan J. Matt, *Homesickness: An American History* (New York: Oxford University Press, 2011), 5–6.

169 *African Americans, Native Americans, and women*: Ibid.

169 *"Nostalgia . . . is the first and most effective aid"*: Ibid.

170 *Working out of his shop in London's East End*: "John Daniel Hamlyn (1858–1922)," St George-in-the-East Church, http://www.stgite.org.uk/media/hamlyn.html (accessed June 15, 2013); "The Avicultural Society," *Avicultural Magazine: For the Study of Foreign and British Birds in Freedom and Captivity* 112 (2006).

170 *Hamlyn was also said to have kept chimpanzees as children in his house*: Bo Beolens, Michael Watkins, and Michael Grayson, *The Eponym Dictionary of Mammals* (Baltimore: Johns Hopkins University Press, 2009), 175; *Hamlyn's Menagerie Magazine* 1, no. 1 (London, 1915), Biodiversity Heritage Library, http://www.biodiversitylibrary.org/bibliography/61908.

170 *When the young gorilla arrived from Gabon*: "John Daniel Hamlyn (1858–1922)"; American Museum of Natural History, "Mammalogy," *Natural History* 21 (1921): 654.

171 *A young woman named Alyse Cunningham*: Alyse Cunningham, "A Gorilla's Life in Civilization," *Zoological Society Bulletin* 24, no. 5 (1921): 118–19.

171 *She was convinced that his fears stemmed*: Ibid.

171 *John was a picky eater*: Ibid.

172 *he had been trying to secure a gorilla*: Bridges, *Gathering of Animals*, 346.

172 *One of the few gorillas to live more than a few months*: "Garner Found Ape That Talked to Him: Waa-hooa, Said the Monkey: Ahoo-ahoo, Replied Professor at Their Meeting," *New York Times*, June 6, 1919; "Zoo's Only Gorilla Dead: Mlle. Ninjo Could Not Endure Our Civilization. Nostalgia Ailed Her," *New York Times*, October 6, 1911; "Death of a Young Gorilla," *New York Times*, January 3, 1888; "Jungle Baby Lolls in Invalid's Luxury," *New York Times*, December 21, 1914.

172 *On a trip to Gabon in 1893*: R. L. Garner, "Among the Gorillas," *Los Angeles Times*, August 27, 1893.

173 *fine health didn't seem to be due to his diet*: William T. Hornaday, "Gorilla a Model for Small Boys: He Always Put Things Back," *Boston Daily Globe*, November 25, 1923.

173 *Just three years earlier Hornaday had proclaimed*: William T. Hornaday, "Gorillas Past and Present," *Zoological Society Bulletin* 18, no. 1 (1915): 1185.

173 *For more than two years Alyse and Rupert encouraged:* Cunningham, "A Gorilla's Life in Civilization," 123.

173 *Occasionally, they brought him to the London Zoo:* Fred D. Pfenig Jr. and Richard J. Reynolds III, "In Ringling Barnum Gorillas and Their Cages," *Bandwagon* (November–December), 6.

173 *"The only way to deal with him":* Ibid.

173 *It's not clear why they weren't able to locate a suitable spot for him:* Ibid.

173 *"sitting quietly in one corner":* "Circus's Gorilla a Bit Homesick," *New York Times,* April 3, 1921.

174 *Soon both circus-goers and the press reported:* "Gorilla Dies of Homesickness," *Los Angeles Times,* 1921; "Grieving Gorilla Dead at Garden," *New York Times,* April 18, 1921.

175 *In the weeks before his death:* "Grieving Gorilla Dead at Garden."

175 *In the three weeks that Ringling Brothers displayed John:* "Gorilla Dies of Homesickness"; "Inflation Calculator," Dollar Times, http://www.dollartimes.com/calculators/inflation.htm.

176 *The famous primatologist Robert Yerkes came to the museum:* Richard J. Reynolds III, circus historian, personal communication, March 14, 2011; Fred D. Pfenig Jr. and Richard J. Reynolds III, "The Ringling-Barnum Gorillas and Their Cages," *Bandwagon* 50, no. 6 (November -December 2006): 4–29; James C. Young, "John Daniel, Gorilla, Sees the Passing Show," *New York Times,* April 13, 1924.

176 *"He offers . . . ocular proof of everything":* Young, "John Daniel, Gorilla, Sees the Passing Show."

176 *and only occasionally bit his mistress:* Richard J. Reynolds III, circus historian, personal communication, March 14, 2011, and his images, featured here http://bucklesw.blogspot.com/2009_04_01_archive.html; Pfenig Jr. and Reynolds, "The Ringling-Barnum Gorillas and Their Cages," 4–29; John C. Young, "John, the Gorilla, Bites His Mistress," *New York Times,* April 8, 1924.

177 *her first gorilla had been mounted and studied:* "Darwinian Theory Given New Boost: Educated Gorilla's Big Toe Became Much Like That of Human," *Los Angeles Times,* April 17, 1922; "Gorilla Most Like Us, Say Scientists: Nearer to Man in 'Dictatorial Egoism' than Other Primates, Neurologist Finds. Comparison with a Child Surgeons Report Study of 'John Daniel,' Dead Circus Gorilla, to Society of Mammalogists. No Mention of Bryan. Chimpanzees Got Drunk. Darwin's Theory

Discussed," *New York Times*, May 18, 1922; "Specialists Study John Daniel's Body," *New York Times*, April 25, 1921.

177 *Almost a hundred years after his death:* John and Meshie are only two of the many human-raised ape-children whose responses to their domestic environment shifted perceptions of apes' emotional lives and intelligence and reflected back, often uncomfortably, the desires of those who raised them. Henry Cushier Raven, "Meshie: The Child of a Chimpanzee. A Creature of the African Jungle Emigrates to America," *Natural History Magazine*, April 1932; Joyce Wadler, "Reunion with a Childhood Bully, Taxidermied," *New York Times*, June 6, 2009. See also the case of J. T. Junior, monkey captured and raised by the Akeley family first in Africa and then in their apartment in New York City, until she was sent to the National Zoo. Delia J. Akeley, *"J. T. Jr.": The Biography of an African Monkey* (New York: Macmillan, 1928). Toto the gorilla was raised by an American woman in her home and then given to the circus. Augusta Maria Daurer Hoyt, *Toto and I: A Gorilla in the Family* (Philadelphia: J. B. Lippincott, 1941). Lucy the signing chimpanzee was sent from the human home in America where she was raised to Africa, where she was eventually killed. Maurice K. Temerlin, *Lucy: Growing Up Human. A Chimpanzee Daughter in a Psychotherapist's Family* (Palo Alto, CA: Science and Behavior Books, 1976); Eugene Linden, *Silent Partners: The Legacy of the Ape Language Experiments* (New York: Times Books, 1986). On Nim Chimpsky the chimpanzee, see Elizabeth Hess, *Nim Chimpsky: The Chimp Who Would Be Human* (New York: Bantam Books, 2009).

177 *The psychological effects of the fighting:* Matt, *Homesickness*, 178–83.

177 *During the war era and for some time afterward:* Hayden Church, "American Women in London Minister to Homesick Yankees in British Hospitals," *San Francisco Chronicle*, December 16, 1917; Helen Dare, "Seeing to It That Soldier Boy Won't Feel Homesick: Even Has Society Organized for Keeping His Mind Off the Girl He Left behind Him," *San Francisco Chronicle*, September 12, 1917; "Our Men in France Often Feel Homesick," *New York Times*, June 9, 1918; "Need Musical Instruments: Appeal by Dr. Rouland in Behalf of Homesick Soldiers," *New York Times*, November 24, 1918; "Recipe for Fried Chicken Gives Soldier Nostalgia," *San Francisco Chronicle*, May 11, 1919; "Doty Was Homesick, and Denies Cowardice: Explains Desertion from French

Foreign Legion. Will Be Tried but Not Shot," *New York Times*, June 18, 1926; "Nostalgia," *New York Times*, June 27, 1929.

177 *Away from the front, war brides:* "War Bride Takes Gas: German Girl Who Married American Soldier Was Homesick," *New York Times*, July 2, 1921; "Woman Jumps into Bay with Child Rescued."

178 *Country boys new to cities:* "Geisha Girls Are Homesick: Japanese World's Fair Commissioner Resorts to Courts to Secure Return of Maids to Japan," *Chicago Daily Tribune*, October 9, 1904; "Boy Coming into a City Finds It Hard to Save: Exaggerate Value of Salary. Adopts More Economical Plan. Country Dollar Is 50c in City. 'Blues' Bred in Hall Bedrooms," *Chicago Daily Tribune*, June 4, 1905; "Glass Eye Blocks Suicide: Deflects Bullet Fired by Owner Who Is Ill and Homesick in New York," *Chicago Daily Tribune*, August 6, 1910; "Homesick: Ends Life. Irish Girl, Unable to Get Back to Erin to See Her Mother, Takes Gas," *Chicago Daily Tribune*, September 8, 1916; "English Writer, Ill, Ends His Life Here: Bertram Forsyth, Homesick and Depressed, Dies by Gas in Apartment. Left Letter to His Wife. 'Life of Little Account,' He Wrote, Expressing Hope His Son Had Not Inherited His Pessimism," *New York Times*, September 17, 1927; "Homesick Stranger Steals Parrot That Welcomes Him: Heart of Albert Schwartz So Touched by Bird's Greeting He Commits Theft, but Capture Follows," *Chicago Daily Tribune*, January 2, 1908.

178 *One case in 1892 concerned a mule:* "Bill Zack and His Knowing Mule: Following His Master, He Walked from Louisiana to Tennessee," *Chicago Daily Tribune*, July 10, 1892.

178 *Dogs who whined with nostalgia:* "Care for Sick Pets: Chicago Sanitariums for Birds, Cats, and Dogs. Methods of Treatment. Queer Incident Recounted at the Animal Hospitals. Teach Parrots to Speak. School Where the Birds Learn to Repeat Catching Phrases. Swearing Is a Special Course. Died of a Broken Heart. School for Parrots. Hospital for Sick Cats. Sanitarium for Dogs. Canine Victim to Alcohol," *Chicago Daily Tribune*, May 9, 1897.

178 *Jocko, a monkey mascot:* "Jocko, Homesick, Tries to Die: Sailors' Singing Awakens Fond Memories. Waving Farewell a Naval Mascot Swallows Poison," *Chicago Daily Tribune*, July 19, 1903.

178 *an African elephant named Jingo:* What killed him may have been seasickness, disease, his strict confinements, or any number of other stressors. "Jingo, Rival of Jumbo, Is Dead: Tallest Elephant Ever in Captivity

Unable to Stand Ocean Voyage. Could Not Be Consoled. Refuses to Eat for Several Days and Is Thought to Have Died of Homesickness. Big Beast Is Homesick. Second Officer Tells His Story. Varying Views of Jingo's Value," *Chicago Daily Tribune,* March 19, 1903.

178 *a young female mountain gorilla named Congo:* "Only One Gorilla Now in Captivity," *New York Times,* July 18, 1926.

178 *"What caused her death is unknown":* "Miss Congo, the Lonely Young Gorilla, Dies at Her Shrine on Ringling's Florida Estate," *New York Times,* April 25, 1928.

178 *the family of a young San Francisco boy:* "Broken Heart or Nostalgia Causes Pet Duck's Death: Mandarin Gives Up Ghost at Park after Fight with Mudhens," *San Francisco Chronicle,* February 10, 1919.

179 *Ota Benga, an African pygmy man:* Phillips Verner Bradford, *Ota: The Pygmy in the Zoo* (New York: St. Martin's Press, 1992); Elwin R. Sanborn, ed., "Suicide of Ota Benga, the African Pygmy," *Zoological Society Bulletin* 19, no. 3 (1916): 1356; Samuel P. Verner, "The Story of Ota Benga, the Pygmy," *Zoological Society Bulletin* 19, no. 4 (1916): 1377–79.

179 *"The asylums of this and every country":* "A Broken Heart: Often Said to Be a Cause of Death. What the Term Means. A Common Figure of Speech That Has Some Foundation in Fact," *San Francisco Chronicle,* February 27, 1888.

179 *many deaths attributed to heartbreak:* Georges Minois, *History of Suicide: Voluntary Death in Western Culture* (Baltimore: Johns Hopkins University Press, 1999), 316.

180 *There were lovers' hearts that gave out:* "Died Before His Wife Did: Husband, Who Had Said He Would Go before Her, Stricken as She Lay Dying," *New York Times,* July 5, 1901; "Died of a Broken Heart," *New York Times,* December 22, 1894; "Died of a Broken Heart," *New York Times,* July 5, 1883; "Widow Killed by Grief: Dies of a Broken Heart Following Loss of Her Husband," *New York Times,* January 9, 1910; "Veteran Dies of a Broken Heart: Fails to Rally after Wife Passes Beyond: He Soon Follows Her," *Los Angeles Times,* September 4, 1915; "Died of a Broken Heart," *New York Times,* June 27, 1884; "Died of a Broken Heart: Sad Ending of the Life of an Intelligent Girl," *New York Times,* September 3, 1884; "Died of a Broken Heart: Sad End of Michigan Man Whose Wife Deserted Him for Love of Younger Man," *Los Angeles Times,* October

21, 1910; "Brigham Young's Heirs," *New York Times*, March 10, 1879; "Died of a Broken Heart," *New York Times*, January 9, 1886; "Broken Heart Kills Mother: Burden of Her Grief Too Heavy to Bear: Mrs. Franklin, Whose Son Met Cruel Fate in Santa Fe Train Collision on River Bridge, Ages in Few Months, Pines Away and Dies Pitifully," *Los Angeles Times*, June 1, 1907; "Drowned Lad's Mother Dies: News of His Death While Skating on Thin Ice Crushed Her," *New York Times*, January 16, 1903; "Prostrated by His Son's Death," *New York Times*, November 13, 1893; "Kidnapped Children Recovered," *New York Times*, July 22, 1879; "Died of a Broken Heart," *New York Times*, December 8, 1897; "Girl Dies of a Broken Heart," *Los Angeles Times*, April 11, 1905; "General G. K. Warren Obituary," *New York Times*, August 9, 1882; "Spotted Tail's Daughter: How the Princess Monica Died of a Broken Heart from Unrequited Love for a Pale-face Soldier. The Chaplain's Story," *New York Times*, July 15, 1877; *Fort Laramie: Historical Handbook Number Twenty*, National Park Service, 1954, http://www.cr.nps.gov/history/online_books/hh/20/hh20m.htm.

180 *Loyal hounds who died of heartbreak and sadness*: Marjorie Garber, *Dog Love* (New York: Touchstone, 1997), 241–42, 249–52, 257–58.

180 *Greyfriars Bobby*: Ibid., *Dog Love*, 255–56. Recently a researcher in Cardiff has argued that Bobby was actually two dogs. "Greyfriars Bobby Was Just a Victorian Publicity Stunt, Claims Academic," *Telegraph*, August 3, 2011, http://www.telegraph.co.uk/news/newstopics/howaboutthat/8678875/Greyfriars-Bobby-was-just-a-Victorian-publicity-stunt-claims-academic.html (accessed August 3, 2011).

180 *a German Shepherd named Teddy stopped eating*: "Horse Dies, Dog Follows: Shepherd Refused Food, Grieving over Passing of Friend," *New York Times*, September 19, 1937.

180 *Horses were also supposedly done in by heartbreak*: "An Affectionate Horse," *New York Times*, January 14, 1887.

181 *"will strive and struggle to get out"*: "Army Mule Aristocrat of Allied Armies' Transport: Humble American Has Won the Heart of the British Army. Seldom Sick, and Never Afraid, He Survives Where Horses Succumb," *Boston Daily Globe*, February 25, 1917.

181 *Besides faithful dogs*: "Rhinoceros Bomby Is Dead: New-York Climate and Isolation from His Sweetheart Killed Him," *New York Times*, June 27, 1886; "Grieving Sea Lion Dies at Aquarium: Trudy Had Refused to Touch Food Since Death of Mate Ten Days Ago. Many

Children Knew Her. Bought from California for a Circus Career. Blind Eye Kept Her from Learning Tricks," *New York Times*, September 10, 1928; "Berlin Sea Elephant Dies of a Broken Heart," *New York Times*, December 31, 1935; "Zoo Penguin Dies of Broken Heart Mourning Mate," *Los Angeles Times*, August 4, 1947.

181 *Wild animals were occasionally thought to suffer*: See, for example, Martin Johnston, "Helpless, but Unafraid, the Giraffe Thrives on Persecution," *Daily Boston Globe*, December 2, 1928.

181 *the inability to keep many animals, from lions to songbirds, alive*: "The Story of a Lion's Love: Wynant Hubbard's Account of Moving Jungle Romance Wherein King of Beasts Enters Captivity out of Affection for His 'Wife,'" *Daily Boston Globe*, May 19, 1929; "Most Fastidious of Wild Beasts Are the Leopards: So Says Mme. Morelli, and She Ought to Know, for They've Tried to Eat Her Several Times. How Bostock Saved the Plucky Woman Trainer. One of Her Pets Died from a Broken Heart," *New York Times*, June 18, 1905; "Birds I Know," *Daily Boston Globe*, July 23, 1946.

181 *In 1966 a killer whale named Namu*: "Lovelorn Killer Whale Dies in Frantic Dash for Freedom," *Los Angeles Times*, July 11, 1966; David Kirby, *Death at SeaWorld: Shamu and the Dark Side of Killer Whales in Captivity* (New York: St. Martin's Press, 2012), 151–52.

181 *Belle Benchley, director of the San Diego Zoo*: Belle J. Benchley, "'Zoo-Man' Beings," *Los Angeles Times* (1923—Current File), August 14, 1932; Belle J. Benchley, "The Story of Two Magnificent Gorillas," *Bulletin of the New York Zoological Society* 43, no. 4 (1940): 105–16; Belle Jennings Benchley, *My Animal Babies* (London: Faber and Faber, 1946); Gerald B. Burtnett, "The Low Down on Animal Land," *Los Angeles Times* (1923—Current File), June 2, 1935.

182 *One of these friendships, at the Berlin Zoo*: "Ape Porcupine Firm Friends: Melancholia Banished When Huge Monkey Romps with Strange Playmate," *Los Angeles Times* (1923—Current File), November 27, 1924. Almost sixty years later my email inbox is full of friendships like this. Not a week goes by that someone doesn't forward me photos of a dog and an orangutan splashing around in a plastic tub, or baby pigs cuddling up to a tiger, or a snake and a rabbit curled around each other in some sort of multispecies friendship ring. A number of these little photo essays, as anthropomorphic, cutesy, and staged as many of them are, also discuss loneliness and depression as reasons for the surprising relationships.

182 *Monarch's mounted body served as one of the models:* Tracy I. Storer and Lloyd P. Tevis, *California Grizzly* (Berkeley: University of California Press, 1996), 276.

182 *few people know the bear was an actual:* "Grizzly Comes as Mate to Monarch: Young Silver Tip Shipped from Idaho Arrives Safely and Is Now in Park Bear Pit," *San Francisco Chronicle*, February 10, 1903.

183 *Monarch was an icon of a recently neutered wilderness:* There has been much good scholarship on the shifting attitudes toward Western wilderness over the course of the late nineteenth and early twentieth centuries, as American Indians were removed from the nation's new national park system, grizzly and wolf populations were decimated, and the federal government established monitoring agencies (U.S. Fish and Wildlife Service, the National Park Service, and more) to oversee the newly consolidated governmental land holdings—regulating who did or did not have access to resource-rich lands and how this both reflected and helped shape American conceptions of wilderness and the American frontier. William Cronon, "The Trouble with Wilderness," in *Uncommon Ground: Toward Reinventing Nature*, ed. William Cronon (New York: Norton, 1995); Karl Jacoby, *Crimes against Nature: Squatters, Poachers, Thieves, and the Hidden History of American Conservation* (Berkeley: University of California Press, 2003); Roderick Frazier Nash, *Wilderness and the American Mind* (New Haven, CT: Yale University Press, 2001); Philip Shabecoff, *A Fierce Green Fire: The American Environmental Movement* (Washington, DC: Island Press, 2003); Louis S. Warren, *The Hunter's Game: Poachers and Conservationists in Twentieth-Century America* (New Haven, CT: Yale University Press, 1999); Richard White, Patricia Nelson Limerick, and James R. Grossman, *The Frontier in American Culture: An Exhibition at the Newberry Library, August 26, 1994–January 7, 1995* (Chicago: Newberry Library, 1994).

183 *In 1858 a sheriff in Sacramento:* Susan Snyder, *Bear in Mind: The California Grizzly* (Berkeley, CA: Heyday Books, 2003), 117–40; Storer and Tevis, *California Grizzly*, 249.

183 *the bears "were everywhere":* Snyder, *Bear in Mind*, 65.

183 *Grizzly Adams, the famous bear hunter:* Storer and Tevis, *California Grizzly*, 244–49.

183 *Well into the 1860s captive bears:* Snyder, *Bear in Mind*, 160; Storer and Tevis, *California Grizzly*, 240–41.

183 *Those that hadn't been killed:* Storer and Tevis, *California Grizzly*, 249.

183 *William Randolph Hearst, the eccentric California newspaper magnate:* "The New Bear Flag Is Grizzlier," *San Francisco Chronicle*, September 19, 1953; Storer and Tevis, *California Grizzly*, 249–50.

184 *A few weeks became months:* How verifiable those details are is a matter of interpretation, since no other account exists. While not everything that happened in the tale of Monarch may have happened to Monarch the individual, it probably happened at some point with some bear of Kelly's. Ernest Thompson Seton questioned Kelly in 1889 about the authenticity of the Monarch account published in the *Examiner.* Seton wrote, "It is safe to say that many adventures ascribed to this bear belonged to various and different bears." Storer and Tevis, *California Grizzly*, 250; Allen Kelly, *Bears I Have Met—and Others* (Philadelphia: Drexel Biddle, 1903), http://www.gutenberg.org/files/15276/15276-h/15276-h.htm.

184 *a Mexican man trapped a large grizzly:* Kelly, *Bears I Have Met.*

184 *For a full week he raged and refused to touch food:* Ibid.

184 *Egged on by wildly embellished tales:* Board of Parks Commissioners, *Annual Report of the Board of Parks Commissioners 1895*, June 30, 1895, California Academy of Sciences, 7, 15.

185 *he had taken to spending all day inside a hole:* "Grizzly Comes as Mate to Monarch."

185 *They also claimed that he might be grieving:* Ibid.

186 *These vast changes:* Nash, *Wilderness and the American Mind*; Shabecoff, *A Fierce Green Fire*, 21–31, 61–67.

186 *In 1896, seven years into Monarch's stint in San Francisco:* Nash, *Wilderness and the American Mind*, 146.

186 *The country's wildlands and wildlife:* Ibid., 76.

187 *Places like Yosemite and Yellowstone could now be seen as antidotes:* The historian William Cronon has argued that when Turner announced the closing of the frontier, Americans who felt a sense of loss were already looking backward, mourning an older, simpler, and more peaceable country. The problem is that such a country never actually existed: the extirpation of native peoples, the decimation of buffalo, wolf, and grizzly populations, and the drastic ecological damage brought on by mining and massive deforestation were the kinds of violence that did not figure into romantic ideas of the nation's wilderness ideals. Cronon, "The Trouble with Wilderness."

187 *Efforts to protect and celebrate these places:* Ibid., 76, 78.

188 *the new female "typified the saying":* "Grizzly Comes as Mate to Monarch."

188 *"The poor little cub died"*: "Grizzly Bear Cub Is Dead," *San Francisco Chronicle*, January 18, 1904. A subsequent article claimed the cub died of a head injury after Montana dropped it the day it was born. A second serving of disdain for Montana's mothering skills was served up for San Franciscans: "Monarch and his diffident spouse Montana, the two grizzlies who live in an iron cage near the buffalo paddock, held an informal reception yesterday for many visitors who called to offer their sympathy for the untimely death of their offspring. Considering that Mrs. Monarch partially ate one of her twins and ignored the other almost entirely, the condolences of her visitors seemed ill-placed." "Great Crowd Visits Park: Police Estimate Attendance of Fully Forty Thousand People at the Recreation Grounds," *San Francisco Chronicle*, January 25, 1904.

188 *Four years later Monarch was confirmed:* Board of Parks Commissioners, *Annual Report of the Board of Parks Commissioners 1910*, June 30, 1910, 40–43.

188 *after twenty-two years in captivity*: "Park Museum Has New Attractions: Grizzly Monarch Will Be Put on Exhibition for the Labor Day Crowds," *San Francisco Chronicle* (1869—Current File), August 28, 1911.

189 *"Templeton," said Wilbur in desperation:* E. B. White, *Charlotte's Web*, (New York: Harper Brothers, 1952).

190 *two elderly male otters:* " 'Heartbroken' Male Otters Die within an Hour of Each Other," April 1, 2010, Advocate.com, http://www.advocate.com/News/Daily_News/2010/04/01/Heartbroken_Male_Otters_Die_Within_An_Hour_of_Each_Other/ (accessed November 5, 2012).

190 *the story of a Miniature Schnauzer named Pepsi*: Bekoff, *The Emotional Lives of Animals*, 66.

190 *In March 2011 another heartbreak story:* Jill Lawless, "Hours after Soldier Killed in Action, His Faithful Dog Suffers Seizure," *Toronto Star*, March 10, 2011.

191 *Japanese cardiologists named the syndrome:* Salim S. Virani, A. Nasser Khan, Cesar E. Mendoza, Alexandre C. Ferreira, and Eduardo de Marchena, "Takotsubo Cardiomyopathy, or Broken-Heart Syndrome," *Texas Heart Institute Journal* 34, no. 1 (2007): 76–79.

192 *proof of the powerful connection:* Natterson-Horowitz and Bowers, *Zoobiquity*, 5.

192 *She and Bowers point to a few fascinating public health statistics:* Ibid., 110–13.

192 *games that end in "sudden death" shootouts:* Ibid., 112–13.

193 *Since then, sudden death among terrified animals:* Ibid., 118–20.

Chapter Five: Animal Pharm

194 *When the reality-TV star Anna Nicole Smith:* "Eternal Sunshine," *Guardian,* May 13, 2007, http://www.guardian.co.uk/society/2007/may/13/social care.medicineandhealth (accessed March 10, 2009).

194 *Maltese and Bichon Frise mix, Sumo:* Stanley Coren, "The Former French President's Depressed Dog: Jacques Chirac and Sumo," *Psychology Today,* October 5, 2009, http://www.psychologytoday.com/blog/canine-corner/200910/the-former-french-president-s-depressed-dog-jacques-chirac-and-sumo (accessed April 5, 2012); Ian Sparks, "Former French President Chirac Hospitalised after Mauling by His Clinically Depressed Poodle," *Mail Online,* January 21, 2009, http://www.dailymail.co.uk/news/article-1126136/Former-French-President-Chirac-hospitalised-mauling-clinically-depressed-poodle.html (accessed April 5, 2012).

194 *Fluoxetine, or generic Prozac, is available:* "Fluoxetine (AS HCL): Oral Suspension," Wedgewood Pharmacy, http://www.wedgewoodpetrx.com/items/fluoxetine-as-hcl-oral-suspension.html (accessed April 6, 2012).

195 *Most small-animal doctoring, until the turn of the twentieth century:* Nineteenth-century homeopathic practices also included pets. Homeopathic manuals for pet health were published well into the 1930s, and you could buy veterinary kits from homeopathic doctors. Katherine C. Grier, *Pets in America: A History* (Chapel Hill: University of North Carolina Press, 2006), 90–96.

196 *In May 1950 Henry Hoyt and Frank Berger:* Andrea Tone, *The Age of Anxiety: A History of America's Turbulent Affair with Tranquilizers* (New York: Basic Books, 2008), 43–51.

196 *another drug was relaxing rats:* Shorter, *A Historical Dictionary of Psychiatry,* 54.

197 *In 1951, a company pharmacist:* David Healy, *The Creation of Psychopharmacology* (Cambridge, MA: Harvard University Press, 2002), 78.

197 *Rats were given the antihistamine:* Ibid., 80–81.

197 *The rats' indifference piqued the curiosity:* Ibid., 46, 80–81, 84. Other uses were tested in dogs. It turned out that the compound made nauseated dogs stop vomiting, even in one experimental group that was swung incessantly in hammocks.

197 *The drug also caused certain patients at Sainte-Anne:* Ibid., 90–91.

197 *A barber from Lyon, France, was a typical case:* Ibid.

198 *In 1954 Rhône-Poulenc sold the U.S. chlorpromazine license:* Shorter, *A Historical Dictionary of Psychiatry*, 55.

198 *It was marketed as an antinausea agent:* Healy, *The Creation of Psychopharmacology*, 98–99.

198 *One 1968 journal article summed up the veterinary use:* J. W. Kakolewski, "Psychopharmacology: Clinical and Experimental Subjects," in *Abnormal Behavior in Animals*, ed. Michael W. Fox (Philadelphia: Saunders, 1968), 527.

198 *"Litter-savaging" pigs:* A. F. Fraser, "Behavior Disorders in Domestic Animals," in *Abnormal Behavior in Animals*, ed. Michael W. Fox (Philadelphia: Saunders, 1968), 184.

198 *Soon other new antipsychotic drugs:* W. Ferguson, "Abnormal Behavior in Domestic Birds," in *Abnormal Behavior in Animals*, ed. Michael W. Fox (Philadelphia: Saunders, 1968), 195.

198 *A year after Smith Kline bought the chlorpromazine license:* Tone, *The Age of Anxiety*, 43–52.

199 *Meanwhile scientists at Walter Reed:* Ibid., 109–10.

199 *The 1950s was a key decade in the forging of new links:* Jonathan Michel Metzl, *Prozac on the Couch: Prescribing Gender in the Era of Wonder Drugs* (Durham, NC: Duke University Press, 2003), 72, 74–75.

200 *Pharmaceutical marketing pitches:* Ferdinand Lundberg and Marynia Farnham, *Modern Woman: The Lost Sex* (New York: Harper and Brothers, 1947).

200 *When tranquilizers arrived, these dangerous states:* Metzl, *Prozac on the Couch*, 81, 159.

200 *Before the mid-1950s talk therapy, not drugs:* Ibid., 101–2.

200 *Miltown went to market in 1955:* Tone, *The Age of Anxiety*, 109–10.

200 *Physicians' reference manuals published by pharma companies:* Roche Laboratories, *Aspects of Anxiety* (Philadelphia: Lippincott, 1968).

201 *two years after Miltown was released:* Tone, *The Age of Anxiety*, 57, 108–10, 113.

201 *Children were given the drug:* Metzl, *Prozac on the Couch*, 100; R. Huebner, "Meprobamate in Canine Medicine: A Summary of 77 Cases," *Veterinary Medicine* 51 (October 1956): 488.

201 *If a drug could cure anxiety, Berger later argued:* Metzl, *Prozac on the Couch*, 73.

201 *the addictive nature of Miltown came to light:* Tone, *The Age of Anxiety*, 144–47.

202 *In 1967 it was placed under abuse control amendments:* "Tranquilizer Is Put under U.S. Curbs," *New York Times,* December 6, 1967.

202 *The industry's success:* Ibid., 144–47, David Healy, *Let Them Eat Prozac: The Unhealthy Relationship between the Pharmaceutical Industry and Depression* (New York: New York University Press, 2004).

202 *Other psychopharmaceutical drugs were also in development:* Tone, *The Age of Anxiety,* 129.

202 *The new drug also passed the industry-wide "cat test":* Ibid., 129, 135.

203 *This cat-mouse-human relaxant:* Ibid., 153–56.

203 *Sometime in the 1960s he was captured in Congo:* Rebecca Burns, "11 Years Ago This Month: Willie B.'s Memorial," Atlantamagazine.com, February 2000, http://www.atlantamagazine.com/flashback/Story.aspx?id=1353208 (accessed July 20, 2012); Dorie Turner, "Famed Atlanta Resident Who Ate Bananas Comes to TV," *USA Today,* August 5, 2008, http://www.usatoday.com/news/nation/2008–08–05–2238724203_x.htm (accessed July 20, 2012).

204 *The drugs are used to overcome phobias in birds:* Liz Wilson and Andrew Luescher, "Parrots and Fear," in *Manual of Parrot Behavior,* ed. Andrew Luescher (Ames, IA: Blackwell, 2006), 227; Peter Holz and James E. F. Barnett, "Long-Acting Tranquilizers: Their Use as a Management Tool in the Confinement of Free-Ranging Red-Necked Wallabies *(Macropus rufogriseus)," Journal of Zoo and Wildlife Medicine* 27, no. 1 (1996): 54–60; Y. Uchida, N. Dodman, and D. DeGhetto, "Animal Behavior Case of the Month: A Captive Bear Was Observed to Exhibit Signs of Separation Anxiety," *Journal of the American Veterinary Medical Association* 212, no. 3 (1998): 354–55; Thomas H. Reidarson, Jim McBain, and Judy St. Leger, "Side Effects of Haloperidol (Haldol(r)) to Treat Chronic Regurgitation in California Sea Lions," *IAAAM Conference Proceedings* (2004): 124–25, http://www.vin.com/Proceedings/Proceedings.plx?&CID=IAAAM2004&PID=pr50067&O=Generic; Leslie M. Dalton and Todd R. Robeck, "Aberrant Behavior in a California Sea Lion *(Zalophus californianus)," IAAAM Conference Proceedings* (1997): 145–46, http://www.vin.com/Proceedings/Proceedings.plx?&CID=IAAAM1997&PID=pr49310&O=Generic; Larry Gage et al., "Medical and Behavioral Management of Chronic Regurgitation in a Pacific Walrus *(Odobenus rosmarus divergens)," IAAAM Conference Proceedings* (2000): 341–42, http://www.vin.com/Proceedings/Proceedings.plx?&CID=IAAAM2000&PID=pr49633&O=Generic.

204 *At the Toledo Zoo Haldol was used:* Jenny Laidman, "Zoos Using Drugs to Help Manage Anxious Animals," *Toledo Blade,* September 14, 2005.

205 *When the antipsychiatry movement:* Healy, *The Creation of Psychopharmacology,* 5; Michel Foucault, *Madness and Civilization: A History of Insanity in the Age of Reason* (New York: Vintage Books, 1973); Michael E. Staub, *Madness Is Civilization: When the Diagnosis Was Social, 1948–1980* (Chicago: University of Chicago Press, 2011), 6, 139–40, 181–83; Roy W. Menninger and John C. Nemiah, eds., *American Psychiatry after World War II, 1944–1994* (Arlington, VA: American Psychiatric Press, 2000), 281–89.

205 *Ken Kesey portrayed the psych ward:* Ken Kesey, *One Flew Over the Cuckoo's Nest* (New York: Signet, 1963).

205 *As the historian David Healy points out:* Healy, *The Creation of Psychopharmacology,* 5, 148–56, 162–63.

205 *their doctor-focused ad campaigns:* Ibid., 237.

205 *More than thirty years later:* Lorna A. Rhodes, *Total Confinement: Madness and Reason in the Maximum Security Prison* (Berkeley: University of California Press, 2004), 126–28. See also Laura Calkins, "Detained and Drugged: A Brief Overview of the Use of Pharmaceuticals for the Interrogation of Suspects, Prisoners, Patients, and POWs in the U.S.," *Bioethics* 24, no. 1 (2010): 27–34; Charles Pillar, "California Prison Behavior Units Aim to Control Troublesome Inmates," *Sacramento Bee,* May 10, 2010; Kenneth Adams and Joseph Ferrandino, "Managing Mentally Ill Inmates in Prisons," *Criminal Justice and Behavior* 35, no. 8 (2008): 913–27; David Jones, A. Bernard Ackerman Professor of the Culture of Medicine, Harvard University, and psychiatrist, personal correspondence, July 29, 2013.

205 *Antipsychotics, antidepressants, and antianxiety medications have, for example:* M. Babette Fontenot et al., "Dose-Finding Study of Fluoxetine and Venlafaxine for the Treatment of Self-Injurious and Stereotypic Behavior in Rhesus Macaques *(Macaca mulatta),*" *Journal of the American Association for Laboratory Animal Science* 48, no. 2 (2009): 176–84; M. Babette Fontenot et al., "The Effects of Fluoxetine and Buspirone on Self-Injurious and Stereotypic Behavior in Adult Male Rhesus Macaques," *Comparative Medicine* 55, no. 1 (2005): 67–74; H. W. Murphy and R. Chafel, "The Use of Psychoactive Drugs in Great Apes: Survey Results," *Proceedings of the American Association of Zoo Veterinarians, American Association of Wildlife Veterinarians,*

Association of Reptile and Amphibian Veterinarians, and National Association of Zoo and Wildlife Veterinarians Joint Conference (September 18, 2001): 244–49.

205 *an easily agitated male gorilla in Ohio:* Laidman, "Zoos Using Drugs to Help Manage Anxious Animals."

206 *At the Guadalajara Zoo in Mexico:* D. Espinosa-Avilés et al., "Treatment of Acute Self Aggressive Behaviour in a Captive Gorilla *(Gorilla gorilla gorilla),*" *Veterinary Record* 154, no. 13 (2004): 401–2.

206 *After their experiences treating Gigi:* H. W. Murphy and R. Chafel, "The Use of Psychoactive Drugs in Great Apes: Survey Results," *Proceedings of the American Association of Zoo Veterinarians, American Association of Wildlife Veterinarians, Association of Reptile and Amphibian Veterinarians, and National Association of Zoo and Wildlife Veterinarians Joint Conference* (September 18, 2001): 244–49; Murphy and Mufson, "The Use of Psychopharmaceuticals to Control Aggressive Behaviors in Captive Gorillas."

207 *gorilla and his keeper were unloaded on the tarmac:* Paul Luther, personal communication, June 2009; Also Darrel Glover, "Cranky Ape Puts His Foot Down, So Pilot Boots Him off Jet," *Seattle Post-Intelligencer,* October 17, 1996; Elizabeth Morell, "Transporting Wild Animals," *Risk Management* (July 1998).

207 *Dolphins, whales, sea lions, walruses:* Dalton and Robeck, "Aberrant Behavior in a California Sea Lion *(Zalophus Californianus),*" 145–46; Reidarson et al., "Side Effects of Haloperidol (Haldol(r)) to Treat Chronic Regurgitation in California Sea Lions," 124–25; Chen et al., "Diagnosis and Treatment of Abnormal Food Regurgitation in a California Sea Lion *(Zalophus californianus).*" Justin Carissimo, "SeaWorld puts Whales on Valium-like Drug, Documents Show," *Buzzfeed,* March 31, 2014. http://www.buzzfeed.com/justincarissimo/seaworld-puts-its-whales-on-valium-like-drug-documents-show#.ibKNLJlGq.

207 *There are incentives at these facilities:* Kirby, *Death at SeaWorld,* 317–34.

208 *SeaWorld treated two of Tilikum's sons:* Melissa Cronin, "SeaWorld Gave Nursing Orca Valium," *The Dodo,* April 2, 2014, https://www.thedodo.com/seaworld-gave-nursing-orca-val-493887337.html.

208 *Other mother orcas may be given:* Gabriela Cowperthwaite, "Exclusive: 'Blackfish,' 'The Cove' creators challenge SeaWorld to a debate," *The Dodo,* January 22, 2014, https://www.thedodo.com/exclusive-blackfish-the-cove-c-399531056.html.

208 *A few of these published cases include:* William Van Bonn, "Medical Management of Chronic Emesis in a Juvenile White Whale *(Delphinapterus leucas)*," *IAAAM Conference Proceedings* (2006): 150–52, http://www.vin.com/Proceedings/Proceedings.plx?CID=IAAAM2006&Category=7556&PID=50364&O=Generic.

209 *The word* antidepressant *was coined:* Edward Shorter, *Before Prozac: The Troubled History of Mood Disorders in Psychiatry* (New York: Oxford University Press, 2009), 2.

209 *From roughly 1900 through 1980:* Ibid., 4.

209 *In Europe before the 1950s:* Healy, *The Creation of Psychopharmacology*, 57.

209 *Shorter has argued that the cause:* Shorter, *Before Prozac*, 2.

209 *Antidepressants, particularly Prozac:* Peter D. Kramer, *Listening to Prozac* (New York: Viking, 1993); Healy, *Let Them Eat Prozac*, 264.

209 *A human psychiatrist prescribed Remeron:* Carla Hall, "Fido's Little Helper," *Los Angeles Times*, January 10, 2007, http://articles.latimes.com/2007/jan/10/local/me-animalmeds10 (accessed September 15, 2010).

209 *Johari, a female gorilla at the Toledo Zoo:* Laidman, "Zoos Using Drugs to Help Manage Anxious Animals."

210 *When tabloids broke the story:* Tad Friend, "It's a Jungle in Here," *New York Magazine*, April 24, 1995.

210 *The bear was on the cover of* Newsday: Will Nixon, "Gus the Neurotic Bear: Polar Bear in New York City Central Park Zoo," *E the Environmental Magazine*, December 1994.

210 *Bipolar disorder came into vogue:* David Healy, "Folie to Folly: The Modern Mania for Bipolar Disorders," in *Medicating Modern America: Prescription Drugs in History*, ed. Andrea Tone and Elizabeth Siegel Watkins (New York: New York University Press, 2007), 43; Emily Martin, *Bipolar Expeditions: Mania and Depression in American Culture* (Princeton, NJ: Princeton University Press, 2007), 223–27.

210 *The zoo's public affairs manager:* Friend, "It's a Jungle in Here." A related sentiment is touched upon in Emily Martin's *Bipolar Expeditions*, 225, in which she suggests that bipolarity was seen, at least for a time and by certain New Yorkers, as a New York City phenomenon—the frenetic pace of the city possibly attracting the already bipolar or inciting bipolarity in those prone to it, influenced the types of disorders New Yorkers recognized in their zoo animals.

210 *In the wake of Gus's news coverage*: Friend, "It's a Jungle in Here."

210 *In fact when Gus first arrived at the zoo from Ohio in 1988*: Nixon, "Gus the Neurotic Bear"; Friend, "It's a Jungle in Here."

211 *Hoping to curb the neurotic behavior*: Ingrid Newkirk, *The PETA Practical Guide to Animal Rights: Simple Acts of Kindness to Help Animals in Trouble* (New York: Macmillan, 2009); Julia Naylor Rodriguez, "Experts Say Prozac for Pets Is a Pretty Depressing Idea," *Forth Worth Star*, September 2, 1994; "Dogs Feeling Wuff in the City Getting a Boost from Prozac," *New York Daily News*, January 11, 2007; June Naylor Rodriguez, "Prozac for Fido? Don't Get Too Anxious for It, Vets Say," *Fort Worth Star*, September 3, 1994.

211 *The zoo also redesigned his exhibit*: N. R. Kleinfeld, "Farewell to Gus, Whose Issues Made Him a Star," *New York Times*, August 28, 2013; http://www.nytimes.com/2013/08/29/nyregion/gus-new-yorks-most-famous-polar-bear-dies-at-27.html?_r=0.

211 *In August 2013, Gus was euthanized*: Ibid.

211 *Because it's impossible to replicate*: E. M. Poulsen et al., "Use of Fluoxetine for the Treatment of Stereotypical Pacing Behavior in a Captive Polar Bear," *Journal of the American Veterinary Medical Association* 209, no. 8 (1996): 1470–74. The study was paid for by Eli Lilly.

211 *Abdi is a male brown bear*: Yalcin and N. Aytug, "Use of Fluoxetine to Treat Stereotypical Pacing Behavior in a Brown Bear *(Ursus arctos)*," *Journal of Veterinary Behavior Clinical Applications and Research* 2, no. 3 (2007): 73–76.

212 *Abdi is doing very well*: Dr. Prof. Nilufer Aytug, Karacabey Bear Sanctuary, personal communication, February 5, 2012.

213 *among 2.5 million insured Americans from 2001 to 2010*: "America's State of Mind," Medco, 2011, http://apps.who.int/medicinedocs/documents/s19032en/s19032en.pdf.

213 *Americans spent more than $16 billion on antipsychotics*: Brendan Smith, "Inappropriate Prescribing," *Monitor on Psychology*, American Psychological Association, June 2012, Vol. 43, No. 6, 36. http://www.apa.org/monitor/2012/06/prescribing.aspx.

213 *According to a recent study by the Centers for Disease Control*: National Ambulatory Medical Care Survey, Factsheet, Psychiatry, CDC, http://www.cdc.gov/nchs/data/ahcd/NAMCS_Factsheet_PSY_2009.pdf (accessed September 1, 2010); Laura A. Pratt, Debra J. Brody, and

Qiuping Gu, "Antidepressant Use in Persons Ages 12 and Over: United States, 2005–2008," CDC, http://www.cdc.gov/nchs/data/databriefs/db76.htm (accessed September 1, 2010).

213 *The U.S. market for pet pharmaceuticals:* Matt Wickenheiser, "Vet Biotech Aims at Generic Pet Medicine Market," *Bangalore Daily News*, February 3, 2012, http://bangordailynews.com/2012/02/03/business/vet-biotech-aims-at-generic-pet-medicine-market/. "Pet Industry Market Size and Ownership Statistics," American Pet Products Manufacturers Association, 2011–12, National Survey, http://www.americanpetproducts.org/press_industrytreras.asp.

213 *raised $2.2 billion in its initial public offering:* Chris Dietrich, "Zoetis Raises $2.2 Billion in IPO," *Wall Street Journal*, January 31, 2013.

213 *Elanco, a pet pharma company owned by Eli Lilly:* "Eli Lilly: Offsetting Generic Erosion through Janssen's Animal Health Business," *Comment-Wire*, March 17, 2011.

213 *Yearly sales of Pfizer's animal pharmaceuticals:* Susan Todd, "Retailers Shaking Up Pet Medicines Market, but Consumers Continue to Rely on Vets for Serious Remedies and Care," *Star-Ledger (NJ)*, October 2, 2011, http://www.nj.com/business/index.ssf/2011/10/retailers_shaking_up_pet_medic.html (accessed October 3, 2011).

214 *The pet pharmaceutical industry:* KPMG, Bureau of Economic Analysis, Packaged Facts, and William Blair and Co., Veterinary Economics, April 2008; and Allison Grant, "Veterinarians Scramble as Retailers Jump Into Pet Meds Market," *The Plain Dealer*, January 9, 2012.

214 *One market research firm recently claimed:* David Lummis, "Human/Animal Bond and 'Pet Parent' Spending Insulate $53 Billion U.S. Pet Market against Downturn, Forecast to Drive Post-Recession Growth," *Packaged Facts*, March 2, 2010, http://www.packagedfacts.com/Pet-Outlook-2553713/.

214 *This has proven to be true:* Susan Jones, *Valuing Animals: Veterinarians and Their Patients in Modern America* (Baltimore: Johns Hopkins University Press, 2003), 119; National Ambulatory Medical Care Survey.

214 *The most lucrative human drugs in 2012:* David Healy, *Pharmageddon* (Berkeley: University of California Press, 2012), 10–11.

214 *The scale of investment in the development:* Adriana Petryna, Andrew Lakoff, and Arthur Kleinman, eds., *Global Pharmaceuticals: Ethics, Markets, Practices* (Durham, NC: Duke University Press, 2006) 9; See also Shorter, *Before Prozac*, 11–33.

214 *Two key historical decisions:* Healy, *The Creation of Psychopharmacology*, 35.

215 *A second FDA decision in 1997:* Shorter, *Before Prozac*, 194–96.

215 *"One of the things that people called me was the Timothy Leary":* "Pet Pharm," CBC Documentaries, September 10, 2010, http://www.cbc.ca/documentaries/doczone/2010/petpharmacy/index.html.

215 *Like Leary, Dodman acted as a sort of pied piper:* See, for example, Nicholas H. Dodman and Louis Shuster, eds., *Psychopharmacology of Animal Behaviour Disorders* (Malden, MA: Blackwell Science, 1998); Dodman et al., "Equine Self-Mutilation Syndrome (57 Cases)"; N. H. Dodman et al., "Investigation into the Use of Narcotic Antagonists in the Treatment of a Stereotypic Behavior Pattern (Crib-Biting) in the Horse," *American Journal of Veterinary Research* 48, no. 2 (1987): 311–19; N. H. Dodman et al., "Use of Narcotic Antagonists to Modify Stereotypic Self-Licking, Self-Chewing, and Scratching Behavior in Dogs," *Journal of the American Veterinary Medical Association* 193, no. 7 (1988): 815–19; N. H. Dodman et al., "Use of Fluoxetine to Treat Dominance Aggression in Dogs," *Journal of the American Veterinary Medical Association* 209, no. 9 (1996): 1585–87; "Dodman to Hold Behavior Workshops in Northern Calif.," *Veterinary Practice News*, April 19, 2011, http://www.veterinarypracticenews.com/vet-breaking-news/2011/04/19/dodman-to-hold-behavior-workshops-in-northern-calif.aspx; Nicholas H. Dodman, "The Well Adjusted Cat—One Day Workshop: Secrets to Understanding Feline Behavior," Pet Docs, http://www.thepetdocs.com/events.html.

215 *He has published research:* See, for example, Dodman et al., "Use of Narcotic Antagonists to Modify Stereotypic Self-Licking, Self-Chewing, and Scratching Behavior in Dogs"; Dodman et al., "Investigation into the Use of Narcotic Antagonists in the Treatment of a Stereotypic Behavior Pattern (Crib-Biting) in the Horse"; B. L. Hart et al., "Effectiveness of Buspirone on Urine Spraying and Inappropriate Urination in Cats," *Journal of the American Veterinary Medical Association* 203, no. 2 (1993): 254–58; A. A. Moon-Fanelli and N. H. Dodman, "Description and Development of Compulsive Tail Chasing in Terriers and Response to Clomipramine Treatment," *Journal of the American Veterinary Medical Association* 212, no. 8 (1998): 1252–57; Raphael Wald, Nicholas Dodman, and Louis Shuster, "The Combined Effects of Memantine and Fluoxetine on an Animal Model of Obsessive Compulsive Disorder,"

Experimental and Clinical Psychopharmacology 17, no. 3 (2009): 191–97; L. S. Sawyer, A. A. Moon-Fanelli, and N. H. Dodman, "Psychogenic Alopecia in Cats: 11 Cases (1993–1996)," *Journal of the American Veterinary Medical Association* 214, no. 1 (1999): 71–74.

215 *In his book* The Well-Adjusted Dog: Nicholas H. Dodman, *The Well-Adjusted Dog: Dr. Dodman's Seven Steps to Lifelong Health and Happiness for Your Best Friend* (Boston: Houghton Mifflin Harcourt, 2008), 212.

215 *Dodman prescribes a wide variety of psychopharmaceuticals:* Buspar, another drug he uses, was first tested in the late 1980s in dogs with storm phobias. Dogs given the drug acted calmer when the thunder was far away but still got extremely anxious when it passed overhead. He has also used it for fear aggression and social anxiety and says it's really effective for treating dogs who like to pee in the house and for canine car sickness. Dodman, *The Well-Adjusted Dog*, 233–34.

216 *Dodman shared his ideas for the first time:* James Vlahos, "Pill-Popping Pets," *New York Times Magazine*, July 13, 2008.

216 *Dodman remembers hearing the former dean of the Tufts vet school:* Dodman, *Well-Adjusted Dog*, 232.

219 *Dodman argues that the great salvation:* "Pet Pharm."

219 *According to the ASPCA, 3.7 million of them:* "Animal Shelter Euthanasia," American Humane Asssociation, www.americanhumane.org/animals/stop-animal-abuse/fact-sheets/animal-shelter-euthanasia.htm (accessed December 20, 2013).

219 *psychopharm for pets can be a useful way station:* See, for example, D. A. Babcock et al., "Effects of Imipramine, Chlorimipramine, and Fluoxetine on Cataplexy in Dogs," *Pharmacology, Biochemistry, and Behavior* 5, no. 6 (1976): 599; Sharon L. Crowell-Davis and Thomas Murray, *Veterinary Psychopharmacology* (Wiley-Blackwell, 2005); Hart et al., "Effectiveness of Buspirone on Urine Spraying and Inappropriate Urination in Cats," 254–58; Charmaine Hugo et al., "Fluoxetine Decreases Stereotypic Behavior in Primates," *Progress in Neuro-Psychopharmacology and Biological Psychiatry* 27, no. 4 (2003): 639–43; Mami Irimajiri et al., "Randomized, Controlled Clinical Trial of the Efficacy of Fluoxetine for Treatment of Compulsive Disorders in Dogs," *Journal of the American Veterinary Medical Association* 235, no. 6 (2009): 705–9; Rapoport et al., "Drug Treatment of Canine Acral Lick," 517; Wald et al., "The

Combined Effects of Memantine and Fluoxetine on an Animal Model of Obsessive Compulsive Disorder."

224 *The humans who own the more than 78 million:* "Pet Industry Market Size and Ownership Statistics," American Pet Products Manufacturers Association, 2011–12, National Survey, http://www.americanpetproducts.org/press_industrytrends.asp.

224 *Simultaneously the company released results:* "Eli Lilly and Company Introduces Reconcile™ for Separation Anxiety in Dogs," *Medical News Today*, April 26, 2007, http://www.medicalnewstoday.com/releases/68990.php (accessed May 1, 2009).

224 *A 2008 study estimated that 14 percent of American dogs:* Vlahos, "Pill-Popping Pets."

224 *Lilly's Reconcile website:* www.reconcile.com (accessed January 15, 2012).

224 *An older version of the site:* www.reconcile.com/downloads (accessed June 15, 2009).

225 *published in* Veterinary Therapeutics *in 2007:* Barbara Sherman Simpson et al., "Effects of Reconcile (Fluoxetine) Chewable Tablets Plus Behavior Management for Canine Separation Anxiety," *Veterinary Therapeutics: Research in Applied Veterinary Medicine* 8, no. 1 (2007): 18–31. Another study on the drug, also sponsored by Lilly but focused on the effect of fluoxetine on compulsive behaviors in dogs, was equivocal. Irimajiri et al., "Randomized, Controlled Clinical Trial of the Efficacy of Fluoxetine for Treatment of Compulsive Disorders in Dogs," 705–9.

225 *Beagles were sent traveling:* Diane Frank, Audrey Gauthier, and Renée Bergeron, "Placebo-Controlled Double-Blind Clomipramine Trial for the Treatment of Anxiety or Fear in Beagles during Ground Transport," *Canadian Veterinary Journal* 47, no. 11 (2006): 1102–8.

225 *The drug has been more successful:* E. Yalcin, "Comparison of Clomipramine and Fluoxetine Treatment of Dogs with Tail Chasing," *Tierärztliche Praxis: Ausgabe K, Kleintiere/Heimtiere* 38, no. 5 (2010): 295–99; Moon-Fanelli and Dodman, "Description and Development of Compulsive Tail Chasing in Terriers and Response to Clomipramine Treatment," 1252–57; Seibert et al., "Placebo-Controlled Clomipramine Trial for the Treatment of Feather Picking Disorder in Cockatoos"; Dodman and Shuster, "Animal Models of Obsessive-Compulsive Behavior: A Neurobiological and Ethological Perspective."

225 *Medicating a small dog costs roughly:* 1–800PetMeds, http://www.1800petmeds.com/Clomicalm-prod10439.html (accessed February 4, 2012).

226 *He leads training classes and workshops:* "Dr. Ian Dunbar," Sirius Dog Training, http://www.siriuspup.com/about_founder.html (accessed June 3, 2013); "Ian Dunbar Events and Training Courses," https://www.jamesandkenneth.com/store/show_by_tags/Events (accessed June 3, 2013).

226 *"Drugs are simply unnecessary":* "Pet Pharm"; Vlahos, "Pill-Popping Pets"; "About Founder," *Sirius Dog Training,* http://www.siriuspup.com/about_founder.html (accessed June 3, 2013); "Ian Dunbar Events and Training Courses," https://www.jamesandkenneth.com/store/show_by_tags/Events (accessed June 3, 2013).

226 *Dunbar argues that pet owners:* "Pet Pharm."

227 *Debating whether dosing other animals:* Nigel Rothfels, *Savages and Beasts: The Birth of the Modern Zoo* (Baltimore: Johns Hopkins University Press, 2002), 81.

227 *demonstrated the presence of a range:* Chris D. Metcalfe et al., "Antidepressants and Their Metabolites in Municipal Wastewater, and Downstream Exposure in an Urban Watershed," *Environmental Toxicology and Chemistry* 29, no. 1 (2010): 79–89.

227 *In one experiment, bass exposed to Prozac:* Ibid.; Janet Raloff, "Environment: Antidepressants Make for Sad Fish. Drugs May Affect Feeding, Swimming and Mate Attracting," *Science News* 174, no. 13 (2008): 15.

227 *Another study looked at the effects of Prozac:* Nina Bai, "Prozac Ocean: Fish Absorb Our Drugs, and Suffer for It," Discover Magazine Blog, December 2, 2008, http://blogs.discovermagazine.com/discoblog/2008/12/02/prozac-ocean-fish-absorb-our-drugs-and-suffer-for-it/ (accessed March 2, 2009); Yasmin Guler and Alex T. Ford, "Anti-Depressants Make Amphipods See the Light," *Aquatic Toxicology* 99, no. 3 (2010): 397–404; Metcalfe et al., "Antidepressants and Their Metabolites in Municipal Wastewater, and Downstream Exposure in an Urban Watershed."

227 *found an array of psychopharmaceuticals in the feathers:* D. C. Love et al., "Feather Meal: A Previously Unrecognized Route for Reentry into the Food Supply of Multiple Pharmaceuticals and Personal Care Products (PPCPs)," *Environmental Science and Technology* 46, no. 7 (2012): 3795–802; Sarah Parsons, "This Is Your Chicken on Drugs: Count the Antibiotics in Your Nuggets," *Good,* April 10, 2012; "Researchers Find Evidence of Banned Antibiotics in Poultry Products," Center for

a Livable Future, Johns Hopkins Bloomberg School of Public Health, April 2012.

228 *According to the journalist Nicolas Kristof*: Nicholas D. Kristof, "Arsenic in Our Chicken?," *New York Times*, April 4, 2012; Love et al., "Feather Meal: A Previously Unrecognized Route for Reentry into the Food Supply of Multiple Pharmaceuticals and Personal Care Products (PPCPs); Sarah Parsons, "This Is Your Chicken on Drugs: Count the Antibiotics in Your Nuggets," *Good*, April 10, 2012; "Researchers Find Evidence of Banned Antibiotics in Poultry Products," Center for a Livable Future, Johns Hopkins Bloomberg School of Public Health, April 2012.

Chapter Six: If Juliet Were a Parrot

230 *According to the study's lead researcher*: Susanne Antonetta, "Language Garden," *Orion*, April 2005, http://www.orionmagazine.org/index.php/articles/article/152/ (accessed August 10, 2011).

232 *Aristotle told the story*: Edmund Ramsden and Duncan Wilson, "The Nature of Suicide: Science and the Self-Destructive Animal," *Endeavor* 34, no. 1 (2010): 21.

232 *Since its first documentation*: "Suicide," Oxford English Dictionary Online, http://www.oed.com/view/Entry/193691?rskey=6JPREr&result=1. Before the eighteenth century, someone could be a "self-destroyer," "self-killer" "self-murderer," or "self-slayer," but not a suicide victim.

232 *The DSM-V does not include suicide*: Nor does it mention people who choose to end their own lives because they have terminal diseases. "Proposed Revision," DSM5.org, http://www.dsm5.org/ProposedRevision/Pages/proposedrevision.aspx?rid=584# (accessed April 1, 2013). See also American Psychiatric Association, *Diagnostic and Statistical Manual of Mental Disorders-IV* (Washington, DC: American Psychiatric Association, 1994) for suicidal ideation and attempts but no suicidal disorder.

232 *Self-destructive behaviors like these among nonhuman animals:* See, for example, Kathryn Bayne and Melinda Novak, "Behavioral Disorders," *Nonhuman Primates in Biomedical Research* (1998): 485–500; P. S. Bordnick, B. A. Thyer, and B. W. Ritchie, "Feather Picking Disorder and Trichotillomania: An Avian Model of Human Psychopathology," *Journal of Behavior Therapy and Experimental Psychiatry* 25, no. 3 (1994): 189–96; John C. Crabbe, John K. Belknap, and Kari J. Buck, "Genetic

Animal Models of Alcohol and Drug Abuse," *Science* 264, no. 5166 (1994): 1715–23; J. N. Crawley, M. E. Sutton, and D. Pickar, "Animal Models of Self-Destructive Behavior and Suicide," *Psychiatric Clinics of North America* 8, no. 2 (1985): 299–310; Cross and Harlow, "Prolonged and Progressive Effects of Partial Isolation on the Behavior of Macaque Monkeys," 39–49; Kalueff et al., "Hair Barbering in Mice"; A. J. Kinnaman, "Mental Life of Two Macacus Rhesus Monkeys in Captivity. I," *American Journal of Psychology* 13, no. 1 (1902): 98–148; Kurien et al., "Barbering in Mice"; O. Malkesman et al., "Animal Models of Suicide-Trait-Related Behaviors," *Trends in Pharmacological Sciences* 30, no. 4 (2009): 165–73; Melinda A. Novak and Stephen J. Suomi, "Abnormal Behavior in Nonhuman Primates and Models of Development," *Primate Models of Children's Health and Developmental Disabilities* (2008): 141–60; Overall, "Natural Animal Models of Human Psychiatric Conditions"; J. L. Rapoport, D. H. Ryland, and M. Kriete, "Drug Treatment of Canine Acral Lick: An Animal Model of Obsessive-Compulsive Disorder," *Archives of General Psychiatry* 49, no. 7 (1992): 517–21; Richard E. Tessel et al., "Rodent Models of Mental Retardation: Self-Injury, Aberrant Behavior, and Stress," *Mental Retardation and Developmental Disabilities Research Reviews* 1, no. 2 (1995): 99–103.

233 *"self-destructive and suicidal behaviors"*: Crawley, Sutton, and Pickar, "Animal Models of Self-Destructive Behavior and Suicide."

233 *In the twenty-five years since this study was published:* For example, Nicholas H. Dodman and Louis Shuster, "Animal Models of Obsessive-Compulsive Behavior: A Neurobiological and Ethological Perspective," in *Concepts and Controversies in Obsessive-Compulsive Disorder*, ed. Jonathan S. Abramowitz and Arthur C. Houts (New York: Springer, 2005), 53–71; Garner et al., "Barbering (Fur and Whisker Trimming) by Laboratory Mice as a Model of Human Trichotillomania and Obsessive-compulsive Spectrum Disorders"; Kurien et al., "Barbering in Mice" Rapoport et al., "Drug Treatment of Canine Acral Lick"; Bordnick et al., "Feather Picking Disorder and Trichotillomania"; Crabbe et al., "Genetic Animal Models of Alcohol and Drug Abuse"; Overall, "Natural Animal Models of Human Psychiatric Conditions"; Tessel et al., "Rodent Models of Mental Retardation."

233 *"Suicide is a complex behavior"*: Malkesman et al., "Animal Models of Suicide-Trait-Related Behaviors."

233 *The traits and behaviors in lab animals:* Ibid.

235 *According to the American Association of Suicidology:* A suicide note demonstrates that the person intended to kill themselves but the notes often stop short of explaining their motivations. There are clear-eyed exceptions, of course, full of insight or directions on how the person wants to be disposed of, or what they want in a memorial service, but more often than not the notes simply illustrate the range of emotions that the people writing them were feeling at the time. Eric Marcus, *Why Suicide?: Answers to 200 of the Most Frequently Asked Questions about Suicide, Attempted Suicide, and Assisted Suicide* (San Francisco: HarperOne, 1996), 14.

236 *Their paper, "The Nature of Suicide":* Justin Nobel, "Do Animals Commit Suicide? A Scientific Debate," *Time,* March 19, 2010; Larry O'Hanlon, "Animal Suicide Sheds Light on Human Behavior," *Discovery News,* March 10, 2010, http://news.discovery.com/animals/animal-suicide-behavior.html?print=true; Rowan Hooper, "Animals Do Not Commit Suicide," *NewScientist,* March 24, 2010, http://www.newscientist.com/blogs/shortsharpscience/2010/03/animals-do-not-commit-suicide.html (accessed April 20, 2010).

236 *"Scientists and social groups":* Ramsden and Wilson, "The Nature of Suicide," 22.

237 *The Victorian period was an extremely interesting time:* Barbara T. Gates, *Victorian Suicide: Mad Crimes and Sad Histories* (Princeton, NJ: Princeton University Press, 1988), 37; Anne Shepherd and David Wright, "Madness, Suicide and the Victorian Asylum: Attempted Self-Murder in the Age of Non-Restraint," *Medical History* 46, no. 2 (2002): 175–96.

237 *In England, the families of suicidal people:* Gates, *Victorian Suicide,* 38.

237 *William Lauder Lindsay devoted an entire chapter:* Lindsay, *Mind in the Lower Animals,* 130–48.

237 *Lindsay's work reflects a larger shift:* Ramsden and Wilson, "The Nature of Suicide."

237 *"sufficiently barbarous . . . to induce":* C. Lloyd Morgan, "Suicide of Scorpions," *Nature* 27 (1883): 313–14. Four years later another member of the Royal Society published a paper, "The Reputed Suicide of Scorpions," arguing that insects were immune to their own venom, and the question of scorpion suicide seemed to be settled. A. G. Bourne, "The Reputed Suicide of Scorpions," *Proceedings of the Royal Society of London* 42 (1887): 17–22.

238 *he published* Mental Evolution in Animals: George John Romanes, *Mental Evolution in Animals* (New York: D. Appleton, 1884).

238 *Romanes, like Lindsay and Darwin, believed that insanity*: Ibid., 148–74; Lorraine Daston has argued that for Romanes, anthropomorphism was a virtue and a necessity since it demonstrated a direct evolutionary relationship between humans and other animals. To Morgan, though, it was faulty human projection. Lorraine Daston, "Intelligences: Angelic, Animal, Human," in *Thinking with Animals: New Perspectives on Anthropomorphism*, ed. Lorraine Daston and Gregg Mitman (New York: Columbia University Press, 2005), 37–58.

238 *Meanwhile, the idea of suicide*: Ramsden and Wilson, "The Nature of Suicide," 24.

239 *people's motivations to do themselves in could have "secret causes"*: Enrico Morselli, *Suicide: An Essay on Comparative Moral Statistics* (London: Kegan Paul, 1881), 8.

239 *Durkheim published his landmark book*: In the introduction he made a point of explaining why he would not be including nonhuman animals: "Our knowledge of animal intelligence does not really allow us to attribute to them an understanding anticipatory of their death nor, especially, of the means to accomplish it. . . . All cases cited at all authentically which might appear true suicides may be quite differently explained. If the irritated scorpion pierces itself with its sting (which is not at all certain), it is probably from an automatic, unreflecting reaction. The motive energy aroused by his irritation is discharged by chance and at random; the creature happens to become its victim, though it cannot be said to have had a preconception of the result of its action. On the other hand, if some dogs refuse to take food on losing their masters, it is because the sadness into which they are thrown has automatically caused lack of hunger; death has resulted, but without having been foreseen. Neither fasting in this case nor the wound in the other have been used as means to a known effect." Emile Durkheim, *Suicide: A Study in Sociology* (Glencoe, IL: Free Press, 1951), 44–45.

239 *in an article for the journal* Mind: Ramsden and Wilson, "The Nature of Suicide," 23.

239 *In 1903 Morgan reiterated his earlier stance*: Daston, "Intelligences," 37–58.

239 *"In no case is an animal activity to be interpreted*: Conwy Lloyd Morgan, *An Introduction to Comparative Psychology* (London: Morgan, 1894), 53.

240 *Accounts of animal suicides appeared*: For fictionalized accounts, see, for example, Claire Goll, *My Sentimental Zoo* (New York: Peter Pauper

Press, 1942). For other accounts, see all examples to follow, as well as "Texas Cattle: Peculiarities of the Long-Horned Beasts," *San Francisco Chronicle*, July 7, 1885; "A Bull's Suicide," *San Francisco Chronicle*, December 22, 1891; Paul Eipper, *Animals Looking at You* (New York: Viking Press, 1929).

240 *One early article, published in the* New York Sun: "Suicide by Animals: Self-Destruction of Scorpion and Star-Fish," *New York Sun*, December 18, 1881.

240 *"You know, lions are as vain as a society woman"*: "The Suicide of a Lion," *San Francisco Chronicle*, August 25, 1901.

240 *The most commonly reported suicidal animals*: For representative samples, see "Suicide of a Dog," *San Francisco Chronicle*, July 29, 1897; "Suicide: Do Animals Seek Their Own Death?," *San Francisco Chronicle*, April 7, 1884; "A Mare's Suicide," *San Francisco Chronicle*, November 7, 1894.

240 *Among the general population*: Ritvo, *The Animal Estate*, 19, 35.

241 *organizations like the Royal Society for the Prevention of Cruelty to Animals*: Ramsden and Wilson, "The Nature of Suicide," 22.

241 *As for horses, popular natural history writers*: Ritvo, *The Animal Estate*, 19, 35.

241 *In February 1905 a superior county court*: "Court Decides That Horse Committed Suicide," *San Francisco Chronicle*, February 2, 1905.

241 *Other horses supposedly jumped*: "Aged Gray Horse, Weary of Life, Commits Suicide," *San Francisco Chronicle*, January 23, 1922; "Horse Fails in Suicide Attempts," *San Francisco Chronicle*, May 22, 1922.

242 *Ric O'Barry is an outspoken former dolphin trainer*: He first discussed it in his book, then in an interview for a *Frontline* documentary. Richard O'Barry and Keith Coulbourn, *Behind the Dolphin Smile: A True Story That Will Touch the Hearts of Animal Lovers Everywhere* (New York: St. Martin's Griffin, 1999), 248–50; "Interview with Richard O'Barry," *Frontline: A Whale of a Business*, PBS, November 11, 1997, http://www.pbs.org/wgbh/pages/frontline/shows/whales/interviews/obarry2.html.

242 *O'Barry writes that the part of Flipper*: O'Barry and Coulbourn, *Behind the Dolphin Smile*, 136.

243 *mounted capture expeditions*: Ibid.; Richard O'Barry, personal communication, June 16, 2009.

243 *Everything changed for O'Barry when*: "Interview with Richard O'Barry."

243 *When O'Barry arrived at Seaquarium, he found Kathy*: O'Barry and Coulbourn, *Behind the Dolphin Smile*, 248–50.

243 *"Kathy died of suicide"*: "Interview with Richard O'Barry."

243 *Twenty million people gathered:* Shabecoff, *A Fierce Green Fire*, 131; "Earth Day: The History of a Movement," Earth Day Network, http://www.earthday.org/earth-day-history-movement (accessed August 20, 2013).

243 *A week later, inspired by this new environmental groundswell:* O'Barry and Coulbourn, *Behind the Dolphin Smile*, 28–35.

244 *She believes that suicide in captive whales and dolphins is possible:* Dr. Naomi Rose, Humane Society of the United States, personal communication, April 13, 2010.

244 *"The whole spectrum of open-ocean species"*: Ibid.

246 *Stranding events—two or more marine animals:* "Three Beached Whales by Jan Wierix," in R. Ellis, *Monsters of the Sea* (Robert Hale, 1994), http://upload.wikimedia.org/wikipedia/commons/9/94/Three_Beached_Whales%2C_1577.jpg; "Stranded Whale at Katwijk in Holland in 1598," in Ellis, http://en.wikipedia.org/wiki/File:Stranded_whale_Katwijk_1598.jpg; "Scenes from Wellfleet Dolphin Stranding," January 19, 2012, http://www.youtube.com/watch?v=AGbdp4saMoI (accessed May 1, 2012); "Raw Video: Mass Stranding of Pilot Whales," May 6, 2011, http://www.youtube.com/watch?v=w636wkpsBBg (accessed May 1, 2012).

246 *Scientific statistics on stranding:* Angela D'Amico et al., "Beaked Whale Strandings and Naval Exercises," *Aquatic Mammals* 35 (December 1, 2009): 452–72.

246 *From the 1930s on:* This discussion is based on searches of historical archives of American newspapers from the nineteenth century through the present, including the *New York Times*, *Washington Post*, *San Francisco Chronicle*, *Boston Globe*, *Los Angles Times*, and other major newspapers.

246 *In one representative account from 1937:* "Enigma of Suicidal Whales," *New York Times*, June 6, 1937.

246 *"deliberately beached themselves"*: "Whales Swim In and Die," *New York Times*, October 8, 1948.

246 *A particularly large mass stranding in Scotland:* This is a rather unlikely explanation, considering such a self-destructive species would not live long. "Scotland's 274 Dead Whales Stir Question," *Los Angeles Times*, July 5, 1950.

247 *cetologists increasingly reacted to reports:* See, for example, Murray D. Dailey and William A. Walker, "Parasitism as a Factor (?) in Single Strandings of Southern California Cetaceans," *Journal of Parasitology* 64, no. 4

(1978): 593–96; Robert D. Everitt et al., *Marine Mammals of Northern Puget Sound and the Strait of Juan de Fuca: A Report on Investigations, November 1, 1977–October 31, 1978*, Environmental Research Laboratories, Marine Ecosystems Analysis Program, 1979; C. H. Fiscus and K. Niggol, "Observations of Cetaceans off California, Oregon, and Washington," *U.S. Fish and Wildlife Service Special Scientific Report* 498 (1965): 1–27; S. Ohsumi, "Interspecies Relationships among Some Biological Parameters in Cetaceans and Estimation of the Natural Mortality Coefficient of the Southern Hemisphere Minke Whale," *Report of the International Whaling Commission* 29 (1979): 397–406; D. E. Sergeant, "Ecological Aspects of Cetacean Strandings," in *Biology of Marine Mammals: Insights through Strandings*, ed. J. R. Geraci and D. J. St. Aubin, Marine Mammal Commission Report No. MMC-77/13, 1979, 94–113.

247 *When twenty-four pilot whales stranded near Charleston:* "Scientists Study Mystery of 24 Pilot Whales That Died after Stranding Themselves on Carolina Island Beach," *New York Times*, October 8, 1973.

247 *Reports of dolphin and whale mass suicides:* See, for example, "Mass Suicide: Whale Beachings Puzzle to Experts," *Observer Reporter*, July 27, 1976. One reason for the increasing numbers of reported strandings may have been, as Norman et al. suggest, the growing establishment of formal stranding responder networks between 1930 and 2002. Spikes of reported strandings also happen to occur during summer months, when more human observers are on beaches and along waterways, where they may witness stranded animals. S. A. Norman, et al., "Cetacean Strandings in Oregon and Washington between 1930 and 2002," *Journal of Cetacean Research and Management* 6 (2004): 87–99.

247 *One reason for the general public's openness:* D. Graham Burnett, "A Mind in the Water," *Orion*, June 2010, http://www.orionmagazine.org/index.php/articles/article/5503/ (accessed July 1, 2010); D. Graham Burnett, *The Sounding of the Whale: Science and Cetaceans in the Twentieth Century* (Chicago: University of Chicago Press, 2012), chapter 6.

247 *The historian Etienne Benson has argued:* Etienne Benson, *Wired Wilderness: Technologies of Tracking and the Making of Modern Wildlife* (Baltimore: Johns Hopkins University Press, 2010), 1–48.

248 *doubt and confusion surrounding the reasons for strandings:* National Research Council (U.S.), Committee on Potential Impacts of Ambient Noise in the Ocean on Marine Mammals, *Ocean Noise and Marine Mammals* (Washington, DC: National Academies Press, 2003);

L. S. Weilgart, "A Brief Review of Known Effects of Noise on Marine Mammals," *International Journal of Comparative Psychology* 20 (2007): 159–68; D'Amico et al., "Beaked Whale Strandings and Naval Exercises"; K. C. Balcomb and D. E. Claridge, "A Mass Stranding of Cetaceans Caused by Naval Sonar in the Bahamas," *Bahamas Journal of Science* 8, no. 2 (2001): 2–12; D. M. Anderson and A. W. White, "Marine Biotoxins at the Top of the Food Chain," *Oceanus* 35, no. 3 (1992): 55–61; R. J. Law, C. R. Allchin, and L. K. Mead, "Brominated Diphenyl Ethers in Twelve Species of Marine Mammals Stranded in the UK," *Marine Pollution Bulletin* 50 (2005): 356–59; R. J. Law et al., "Metals and Organochlorines in Pelagic Cetaceans Stranded on the Coasts of England and Wales," *Marine Pollution Bulletin* 42 (2001): 522–26; R. J. Law et al., "Metals and Organochlorines in Tissues of a Blainville's Beaked Whale *(Mesoplodon densirostris)* and a Killer Whale *(Orcinus orca)* Stranded in the United Kingdom," *Marine Pollution Bulletin* 34 (1997): 208–12; K. Evans et al., "Periodic Variability in Cetacean Strandings: Links to Large-Scale Climate Events," *Biology Letters* 1, no. 2 (2005): 147–50; M. D. Dailey et al., "Prey, Parasites and Pathology Associated with the Mortality of a Juvenile Gray Whale *(Eschrichtius robustus)* Stranded along the Northern California Coast," *Diseases and Aquatic Organisms* 42 (2000): 111–17; J. Geraci et al., "Humpback Whales *(Megaptera novaeanglie)* Fatally Poisoned by Dinoflagellate Toxin," *Canadian Journal of Fisheries and Aquatic Science* 46 (1989): 1895–98; H. Thurston, "The Fatal Shore," *Canadian Geographic*, January–February 1995, 60–68.

248 *These stressors, combined with recent research:* See, for example, Felicity Muth, "Animal Culture: Insights from Whales," Scientific American.com, April 27, 2013, http://blogs.scientificamerican.com/not-bad-science/2013/04/27/animal-culture-insights-from-whales/ (accessed April 28, 2013); Jenny Allen et al., "Network-Based Diffusion Analysis Reveals Cultural Transmission of Lobtail Feeding in Humpback Whales," *Science* 340, no. 6131 (2013): 485–88; John K. B. Ford, "Vocal Traditions among Resident Killer Whales *(Orcinus orca)* in Coastal Waters of British Columbia," *Canadian Journal of Zoology* 69, no. 6 (1991): 1454–83; Luke Rendell et al., "Can Genetic Differences Explain Vocal Dialect Variation in Sperm Whales, *Physeter macrocephalus*?," *Behavior Genetics* 42, no. 2 (2011): 332–43.

248 *predicated on really tight social bonds:* J. R. Geraci and V. J. Lounsbury, *Marine Mammals Ashore: A Field Guide for Strandings* (Galveston: Texas

A&M University Sea Grant College Program, 1993); A. F. González and A. López, "First Recorded Mass Stranding of Short-Finned Pilot Whales *(Globicephala macrorhynchus* Gray, 1846) in the Northeastern Atlantic," *Marine Mammal Science* 16, no. 3 (2000): 640–46.

248 *Attendees are largely unpaid volunteers:* Other nations have their own stranding networks, such as New Zealand's Project Jonah, http://www.projectjonah.org.nz; Indonesia's Whale Strandings Indonesia, http://www.whalestrandingindonesia.com/index.php; and Canada's Marine Mammal Response Society, http://www.marineanimals.ca/.

250 *This may have affected the evolution of their social worlds:* Richard C. Connor, "Group Living in Whales and Dolphins," in *Cetacean Societies: Field Studies of Dolphins and Whales,* ed. Janet Mann (Chicago: University of Chicago Press, 2000), 199–218.

250 *Only one of a pod of nineteen white-sided dolphins:* "Why Do Cetaceans Strand? A Summary of Possible Causes," Hal Whitehead Laboratory Group, http://whitelab.biology.dal.ca/strand/StrandingWebsite.html#social; E. Rogan et al., "A Mass Stranding of White-Sided Dolphins *(Lagenorhynchus acutus)* in Ireland: Biological and Pathological Studies," *Journal of Zoology* 242, no. 2 (1997): 217–27.

251 *Since the Hilo conference in 2010:* These studies built on earlier published research that also suggested causal relationships. David Suzuki, "Sonar and Whales Are a Deadly Mix," *Huffington Post,* February 27, 2013, http://www.huffingtonpost.ca/david-suzuki/sonar-naval-training-kills-whales_b_2769130.html; Tyack et al., "Beaked Whales Respond to Simulated and Actual Navy Sonar"; National Resource Defense Council, "Lethal Sounds: The Use of Military Sonar Poses a Deadly Threat to Whales and Other Marine Mammals," *NRDC,* http://www.nrdc.org/wildlife/marine/sonar.asp (accessed July 5, 2013); Weilgart, "A Brief Review of Known Effects of Noise on Marine Mammals"; National Research Council (U.S.), *Ocean Noise and Marine Mammals.*

251 *The U.S. National Marine Fisheries Service and U.S. Navy released a report:* "U.S. Sued over U.S. Navy Sonar Tests in Whale Waters," *NBC News,* January 26, 2012, http://usnews.nbcnews.com/_news/2012/01/26/10244852-us-sued-over-navy-sonar-tests-in-whale-waters?lite (accessed January 27, 2012); "Marine Mammals and the Navy's 5-Year Plan," *New York Times,* October 11, 2012, http://www.nytimes.com/2012/10/12/opinion/marine-mammals-and-the-navys-5-year-plan.html (accessed October 11, 2012); Natural Resources Defense Center Council, "Navy Training

Blasts Marine Mammals with Harmful Sonar," *National Resource Defense Council Media*, news release, January 26, 2012, http://www.nrdc.org/media/2012/120126a.asp (accessed January 27, 2012).

251 *Other, similar legal battles are now under way:* Lauren Sommer, "Navy Sonar Criticized for Harming Marine Mammals," *All Things Considered*, National Public Radio, April 26, 2013, http://www.npr.org/2013/04/26/179297747/navy-sonar-criticized-for-harming-marine-mammals; Jeremy A. Goldbogen et al., "Blue Whales Respond to Simulated Mid-Frequency Military Sonar," *Proceedings of the Royal Society: Biological Sciences* 280, no. 1765 (2013).

251 *The latest study on these animals and anthropogenic sound:* Goldbogen et al., "Blue Whales Respond to Simulated Mid-Frequency Military Sonar"; "Study: Military Sonar May Affect Endangered Blue Whale Population," *CBS News*, July, 8, 2013, http://seattle.cbslocal.com/2013/07/08/study-military-sonar-may-affect-endangered-blue-whale-population; Suzuki, "Sonar and Whales Are a Deadly Mix" (accessed July 8, 2013); "U.S. Military Sonar May Affect Endangered Blue Whales, Study Suggests," *Washington Post*, July 8, 2013, http://articles.washingtonpost.com/2013–07–08/national/40435944_1_blue-whales-cascadia-research-collective-mid-frequency-sonar (accessed July 8, 2013); Damian Carrington, "Whales Flee from Military Sonar Leading to Mass Strandings, Research Shows," *Guardian*, July 2, 2013; Victoria Gill, "Blue and Beaked Whales Affected by Simulated Navy Sonar," *BBC News*, July 2, 2013, http://www.bbc.co.uk/news/science-environment-23115939 (accessed July 8, 2013); Richard Gray, "Blue Whales Are Disturbed by Military Sonar," *Telegraph*, July 3, 2013, http://www.telegraph.co.uk/earth/wildlife/10158068/Blue-whales-are-disturbed-by-military-sonar.html (accessed July 4, 2013); Megan Gannon, "Military Sonar May Hurt Blue Whales," *Yahoo News*, July 4, 2013, http://news.yahoo.com/military-sonar-may-hurt-blue-whales-141911253.html (accessed July 4, 2013).

252 *"It would be like someone walking in front of a car":* Wynne Parry, "16 Whales Mysteriously Stranded in Florida Keys," *Live Science*, May 6, 2011, http://www.livescience.com/14052-pilot-whale-stranding-florida-pod-noise.html (accessed May 6, 2011).

252 *the hatters didn't just shake:* H. A. Waldron, "Did the Mad Hatter Have Mercury Poisoning?," *British Medical Journal* 287, no. 6409 (1983): 1961.

252 *Today the most widespread source of mercury exposure:* Katherine H. Taber and Robin A. Hurley, "Mercury Exposure: Effects across the Lifespan,"

Journal of Neuropsychiatry and Clinical Neurosciences 20, no. 4 (2008): 384–89; S. Allen Counter and Leo H. Buchanan, "Mercury Exposure in Children: A Review," *Toxicology and Applied Pharmacology* 198, no. 2 (2004): 213.

253 *Almost all of this ingested mercury:* Counter and Buchanan, "Mercury Exposure in Children," 213.

253 *Chronic mercury poisoning can result in anxiety:* Taber and Hurley, "Mercury Exposure," 389.

253 *The effect of mercury on marine mammals:* Wendy Noke Durden et al., "Mercury and Selenium Concentrations in Stranded Bottlenose Dolphins from the Indian River Lagoon System, Florida," *Bulletin of Marine Science* 81, no. 1 (2007): 37–54; H. Gomercic Srebocan and A. Prevendar Crnic, "Mercury Concentrations in the Tissues of Bottlenose Dolphins *(Tursiops truncatus)* and Striped Dolphins *(Stenella coeruloalba)* Stranded on the Croatian Adriatic Coast," *Science and Technology* 2009, no. 12 (2009): 598–604; Dan Ferber, "Sperm Whales Bear Testimony to Ocean Pollution," *Science Now*, August 17, 2005, http://news.sciencemag.org/sciencenow/2005/08/17-02.html (accessed August 9, 2010); "Mercury Levels in Arctic Seals May Be Linked to Global Warming," *Science Daily*, May 4, 2009, http://www.sciencedaily.com/releases/2009/05/090504165950.htm (accessed May 5, 2009); A. Gaden et al., "Mercury Trends in Ringed Seals *(Phoca hispida)* from the Western Canadian Arctic since 1973: Associations with Length of Ice-Free Season," *Environmental Science and Technology* 43 (May 15, 2009): 3646–51.

253 *It has also been possible to work backward:* Shawn Booth and Dirk Zeller, "Mercury, Food Webs, and Marine Mammals: Implications of Diet and Climate Change for Human Health," *Environmental Health Perspectives* 113 (February 2, 2005): 521–26.

253 *toxicologists have shown that the bodies:* Durden et al., "Mercury and Selenium Concentrations in Stranded Bottlenose Dolphins from the Indian River Lagoon System, Florida"; Srebocan and Crnic, "Mercury Concentrations in the Tissues of Bottlenose Dolphins *(Tursiops truncatus)* and Striped Dolphins *(Stenella coeruloalba)* Stranded on the Croatian Adriatic Coast"; Ferber, "Sperm Whales Bear Testimony to Ocean Pollution"; "Mercury Levels in Arctic Seals May Be Linked to Global Warming"; Gaden et al., "Mercury Trends in Ringed Seals *(Phoca hispida)* from the Western Canadian Arctic since 1973."

254 *In harbor seals this contamination:* "Mercury Pollution Causes Immune Damage to Harbor Seals," Science Daily, October 20, 2008, http://www.sciencedaily.com/releases/2008/10/081020191532.htm# (accessed May 30, 2010).

254 *Mercury isn't the only environmental toxin:* Jordan Lite, "What Is Mercury Posioning?," *Scientific American,* December 19, 2008, http://www.scientificamerican.com/article.cfm?id=jeremy-piven-mercury-poisoning (accessed April 20, 2010); "Pollution 'Makes Birds Mate with Each Other,' Say Scientists," *Mail Online,* December 10, 2010, http://www.dailymail.co.uk/sciencetech/article-1334725/Mercury-diet-making-male-birds-gay.html (accessed December 11, 2010); "Fish Consumption Advisories," U.S. Environmental Protection Agency, http://www.epa.gov/hg/advisories.htm (accessed December 11, 2010); Bob Condor, "Living Well: How Much Mercury Is Safe? Go Fishing for Answers," Seattlepi.com, October 19, 2008, http://www.seattlepi.com/lifestyle/health/article/Living-Well-How-much-mercury-is-safe-Go-fishing-1288719.php (accessed November 1, 2009); Francesca Lyman, "How Much Mercury Is in the Fish You Eat? Doctors Recommend Consuming Seafood, but Some Fish Are Tainted," NBCNews.com, April 4, 2003, http://www.nbcnews.com/id/3076632/ns/health-your_environment/t/how-much-mercury-fish-you-eat/ (accessed January 4, 2011); "Weekly Health Tip: Mercury in Fish—How Much Is Too Much?," Huffpost Healthy Living: The Blog, http://www.huffingtonpost.com/deepak-chopra/mercury-fish_b_893631.html; "Mercury Mess: Wild Bird Sex Stifled," *Environmental Health News,* September 22, 2011, http://www.environmentalhealthnews.org/ehs/newscience/2011/08/2011–0920-mercury-messes-with-sex/ (accessed September 22, 2011).

254 *Lead, manganese, arsenic, and organophosphate insecticides:* M. R. Trimble and E. S. Krishnamoorthy, "The Role of Toxins in Disorders of Mood and Affect," *Neurologic Clinics* 18, no. 3 (2000): 649–64; Celia Fischer, Anders Fredriksson, and Per Eriksson, "Coexposure of Neonatal Mice to a Flame Retardant PBDE 99 (2,2',4,4',5-pentabromodiphenyl ether) and Methyl Mercury Enhances Developmental Neurotoxic Defects," *Toxicological Sciences: An Official Journal of the Society of Toxicology* 101, no. 2 (2008): 275–85.

254 *In one human study, factory workers:* Trimble and Krishnamoorthy, "The Role of Toxins in Disorders of Mood and Affect."

254 *exposure to lead, arsenic, mercury:* See, for example, "Some Toxic Effects of Lead, Other Metals and Antibacterial Agents on the Nervous System: Animal Experiment Models," *Acta Neurologica Scandinavica Supplementum* 100 (1984): 77–87; Minoru Yoshida et al., "Neurobehavioral Changes and Alteration of Gene Expression in the Brains of Metallothionein-I/II Null Mice Exposed to Low Levels of Mercury Vapor during Postnatal Development," *Journal of Toxicological Sciences* 36, no. 5 (2011): 539–47; Shuhua Xi et al., "Prenatal and Early Life Arsenic Exposure Induced Oxidative Damage and Altered Activities and mRNA Expressions of Neurotransmitter Metabolic Enzymes in Offspring Rat Brain," *Journal of Biochemical and Molecular Toxicology* 24, no. 6 (2010): 368–78.

255 *the Czech scientist Jaroslav Flegr:* Kathleen McAuliffe, "How Your Cat Is Making You Crazy," *Atlantic*, March 2012.

255 *Flegr puzzled over whether he might be:* Ibid.

255 *Flegr discovered that the French:* Ibid.; "Common Parasite May Trigger Suicide Attempts: Inflammation from T. Gondii Produces Brain-Damaging Metabolites," Science Daily, August 16, 2012, http://www.sciencedaily.com/releases/2012/08/120816170400.htm (accessed August 16, 2012).

256 *the parasite transforms its host animals into cat-delivery systems:* Patrick K. House, Ajai Vyas, and Robert Sapolsky, "Predator Cat Odors Activate Sexual Arousal Pathways in Brains of Toxoplasma Gondii Infected Rats," ed. Georges Chapouthier, *PLoS ONE* 6, no. 8 (2011): e23277; "Toxo: A Conversation with Robert Sapolsky," The Edge.org, December 4, 2009, http://www.edge.org/3rd_culture/sapolsky09/sapolsky09_index.html (accessed August 19, 2012); Jaroslav Flegr, "Effects of Toxoplasma on Human Behavior," *Schizophrenia Bulletin*, 33, no. 3 (2007), http://schizophreniabulletin.oxfordjournals.org/content/33/3/757.full (accessed August 17, 2012).

256 *Oddly, toxo also makes the infected males:* McAuliffe, "How Your Cat Is Making You Crazy"; "Toxo: A Conversation with Robert Sapolsky"; Flegr, "Effects of Toxoplasma on Human Behavior."

256 *For decades we have known that pregnant women:* "Toxoplasmosis," Centers for Disease Control, http://www.cdc.gov/parasites/toxoplasmosis/disease.html (accessed August 17, 2012).

256 *He discovered that people who had been exposed to the parasite:* Flegr, "Effects of Toxoplasma on Human Behavior"; McAuliffe, "How Your Cat Is Making You Crazy"; "Toxo: A Conversation with Robert Sapolsky."

257 *The psychiatric effects in people infected:* Vinita J. Ling, David Lester, Preben Bo Mortensen, Patricia W. Langenberg, and Teodor T. Postolache, "Toxoplasma Gondii Seropositivity and Suicide Rates in Women," *The Journal of Nervous and Mental Disease* 199, no. 7 (July 2011): 440–444; Yuanfen Zhang, Lil Träskman-Bendz, Shorena Janelidze, Patricia Langenberg, Ahmed Saleh, Niel Constantine, Olaoluwa Okusaga, Cecilie Bay-Richter, Lena Brundin, and Teodor T. Postolache, "Toxoplasma Gondii Immunoglobulin G Antibodies and Nonfatal Suicidal Self-Directed Violence," *The Journal of Clinical Psychiatry* 73, no. 8 (August 2012): 1069–1076; David Lester, "Toxoplasma Gondii and Homicide," *Psychological Reports* 111, no. 1 (August 2012): 196–97.

257 *A 2012 Michigan State University study:* "Common Parasite May Trigger Suicide Attempts."

257 *Otters started dying off in larger numbers:* "California Sea Otters Numbers Drop Again," *United States Geological Survey*, August 3, 2010 http://www.usgs.gov/newsroom/article.asp?ID=2560; Miles Grant, "California Sea Otter Population Declining," National Wildlife Federation, March 7, 2011, http://blog.nwf.org/2011/03/california-sea-otter-population-declining (accessed March 8, 2011); "California Sea Otters Mysteriously Disappearing," CBS News, March 3, 2011, http://www.cbsnews.com/news/calif-sea-otters-mysteriously-disappearing (accessed May 4, 2011).

258 *a professor and veterinary parasitologist:* "Study Links Parasites in Freshwater Runoff to Sea Otter Deaths," Science Daily, July 2, 2002, http://www.sciencedaily.com/releases/2002/06/020627004404.htm (accessed October 30, 2010); P. A. Conrad, M. A. Miller, C. Kreuder, E. R. James, J. Mazet, H. Dabritz, D. A. Jessup, Frances Gulland, and M. E. Grigg, "Transmission of Toxoplasma: Clues from the Study of Sea Otters as Sentinels of Toxoplasma Gondii Flow into the Marine Environment," *International Journal for Parasitology* 35, no. 11–12 (October 2005): 1155–1168; M. A. Miller, W. A. Miller, P. A. Conrad, E. R. James, A. C. Melli, C. M. Leutenegger, H. A. Dabritz, et al., "Type X Toxoplasma Gondii in a Wild Mussel and Terrestrial Carnivores from Coastal California: New Linkages between Terrestrial Mammals, Runoff and Toxoplasmosis of Sea Otters," *International Journal for Parasitology* 38, no. 11 (September 2008): 1319–28.

258 *Conrad also found that otters:* Paul Rincon, "Cat Parasite 'Is Killing Otters,' " BBC, February 19, 2006, http://news.bbc.co.uk/2/hi/science/nature/4729810.stm (accessed December 5, 2012); Mariane B. Melo, Kirk D. C.

Jensen, and Jeroen P. J. Saeij, "Toxoplasma Gondii Effectors Are Master Regulators of the Inflammatory Response," *Trends in Parasitology* 27, no. 11 (November 2011): 487–95.

258 *A 2011 study of marine mammals in the Pacific Northwest:* "Dual Parasitic Infections Deadly to Marine Mammals," Science Daily, May 25, 2011, http://www.sciencedaily.com/releases/2011/05/110524171257.htm (accessed May 25, 2013).

260 *Most of these blooms are harmless:* Astrid Schnetzer et al., "Blooms of Pseudo-Nitzschia and Domoic Acid in the San Pedro Channel and Los Angeles Harbor Areas of the Southern California Bight, 2003–2004," *Harmful Algae* 6, no. 3 (2007): 372–87; Frances Gulland, *Domoic Acid Toxicity in California Sea Lions (Zalophus californianus) Stranded along the Central California Coast, May–October 1998*, Report to the National Marine Fisheries Service Working Group on Unusual Marine Mammal Mortality Events, December, 2000; "Domoic Acid Toxicity," Marine Mammal Center, http://www.marinemammalcenter.org/science/top-research-projects/domoic-acid-toxicity.html (accessed June 1, 2012); "Red Tide," Woods Hole Oceanographic Institute, http://www.whoi.edu/redtide (accessed June 1, 2012).

260 *Domoic acid toxicosis was first diagnosed:* Gulland, *Domoic Acid Toxicity in California Sea Lions (Zalophus californianus) Stranded along the Central California Coast, May–October 1998.*

260 *In humans, exposure to the neurotoxin:* "Amnesic Shellfish Poisoning," Woods Hole Oceanographic Institution, http://www.whoi.edu/redtide/page.do?pid=9679&tid=523&cid=27686 (accessed June 1, 2012); D. Baden, L. E. Fleming, and J. A. Bean, "Marine Toxins," in *Handbook of Clinical Neurology: Intoxications of the Nervous System, Part II. Natural Toxins and Drugs*, ed. F. A. de Wolff (Amsterdam: Elsevier Press, 1995), 141–75.

260 *Depending on where their primary hunting grounds are:* Kate Thomas et al., "Movement, Dive Behavior, and Survival of California Sea Lions *(Zalophus californianus)* Posttreatment for Domoic Acid Toxicosis," *Marine Mammal Science* 26, no. 1 (2010): 36–52; E. M. D. Gulland et al., "Domoic Acid Toxicity in Californian Sea Lions *(Zalophus californianus)*: Clinical Signs, Treatment and Survival," *Veterinary Record* 150, no. 15 (2002): 475–80.

261 *Researchers at the Marine Mammal Center and elsewhere have located:* Thomas et al., "Movement, Dive Behavior, and Survival of California

Sea Lions *(Zalophus californianus)* Posttreatment for Domoic Acid Toxicosis."

261 *released a number of sea lions:* Verbal communication with Lee Jackrel, crew leader, Marine Mammal Center, November and December 2010; Thomas et al., "Movement, Dive Behavior, and Survival of California Sea Lions *(Zalophus californianus)* Posttreatment for Domoic Acid Toxicosis."

261 *weirdly overconfident behavior:* Gulland et al., "Domoic Acid Toxicity in Californian Sea Lions (*Zalophus californianus*)"; T. Goldstein et al., "Magnetic Resonance Imaging Quality and Volumes of Brain Structures from Live and Postmortem Imaging of California Sea Lions with Clinical Signs of Domoic Acid Toxicosis," *Diseases of Aquatic Organisms* 91, no. 3 (2010): 243–56; Thomas et al., "Movement, Dive Behavior, and Survival of California Sea Lions *(Zalophus californianus)* Posttreatment for Domoic Acid Toxicosis."

261 *Another sea lion, nicknamed Wilder:* Mark Mullen, "Authorities Remove Sleeping Sea Lion," KRON4 News, December 16, 2002.

262 *In humans, the hippocampus plays an important role:* M. Sala et al., "Stress and Hippocampal Abnormalities in Psychiatric Disorders," *European Neuropsychopharmacology: The Journal of the European College of Neuropsychopharmacology* 14, no. 5 (2004): 393–405; Sapolsky, "Glucocorticoids and Hippocampal Atrophy in Neuropsychiatric Disorders"; Cheryl D. Conrad, "Chronic Stress-Induced Hippocampal Vulnerability: The Glucocorticoid Vulnerability Hypothesis," *Reviews in the Neurosciences* 19, no. 6 (2008): 395–411.

262 *According to the World Health Organization:* J. A. Patz et al., "Climate Change and Infectious Disease," in *Climate Change and Human Health: Risks and Responses,* ed. A. J. McMichael et al. (Geneva: World Health Organization, 2003), 103–32, http://www.who.int/globalchange/publications/climatechangechap6.pdf; "Climate Change and Harmful Algal Blooms," National Oceanic and Atmospheric Administration, http://www.cop.noaa.gov/stressors/extremeevents/hab/current/CC_habs.aspx (accessed May 1, 2012).

262 *rising ocean temperatures may be increasing:* Patz et al., "Climate Change and Infectious Disease"; "Climate Change and Harmful Algal Blooms"; T. Goldstein et al., "Novel Symptomatology and Changing Epidemiology of Domoic Acid Toxicosis in California Sea Lions *(Zalophus californianus)*: An Increasing Risk to Marine Mammal Health," *Proceedings*

of the Royal Society B: Biological Sciences 275, no. 1632 (2008): 267–76.

262 *Until the mid-1980s miners carried the birds*: "1986: Coal Mine Canaries Made Redundant," BBC, December 30, 1986; Walter Hines Page and Arthur Wilson Page, *The World's Work* (New York: Doubleday, Page, 1914), 474.

Epilogue: When the Devil Fish Forgive

274 *Frohoff, the same researcher who:* T. G. Frohoff, "Conducing Research on Human-Dolphin Interactions: Captive Dolphins, Free-Ranging Dolphins, Solitary Dolphins, and Dolphin Groups," in *Wild Dolphin Swim Program Workshop*, ed. K. M. Dudzinski, T. G. Frohoff, and T. R. Spradlin (Maui, 1999); T. G. Frohoff and J. Packard, "Human Interactions with Free-Ranging and Captive Bottlenose Dolphins," *Anthrozoos* 8 (1995): 44–53.

275 *California gray whales summer in the Arctic*: "Gray Whale," American Cetacean Society, http://acsonline.org/fact-sheets/gray-whale/; "Gray Whale," Alaska Department of Fish and Game, http://www.adfg.alaska.gov/static/education/wns/gray_whale.pdf (accessed March 1, 2011); "Gray Whale," NOAA Fisheries, http://www.nmfs.noaa.gov/pr/species/mammals/cetaceans/graywhale.htm; "California Gray Whale," Ocean Institute, http://www.ocean-institute.org/visitor/gray_whale.html.

275 *this seasonal congregation of whales was a target:* "California Gray Whale"; "Gray Whale," American Cetacean Society.

275 *Whaling captains like Charles Melville Scammon:* Charles Melville Scammon, *Marine Mammals of the Northwestern Coast of North America: Together with an Account of the American Whale-Fishery* (Berkeley: Heyday, 2007); Charles Siebert, "Watching Whales Watching Us," *New York Times Magazine*, July 8, 2009.

276 *The grays fought back so intensely:* Dick Russell, *Eye of the Whale: Epic Passage from Baja to Siberia* (Island Press, 2004), 20; Joan Druett and Ron Druett, *Petticoat Whalers: Whaling Wives at Sea, 1820–1920* (Lebanon, NH: University Press of New England, 2001), 139.

276 *By the early 1900s there were fewer than two thousand*: Marine Mammal Commission, Annual Report for 2002, http://www.mmc.gov/species/pdf/ar2002graywhale.pdf (accessed April 1, 2009).

276 *Protections were put in place in the 1930s and 1940s:* Ibid.

276 *Pachico's son Ranulfo:* Siebert, "Watching Whales Watching Us."

276 *roughly 10 to 15 percent of the whales in the lagoon:* Personal communication, Jonas Leonardo Meza Otero, March 17–18, 2010; personal communication with Ranulfo Mayoral, March 9, 2010; personal communication with Marcos Sedano, March 18, 2010.

277 *What's particularly startling about all of this:* "Gray Whale," NOAA Fisheries.

279 *As research on whale sociality, communication, and cognition:* Felicity Muth, "Animal Culture: Insights from Whales," *Scientific American*, April 27, 2013, http://blogs.scientificamerican.com/not-bad-science/2013/04/27/animal-culture-insights-from-whales/; Jenny Allen et al., "Network-Based Diffusion Analysis Reveals Cultural Transmission of Lobtail Feeding in Humpback Whales," *Science* 340, no. 6131 (2013): 485–88; John K. B. Ford, "Vocal Traditions among Resident Killer Whales (*Orcinus orca*) in Coastal Waters of British Columbia," *Canadian Journal of Zoology* 69, no. 6 (1991): 1454–83; Luke Rendell et al., "Can Genetic Differences Explain Vocal Dialect Variation in Sperm Whales, Physeter Macrocephalus?," *Behavior Genetics* 42, no. 2 (2011): 332–43.

279 *Mass killings at the hands of humans:* The near extinction of the population affected not only their behavior (how far they had to travel to find mates, for example, and narrowing their choices) but also their biology (constraining the genetic diversity of the population). How it affected them culturally is a mystery.

280 *Adolf Hitler loved his German Shepherd:* One of his last acts before killing himself was poisoning her. Gertraud Junge, *Until the Final Hour* (Arcade, 2003), 38, 181; James Serpell, *In the Company of Animals: A Study of Human-Animal Relationships* (Cambridge, UK: Cambridge University Press, 1996), 26.

280 *And Kim Jong-Il reportedly spent:* "Nothing's Too Good for Kim Jong-il's Pet Dogs," *Chosunilbo*, April 14, 2011; Peter Foster, "Kim Jong-il Reveals Fondness for Dolphins and Fancy Dogs," Telegraph.co.uk/news/worldnews/asia/northkorea/8869192/kim-jong-il-reveals-fondness-for-dolphins-and-fancy-dogs.html, November 4, 2011; Nadia Gilani, "Kim Jong-Il Spends £120,000 on Food for His Dogs, as Six Million North Koreans Starve," *Daily Mail Online*, September 30, 2011, http://www.dailymail.co.uk/news/article-2043868/Kim-Jong-Il-spends-120-000-food-dogs-million-North-Koreans-starve.html (accessed January 10, 2012).

Afterword

285 "*Imagine yourself*": Becky Chung, "The veteran and the labradoodle: How a service dog helped a TEDActive attendee step back out into the world," Ted.com, September 4, 2014, http://blog.ted.com/2014/09/04/the-veteran-and-the-labradoodle/.

Index

About the Author

LAUREL BRAITMAN received her PhD in history and anthropology of science from MIT. Her work has appeared in *The Wall Street Journal*, *The Guardian*, *Wired*, and a variety of other publications. Laurel is a senior TED fellow, performs live for *Pop-Up Magazine*, and lives on a houseboat in Sausalito, California, with her dog.